Advancing Earth Surface Representation via Enhanced Use of Earth Observations in Monitoring and Forecasting Applications

Advancing Earth Surface Representation via Enhanced Use of Earth Observations in Monitoring and Forecasting Applications

Special Issue Editors

Gianpaolo Balsamo
Fatima Karbou
Vanessa M. Escobar
Benjamin Ruston
Susanne Mecklenburg
Matthias Drusch
Isabel F. Trigo

MDPI • Basel • Beijing • Wuhan • Barcelona • Belgrade

Special Issue Editors

Gianpaolo Balsamo
European Centre for
Medium-Range
Weather Forecasts
USA

Fatima Karbou
Centre d'Études de la Neige
France

Vanessa M. Escobar
Science Systems and Applications Inc.
USA

Benjamin Ruston
Naval Research Laboratory
USA

Susanne Mecklenburg
European Space Agency
The Netherlands

Matthias Drusch
European Space Agency
The Netherlands

Isabel F. Trigo
(IPMA) EUMETSAT Land Surface Analysis
Portugal

Editorial Office
MDPI
St. Alban-Anlage 66
4052 Basel, Switzerland

This is a reprint of articles from the Special Issue published online in the open access journal *Remote Sensing* (ISSN 2072-4292) from 2017 to 2019 (available at: https://www.mdpi.com/journal/remotesensing/special_issues/earthsurface_RS)

For citation purposes, cite each article independently as indicated on the article page online and as indicated below:

LastName, A.A.; LastName, B.B.; LastName, C.C. Article Title. *Journal Name* **Year**, *Article Number*, Page Range.

ISBN 978-3-03921-064-0 (Pbk)
ISBN 978-3-03921-065-7 (PDF)

Contents

Preface to "Advancing Earth Surface Representation via Enhanced Use of Earth Observations in Monitoring and Forecasting Applications" . . . . **vii**

Yongchao Wang, Fang Shen, Leonid Sokoletsky and Xuerong Sun
Validation and Calibration of QAA Algorithm for CDOM Absorption Retrieval in the Changjiang (Yangtze) Estuarine and Coastal Waters
Reprinted from: *Remote Sens.* **2017**, *9*, 1192, doi:10.3390/rs9111192 . . . . **1**

Jérôme Vidot, Pascal Brunel, Marie Dumont, Carlo Carmagnola and James Hocking
The VIS/NIR Land and Snow BRDF Atlas for RTTOV: Comparison between MODIS MCD43C1 C5 and C6
Reprinted from: *Remote Sens.* **2018**, *10*, 21, doi:10.3390/rs10010021 . . . . **21**

Shaoning Lv, Yijian Zeng, Jun Wen, Hong Zhao and Zhongbo Su
Estimation of Penetration Depth from Soil Effective Temperature in Microwave Radiometry
Reprinted from: *Remote Sens.* **2018**, *10*, 519, doi:10.3390/rs10040519 . . . . **33**

E. Eva Borbas, Glynn Hulley, Michelle Feltz, Robert Knuteson and Simon Hook
The Combined ASTER MODIS Emissivity over Land (CAMEL) Part 1: Methodology and High Spectral Resolution Application
Reprinted from: *Remote Sens.* **2018**, *10*, 643, doi:10.3390/rs10050664 . . . . **52**

Michelle Feltz, Eva Borbas, Robert Knuteson, Glynn Hulley and Simon Hook
The Combined ASTER MODIS Emissivity over Land (CAMEL) Part 2: Uncertainty and Validation
Reprinted from: *Remote Sens.* **2018**, *10*, 664, doi:10.3390/rs10040643 . . . . **74**

Sid-Ahmed Boukabara, Kevin Garrett and Christopher Grassotti
Dynamic Inversion of Global Surface Microwave Emissivity Using a 1DVAR Approach
Reprinted from: *Remote Sens.* **2018**, *10*, 679, doi:10.3390/rs10050679 . . . . **95**

Michelle Feltz, Eva Borbas, Robert Knuteson, Glynn Hulley and Simon Hook
The Combined ASTER and MODIS Emissivity over Land (CAMEL) Global Broadband Infrared Emissivity Product
Reprinted from: *Remote Sens.* **2018**, *10*, 1027, doi:10.3390/rs10071027 . . . . **113**

Margaret Wambui Kimani, Joost C. B. Hoedjes and Zhongbo Su
Bayesian Bias Correction of Satellite Rainfall Estimates for Climate Studies
Reprinted from: *Remote Sens.* **2018**, *10*, 1074, doi:10.3390/rs10071074 . . . . **129**

Nina Raoult, Bertrand Delorme, Catherine Ottlé, Philippe Peylin, Vladislav Bastrikov, Pascal Maugis, and Jan Polcher
Confronting Soil Moisture Dynamics from the ORCHIDEE Land Surface Model With the ESA-CCI Product: Perspectives for Data Assimilation
Reprinted from: *Remote Sens.* **2018**, *10*, 1786, doi:10.3390/rs10111786 . . . . **147**

Gianpaolo Balsamo, Anna Agusti-Panareda, Clement Albergel, Gabriele Arduini, Anton Beljaars, Jean Bidlot, Nicolas Bousserez, Souhail Boussetta, Andy Brown, Roberto Buizza, Carlo Buontempo, Frederic Chevallier, Margarita Choulga, Hannah Cloke, Meghan F. Cronin, Mohamed Dahoui, Patricia De Rosnay, Paul A. Dirmeyer, Matthias Drusch, Emanuel Dutra, Michael B. Ek, Pierre Gentine, Helene Hewitt, Sarah P. E. Keeley, Yann Kerr, Sujay Kumar, Cristina Lupu, Jean-Francois Mahfouf, Joe McNorton, Susanne Mecklenburg, Kristian Mogensen, Joaquín Muñoz-Sabater, Rene Orth, Florence Rabier, Rolf Reichle, Ben Ruston, Florian Pappenberger, Irina Sandu, Sonia I. Seneviratne, Steffen Tietsche, Isabel F. Trigo, Remko Uijlenhoet, Nils Wedi, R. Iestyn Woolway and Xubin Zeng
Satellite and In Situ Observations for Advancing Global Earth Surface Modelling: A Review
Reprinted from: *Remote Sens.* **2018**, *10*, 2038, doi:10.3390/rs10122038 . . . . **176**

Gianpaolo Balsamo, Anna Agusti-Panareda, Clement Albergel, Gabriele Arduini, Anton Beljaars, Jean Bidlot, Eleanor Blyth, Nicolas Bousserez, Souhail Boussetta, Andy Brown, Roberto Buizza, Carlo Buontempo, Frédéric Chevallier, Margarita Choulga, Hannah Cloke, Meghan F. Cronin, Mohamed Dahoui, Patricia De Rosnay, Paul A. Dirmeyer, Matthias Drusch, Emanuel Dutra, Michael B. Ek, Pierre Gentine, Helene Hewitt, Sarah P.E. Keeley, Yann Kerr, Sujay Kumar, Cristina Lupu, Jean-François Mahfouf, Joe McNorton, Susanne Mecklenburg, Kristian Mogensen, Joaquín Muñoz-Sabater, Rene Orth, Florence Rabier, Rolf Reichle, Ben Ruston, Florian Pappenberger, Irina Sandu, Sonia I. Seneviratne, Steffen Tietsche, Isabel F. Trigo, Remko Uijlenhoet, Nils Wedi, R. Iestyn Woolway and Xubin Zeng
Correction: Balsamo, G., *et al.* Satellite and In Situ Observations for Advancing Global Earth Surface Modelling: A Review. *Remote Sensing* 2018, *10*, 2038
Reprinted from: *Remote Sens.* **2019**, *11*, 941, doi:10.3390/rs11080941 . . . . **248**

Preface to "Advancing Earth Surface Representation via Enhanced Use of Earth Observations in Monitoring and Forecasting Applications"

The Earth surface modelling community has recognized the need for enhancing the use of Earth observation (EO), in particular from remote sensing global observing satellites, to support and improve the monitoring and understanding of surface processes. These processes include complex and intertwined components of the Earth system such as land, vegetation, snow, ice, coasts and open waters. A review paper is proposed in place of an editorial introduction.

Gianpaolo Balsamo, Fatima Karbou, Vanessa M. Escobar,
Benjamin Ruston, Susanne Mecklenburg, Matthias Drusch, Isabel F. Trigo
Special Issue Editors

 *remote sensing*

Article

Validation and Calibration of QAA Algorithm for CDOM Absorption Retrieval in the Changjiang (Yangtze) Estuarine and Coastal Waters

Yongchao Wang, Fang Shen *, Leonid Sokoletsky and Xuerong Sun

State Key Laboratory of Estuarine and Coastal Research, East China Normal University,
3663 Zhongshan N. Road, Shanghai 200062, China; 51152601015@ecnu.cn (Y.W.);
sokoletsky.leonid@gmail.com (L.S.); 52152601003@ecnu.cn (X.S.)
* Correspondence: fshen@sklec.ecnu.edu.cn; Tel.: +86-021-6223-3467

Received: 29 August 2017; Accepted: 18 November 2017; Published: 21 November 2017

Abstract: Distribution, migration and transformation of chromophoric dissolved organic matter (CDOM) in coastal waters are closely related to marine biogeochemical cycle. Ocean color remote sensing retrieval of CDOM absorption coefficient ($a_g(\lambda)$) can be used as an indicator to trace the distribution and variation characteristics of the Changjiang diluted water, and further to help understand estuarine and coastal biogeochemical processes in large spatial and temporal scales. The quasi-analytical algorithm (QAA) has been widely applied to remote sensing inversions of optical and biogeochemical parameters in water bodies such as oceanic and coastal waters, however, whether the algorithm can be applicable to highly turbid waters (i.e., Changjiang estuarine and coastal waters) is still unknown. In this study, large amounts of in situ data accumulated in the Changjiang estuarine and coastal waters from 9 cruise campaigns during 2011 and 2015 are used to verify and calibrate the QAA. Furthermore, the QAA is remodified for CDOM retrieval by employing a CDOM algorithm (QAA_CDOM). Consequently, based on the QAA and the QAA_CDOM, we developed a new version of algorithm, named QAA_cj, which is more suitable for highly turbid waters, e.g., Changjiang estuarine and coastal waters, to decompose a_g from a_{dg} (CDOM and non-pigmented particles absorption coefficient). By comparison of matchups between Geostationary Ocean Color Imager (GOCI) retrievals and in situ data, it reveals that the accuracy of retrievals from calibrated QAA is significantly improved. The root mean square error (RMSE), mean absolute relative error (MARE) and bias of total absorption coefficients ($a(\lambda)$) are lower than 1.17, 0.52 and 0.66 m^{-1}, and $a_g(\lambda)$ at 443 nm are lower than 0.07, 0.42 and 0.018 m^{-1}. These results indicate that the calibrated algorithm has a better applicability and prospect for highly turbid coastal waters with extremely complicated optical properties. Thus, reliable CDOM products from the improved QAA_cj can advance our understanding of the land-ocean interaction process by earth observations in monitoring spatial-temporal distribution of the river plume into sea.

Keywords: CDOM; absorption coefficient; QAA inversion; GOCI; Changjiang (Yangtze) estuary

---

1. Introduction

Remote sensing of oceans and coastal zones is a key technology for monitoring spatial-temporal distribution of the river plume into sea and understanding of the land-ocean interaction processes. Satellite retrievals of inherent optical parameters (IOPs) of waters such as absorption and scattering characteristics is one of the most important applications of ocean color remote sensing [1]. Furthermore, chlorophyll, chromophoric dissolved organic matter (CDOM), suspended sediment and other water component concentrations can be derived from IOPs, which leads to further estimations of phytoplankton biomass, primary productivity and heat flux [2]. The IOPs mainly include absorption coefficient ($a(\lambda)$, see Table 1 for symbols and definitions) and backscattering coefficient. They are the source of the satellite remote sensing to quantify ocean water information [3], and key parameters of bio-optical models [4]. CDOM absorption coefficient is often used as a tracer to evaluate the amount of nutrients carried by Changjiang diluted water. Numerous applications of remote sensing have allowed to retrieve main components of water (i.e., phytoplankton, non-pigmented particles and CDOM) worldwide [5–7]. In CDOM-rich regions, such as North America and northern Europe, CDOM determines the optical properties of marine waters to a large extent and its existence can affect marine biogeochemical processes by the light absorption. Bricaud et al. [8] pointed out that even low concentrations of CDOM in the open sea may have an effect on absorption and hence on ocean color similar to that of low or moderate algal biomass; Nieke et al. [9] supported the possibility of using light absorption characteristics of CDOM in coastal waters strongly influenced by freshwater runoff in the Estuary and Gulf of St. Lawrence system (Canada); Darecki et al. [10] found a strong influence of CDOM absorption on the quantitative and qualitative features of spectral reflectance of in two different water bodies with similar chlorophyll content in the Baltic Sea; Hu et al. [11] estimated a_g to study occurrence and distribution of red tides in coastal waters off South Florida; Bowers et al. [12] used salinity to determinate a_g in an estuary for exploring the river discharge.

A number of algorithms were proposed to quantify $a_g(\lambda)$ from spectral measurements of ocean water. Empirical algorithms [13–19] were mostly based on spectral reflectance ratios to calculate $a_g(\lambda)$, and these algorithms required adequate data to parameterize the model and may only be valid for specific locations. Algorithms based on statistical modeling [18,20–24], such as optimization (Garver-Siegel-Maritorena, GSM), matrix inversion algorithm, artificial neural network (aNN) and Linear Matrix Inversion (LMI) algorithm, used some semi-analytical methodologies, but required knowledge about specific biochemical parameters [5]. Semi-analytical algorithms [25–27] mostly use $R_{rs}(\lambda)$ to calculate IOPs and further to estimate biochemical parameters, which incorporate both empirical parameters and bio-optical models. The quasi-analytical algorithm (QAA) developed by Lee et al. [28] was widely applied during the last decade. Several updated versions were presented in following years [29–31]. Recent version (QAA_v6) has been presented online by Lee [31].

Although the QAA algorithm is widely used [30,32–35], some researches pointed out that there are still large uncertainties in deriving optical properties for optically complex Case 2 waters [36–38]. Under the joint influences of river runoffs, tidal currents, marine circulations, etc., the hydrodynamic and biogeochemical environment in the Changjiang estuary and its adjacent coastal waters is unique, characterizing by high turbidity and complicated optical properties [39–41]. Among these drivers, the Changjiang diluted water with low salinity and high levels of nutrients and suspended sediment [42–45] makes greatest contribution to the optical complexity in the region of Changjiang River mouth due to its obvious seasonal changes. Compared to the clear oceanic waters, due to the lack of in situ data support, an application of QAA in the Changjiang estuarine and coastal waters is rarely reported in the literature.

Therefore, improvements to QAA over optically complex and highly turbid waters is in great demand considering successful absorption retrieval from ocean color remote sensing would provide a considerable advance in the release of satellite product and estimation of water components in the future. The objective of this study is to improve QAA algorithm and enhance an accuracy of QAA retrieval from satellite data in the Changjiang estuarine and coastal waters. The applicability of QAA is tested in the study area at first and then we perform a verification and calibration of QAA based on large amounts of in situ data accumulated during 9 voyage surveys from 2011 to 2015. Moreover, we calibrate the QAA_v6 and propose QAA_cj for retrieving both total and CDOM spectral absorption coefficients. Through applying to GOCI level-1 products, QAA_cj is finally validated with in situ data and compared with QAA_v6.

Table 1. Symbols and definitions.

Symbol	Description	Unit
a	Total absorption coefficient, $a_w + a_{ph} + a_g + a_d$	m^{-1}
a_{nw}	Non-water absorption coefficient, $a - a_w = a_{ph} + a_g + a_d$	m^{-1}
a_w	Pure water absorption coefficients	m^{-1}
a_{ph}	Phytoplankton absorption coefficients	m^{-1}
a_g	CDOM absorption coefficients	m^{-1}
a_d	Non-phytoplankton particulate absorption coefficients	m^{-1}
a_p	Particulate absorption coefficients, $a_{ph} + a_d$	m^{-1}
a_{dg}	Combined CDOM and non-pigmented particulate absorption coefficient, $a_g + a_d$	m^{-1}
b_{bp}	Particulate backscattering coefficient	m^{-1}
b_{bw}	Pure seawater backscattering coefficient	m^{-1}
b_b	Total backscattering coefficient, $b_{bw} + b_{bp}$	m^{-1}
Y	Power of the spectral particulate backscattering coefficient	
R_{rs}	Above-surface remote-sensing reflectance	sr^{-1}
r_{rs}	Below-surface remote-sensing reflectance	sr^{-1}
S	Exponential slope of the CDOM spectral absorption coefficient	nm^{-1}
u	Ratio of backscattering coefficient to the sum of absorption and backscattering coefficients, $b_b/(a + b_b)$	
λ_0	Reference wavelength	nm

2. Materials and Methods

2.1. Shipborne Samplings and Measurements

Water samples were collected during nine cruise campaigns in the Changjiang estuarine and coastal waters from 2011 to 2015. Spectral radiometric parameters (i.e., spectral downwelling irradiance, E_d, spectral incident radiance, L_s, total spectral upwelling radiance, L_{tot}) for estimating $R_{rs}(\lambda)$ were measured by Hyperspectral surface acquisition system (HyperSAS, Satlantic Inc.®, Halifax, NS, Canada). A total of 371 surface data samples was collected. Two radiance sensors were pointed to the sea and sky, respectively, at an optimal zenith angle of $40°$, and at an optimal azimuth angle of $135°$ away from the sun, in order to maximally avoid the wind speed impact and minimize solar glitter effects [46].

Spectral absorption coefficients $a(\lambda, z)$ and attenuation coefficients $c(\lambda, z)$ (z is the depth in meters) were measured in situ by WETLabs® absorption and attenuation meter (ac-s) during downcasts and upcasts as water flowed through the ac-s meter. A total of 479 data samples of a was obtained. $b_b(\lambda, z)$ values were measured simultaneously by WETLabs® ECO-BB9 backscattering sensors (at wavelengths of 412, 440, 488, 510, 532, 595, 650, 676, and 715 nm, and at a scattering angle of $117°$). 515 data samples of b_{bp} were collected.

CDOM water samples were obtained through filtration on shipboard using a 0.22 μm polycarbonate membrane (Millipore, 47 mm diameter, MilliporeSigma, Burlington, MA, USA) under low vacuum immediately after sampling. The membranes were soaked in 10% HCl for 15 min and then rinsed by Milli-Q water three times before filtration. The filtered CDOM samples were collected in borosilicate glass vials, and then stored in a $-40\ °C$ refrigerator. All vials were pre-soaked in 10% HCl for 24 h, rinsed by Milli-Q water for three times, and pre-combusted at $450\ °C$ for 5 h. A total of 551 data samples were obtained.

Data processing methods were detailed in Section 2.2. A total of 181 matchups containing simultaneous data of R_{rs}, a and b_b were obtained. Furthermore, SPSS software (IBM®, version 22.0) was used to control data quality. Excluding the sampling data deviating from the mean values more than $\pm3\sigma$, 144 matchup data were reserved for the analysis. In addition, 159 matchup data of a_g and R_{rs} were collected. All matchup locations are shown in Figure 1a. Meanwhile, two sets of matchup data were randomly divided into two parts, of which 70% were used to calibrate algorithm, and 30% to validate algorithm.

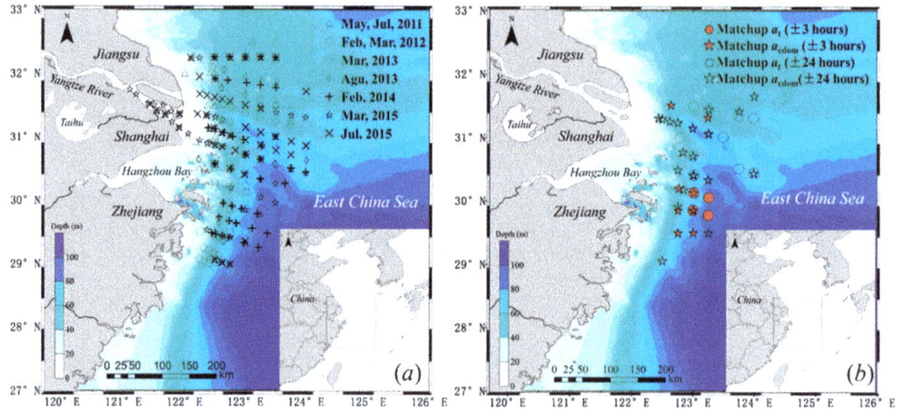

Figure 1. Location of sampling stations in the Changjiang estuarine and coastal waters. (**a**) Samples were collected from 9 cruises in summer (May, July 2011, August 2013 and July 2015) and winter (February, March 2012, March 2013, February 2014 and March 2015); (**b**) Matchup stations selected for in situ and GOCI images (empty circles and stars represent the matchup time windows within ±24 h, filled circles and stars within ±3 h).

Typical spectra of remote-sensing reflectance collected in the Changjiang estuarine and its adjacent coastal waters are shown in Figure 2a,b shows the sun zenith angle and weather conditions of in situ $R_{rs}(\lambda)$ validation data.

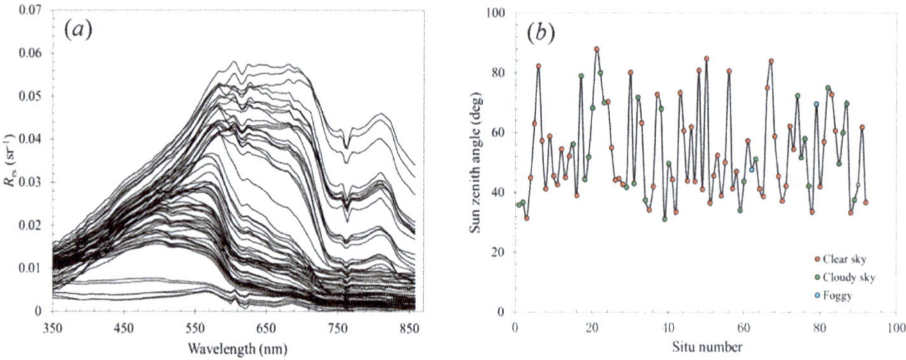

Figure 2. (**a**) Typical spectra of remote-sensing reflectance collected in the Changjiang estuarine and its adjacent coastal waters; (**b**) The sun zenith angle and weather conditions of in situ $R_{rs}(\lambda)$ validation data.

2.2. Data Processing

Through in situ measurements of downwelling spectral irradiance, $E_d(\lambda)$, incident spectral radiance, $L_s(\lambda)$, total upwelling spectral radiance, $L_{tot}(\lambda)$, $R_{rs}(\lambda)$ is estimated by Sokoletsky and Shen [47]:

$$R_{rs}(\lambda) = \frac{L_{tot}(\lambda) - \rho_{sky}(\lambda)L_{s,sky}(\lambda)}{E_d(\lambda)} \tag{1}$$

where $\rho_{sky}(\lambda)$ stands for a ratio of spectral reflected sky radiance, and $L_{s,sky}(\lambda)$ for incident spectral sky radiance. The estimations of $\rho_{sky}(\lambda)$, $L_{s,sky}(\lambda)$ and $R_{rs}(\lambda)$ were detailed in Sokoletsky and Shen [47].

Because $a(\lambda)$ is affected by temperature, salinity, pure water absorption and total scattering, it must be corrected [47–49]. For the most important and simultaneously the most difficult scattering correction procedure for the $a_{nw}(\lambda)$, we have used a modified Boss's method (MBM) which was described in Sokoletsky and Shen [47]. The average of $a(\lambda, z)$ in depths of 0.5~1.5 m is adopted for the surface data. $b_{bp}(\lambda)$ is calculated using a scale factor supplied by the WETLabs Inc. [48]. The specific correction method of b_{bp} is described in Sokoletsky and Shen [47]. In turbid waters, b_{bp} at $\lambda < 488$ nm measured by ECO-BB9 is generally too low due to absorption effects. Therefore, a spectral power function fitting was conducted, based on $b_{bp}(\lambda, z)$ values measured at $\lambda \geq 488$ nm [47]. In this study, the average of $b_{bp}(\lambda, z)$ in depths of 0.5 to 1.5 m is adopted for the surface values.

In laboratory, CDOM samples were unfrozen and warmed to room temperature under dark light conditions. CDOM absorbance spectra, $D(\lambda)$, were measured by PerkinElmer Lambda 1050 UV/VIS spectrophotometer. $a_g(\lambda)$ was derived as follows [39]:

$$a'_g = 2.303 \times \frac{D(\lambda)}{l} \tag{2}$$

where $a'_g(\lambda)$ represent uncorrected values of $a_g(\lambda)$ at wavelength λ, λ in nm; l is the length of cuvette, $l = 0.1$ m. Further, these initial values were scattering corrected as follows [8]:

$$a_g(\lambda) = a'_g(\lambda) - a'_g(700) \times \frac{\lambda}{700} \tag{3}$$

where $a_g(\lambda)$ is the final CDOM absorption coefficient.

2.3. Satellite Images

Satellite images were captured by the Geostationary Ocean Color Imager (GOCI) launched by Korean Ocean Satellite Center, which is the world's first geostationary ocean color observation sensor [50]. GOCI image covers 2500 $\times$ 2500 square kilometers, including Bohai Sea, Yellow Sea and East China Sea. GOCI collects eight images per day between 8:00 to 15:00 (Beijing time) at each hour, in a 500 m spatial resolution. GOCI has 6 visible wavebands centered at 412, 443, 490, 555, 660, 680 nm, and two near-infrared wavebands centered at 745 and 865 nm. GOCI data and products can be downloaded from official website (http://kosc.kiost.ac). This study used GOCI level-1 top-of-atmosphere (TOA) radiance data. Through performing atmospheric correction method proposed by Pan et al. [51], which is applicable for the Changjiang estuarine and coastal waters, the TOA radiance data are then inversed into the water surface remote-sensing spectral reflectance $R_{rs}(\lambda)$. Afterwards, GOCI images $R_{rs}(\lambda)$ were used for CDOM retrieval. Quasi-synchronous matchups between GOCI overpass observations and ground samplings were available during 6 March 2012 and 22 March 2015. A time window between in situ and satellite data was set at ± 3 h for the Changjiang estuary, and ± 24 h for the outer oceanic area. A mean value from a 3 $\times$ 3 pixel box centered at each sampling site is used aiming to reduce sensor and algorithm noise. A total of 28 images were obtained. Locations and time intervals of matchup samples are shown in Figure 1b.

2.4. QAA_v6

The QAA_v6 is developed from QAA that is based on the relationship between r_{rs} and IOPs from Gordon et al. [52]:

$$r_{rs}(\lambda) = g_0 \, u(\lambda) + g_1 \left[u(\lambda)\right]^2 \tag{4}$$

$$u = \frac{b_b}{a + b_b} \tag{5}$$

where the values of $g_0 = 0.089$ and $g_1 = 0.1245$ were accepted in this study in accordance with the QAA_v6. $r_{rs}(\lambda)$ has a computable relation with $R_{rs}(\lambda)$ according to the following Equation (1) (QAA_v6, step 0),which can be derived from R_{rs} to obtain IOPs:

$$r_{rs}(\lambda) = \frac{R_{rs}(\lambda)}{0.52 + 1.7R_{rs}(\lambda)} \tag{6}$$

The QAA_v6 algorithm was divided into two parts: in the first part, reference wavelength λ_0 was selected, and then $b_{bp}(\lambda)$ and $a(\lambda)$ were estimated by semi-analytical and analytical algorithms. In this process, $a_{ph}(\lambda)$, $a_g(\lambda)$, and $a_d(\lambda)$ were not taken into account. In the second part, the total absorption coefficient which was derived from the first part was decomposed into absorption coefficients of its major components.

In the first part of the algorithm, two reference wavelengths were used in QAA_v6, which are 55X (here X means any number from 0 to 9; for example, it was 5 in previous versions of the QAA (version 1, 2, 3 and 4)) and 670 nm, designed for oceanic and coastal waters, respectively. The $a(\lambda_0)$ could be estimated from $R_{rs}(443)$, $R_{rs}(490)$, $R_{rs}(55X)$, and $R_{rs}(670)$ according to empirical formula (QAA_v6, step 2):

If $R_{rs}(670) < 0.0015\ \text{sr}^{-1}$:

$$a(\lambda_0 = 55X) = a_w(\lambda_0) + 10^{h_0 + h_1 x + h_2 x^2}\text{ , where}$$
$$x = \log\left[\frac{r_{rs}(443) + r_{rs}(490)}{r_{rs}(55X) + 5r_{rs}(670)\frac{r_{rs}(670)}{r_{rs}(490)}}\right]$$
$$\text{else :} \tag{7}$$
$$a(\lambda_0 = 670) = a_w(670) + 0.39\left[\frac{R_{rs}(670)}{R_{rs}(443) + R_{rs}(490)}\right]^{1.14}$$

where $a(\lambda_0)$ is an empirical coefficient relating to the specific study area, $h_0 = -1.146$, $h_1 = -1.366$, $h_2 = -0.469$. Therefore, it is calibrated by fitting Equation (7) by using in situ data in the study area, which is detailed in Section 3.1.

$b_b(\lambda)$ is expressed by Lee [31] (QAA_v6, step 5):

$$b_b(\lambda) = b_{bw}(\lambda) + b_{bp}(\lambda_0)\left(\frac{\lambda_0}{\lambda}\right)^Y \tag{8}$$

where Y is an empirical coefficient relating to the specific study area. Therefore, it is calibrated by fitting Equation (8) using in situ data in the study area, which is detailed in Section 3.1.

In the second part, $a(\lambda)$, was decomposed into two partial absorption coefficients: $a_{dg}(\lambda)$ and $a_{ph}(\lambda)$. The expression for a_{dg} is given by Lee [31] (QAA_v6, step 9):

$$a_{dg}(\lambda) = a_{dg}(443) \exp\left[-S(443 - \lambda)\right] \tag{9}$$

where S is the exponential slope for $a_{dg}(\lambda)$. According to QAA_v6, S values can be estimated by spectral ratio (QAA_v6, step 8):

$$S = 0.015 + \frac{0.002}{0.6 + \frac{r_{rs}(443)}{r_{rs}(55x)}} \tag{10}$$

Since S values are influenced by CDOM and phytoplankton detritus, they are difficult to estimate accurately. In this study, we reestablished the empirical formula based on in situ data, which is detailed in Section 3.1.

2.5. QAA_CDOM

To retrieve a_g, the mixture variable a_{dg} needs to be further decomposed to a_g and a_d. However, the QAA_v6 cannot separate a_g from a_{dg} so far. In this study, we use QAA_CDOM algorithm proposed by Zhu and Yu [26] and Zhu et al. [27] to separate $a_g(443)$ and $a_d(443)$ by:

$$a_p(443) = j_1 b_{bp}(555)^{j_2} \tag{11}$$

$$a_g(443) = a(443) - a_p(443) - a_w(443) \tag{12}$$

$$a_d(443) = a_{dg}(443) - a_g(443) \tag{13}$$

where j_1 and j_2 are calculated by fitting Equation (11) by using in situ data in the study area, which is detailed in Section 3.1.

Zhu and Yu [26] and Zhu et al. [27] used the in situ data (water types vary from clear Case 1 to turbid Case 2) to prove the effectiveness of this algorithm. The algorithm takes an advantage of $b_{bp}(555)$ to estimate $a_p(443)$. Therefore, $a_g(443)$ can eventually be obtained by subtracting $a_p(443)$ and $a_w(443)$ from $a(443)$ estimated by the QAA_v6.

2.6. Accuracy Assessment

The accuracy of calibration algorithm can be evaluated by four statistical indices, root-mean-square-error (RMSE), mean absolute relative error (MARE), bias and the coefficient of determination (R^2). These indices are defined as follows (N is the number of samples):

$$\text{RMSE} = \sqrt{\frac{\sum_{i=1}^{N}(X_{est,i} - X_{mea,i})^2}{N}} \tag{14}$$

$$\text{MARE} = \frac{1}{N}\sum_{i=1}^{N}\frac{|X_{est,i} - X_{mea,i}|}{X_{mea,i}} \tag{15}$$

$$\text{bias} = \frac{1}{N}\sum_{i=1}^{N}(X_{est,i} - X_{mea,i}) \tag{16}$$

where $X_{est,i}$ and $X_{mea,i}$ are predicted and in situ values of optical parameters, respectively.

3. Results

3.1. QAA_cj Calibration

In this work, QAA_cj which is a combination of QAA_v6 and QAA_CDOM, is proposed especially for CDOM retrieval in highly turbid waters, i.e., the Changjiang estuarine and its adjacent coastal waters. Considering the QAA_v6 algorithms contain several empirical formula that depends on datasets (i.e., original QAA, IOCCG and NOMAD datasets), IOPs were obtained mostly from oceanic waters and partly from coastal waters, which is significantly different from IOPs data in the Changjiang estuarine and coastal waters. By comparison, IOPs of the Changjiang estuarine and coastal waters have a large variability (Table 2).

Table 2. Descriptive statistics of water constituent concentrations for the Changjiang estuarine and its adjacent coastal waters (CV is the ratio of standard deviation to the mean).

	Min	Max	Median	Mean	Standard Deviation	CV
$a(443)$ (m^{-1})	0.27	8.58	1.02	1.53	1.46	0.95
$b_{bp}(443)$ (m^{-1})	0.014	6.85	0.14	0.38	0.77	2.05
$a_g(443)$ (m^{-1})	0.029	0.65	0.12	0.17	0.13	0.74
Chl-a (μg·L^{-1})	0.082	20.32	0.95	2.02	3.19	1.58
TSM (mg·L^{-1})	0.61	475	13.27	40.78	63.12	1.55

In order to enhance the applicability of QAA_v6 in highly turbid waters, five empirical equations, namely, Equations (6)–(10) have to be calibrated with our in situ data. More specifically, the following five parameters were estimated: calculated parameters $\alpha(\lambda)$ and $\beta(\lambda)$ instead of 0.52 and 1.7 in Equation (6), $a(\lambda)$ at reference wavelength $\lambda_0 = 680$ nm in Equation(7), Y in Equation (8), $a_{dg}(443)$ in Equation (9) and S in Equation (10). Details of calibration are described as follows.

(1) Calculating new values for spectral parameters $\alpha(\lambda)$ and $\beta(\lambda)$ instead of the spectrally-independent constants 0.52 and 1.7 (step 0 in Table 3). According to Yang et al. [53], $\alpha(\lambda)$ and $\beta(\lambda)$ are wavelength dependent, which are calculated by:

$$r_{rs}(\lambda) = \frac{R_{rs}(\lambda)}{\alpha(\lambda) + \beta(\lambda)R_{rs}(\lambda)} \tag{17}$$

where $\alpha(\lambda) = 0.3638 + 8.776 \times 10^{-4}\lambda - 9.193 \times 10^{-7}\lambda^2 + 3.174 \times 10^{-10}\lambda^3$, $\beta(\lambda) = 1.357 + 8.608 \times 10^{-4}\lambda - 6.347 \times 10^{-7}\lambda^2$, λ in nm. Equation (17) was derived from the Aas-Højerslev radiative transfer model [47,54,55] at solar zenith angle $\theta_0 = 40°$, wind speed $= 5$ m·s^{-1}, and the wavelength range of 400 to 800 nm with $R^2 = 0.9995$ and $R^2 = 0.9903$ for $\alpha(\lambda)$ and $\beta(\lambda)$, respectively.

(2) Calibrating $a(\lambda_0)$ formula (step 2 in Table 3). $a(\lambda)$ was significantly underestimated, when reference wavelength in Equation (7) was accepted as 55X or 670 nm. Through the correlation analysis, it was found that $\lambda_0 = 680$ nm is the optimal reference wavelength. Based on our in situ data, the following equation relating non-water absorption at 680 nm, $a_{nw}(680)$, with the spectral ratio $R_{rs}(680)/R_{rs}(490)$ was derived as follows (Figure 3):

$$a_{nw}(680) = 0.9398\left[\frac{R_{rs}(680)}{R_{rs}(490)}\right]^2 + 0.865\frac{R_{rs}(680)}{R_{rs}(490)} - 0.0852 \tag{18}$$

(3) Step 3 is from QAA_v6 (step 3 in Table 3).

(4) Calibrating the Y in Equation (8) (step 4 in Table 3). The unknown parameters m and n were obtained by fitting a power regression with 466 sets of individual in situ measured data including $b_{bp}(680)$ and its corresponding Y values derived from Equation (8). The power regression is:

$$Y = mb_{bp}(680)^n \tag{19}$$

where it was found that $m = 1.75$ and $n = -0.05$, with the significance level $p < 0.001$ (Figure 4).

(5) Step 5 is from QAA_v6 (step 5 in Table 3).

(6) Step 6 is from QAA_v6 (step 6 in Table 3).

(7) Establishing $a_g(\lambda)$ formula (step 7 in Table 3). Though regression analysis, Equation (11) is fitted by in situ data, where it was found that $j_1 = 4.802$ and $j_2 = 0.8055$, with the significance level $p < 0.001$ (Figure 5).

(8) Step 8 is from QAA_CDOM (step 8 in Table 3). The unknown parameters p and q were obtained by fitting a power regression with in situ measured data including spectral ratio $R_{rs}(555)/R_{rs}(490)$ and its corresponding S values derived from Equation (10). The power regression is:

$$S = p[R_{rs}(555)/R_{rs}(490)]^q \tag{20}$$

where it was found that p = 0.0112 and q = 1.0401, with the significance level $p < 0.001$.

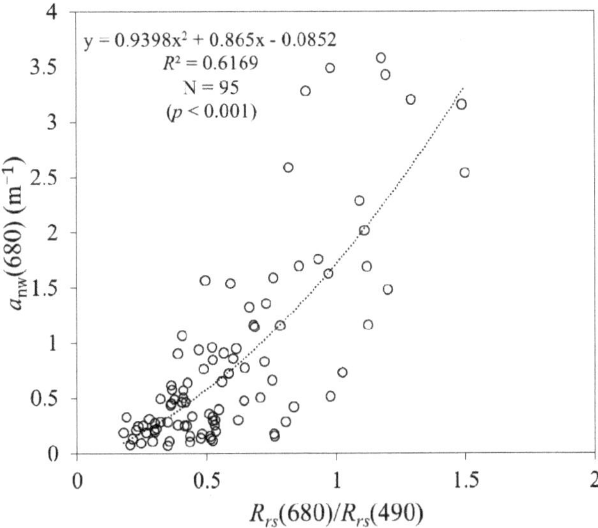

Figure 3. Scatter plot of in situ measured $a_{nw}(680)$ values versus $R_{rs}(680)/R_{rs}(490)$ spectral ratios.

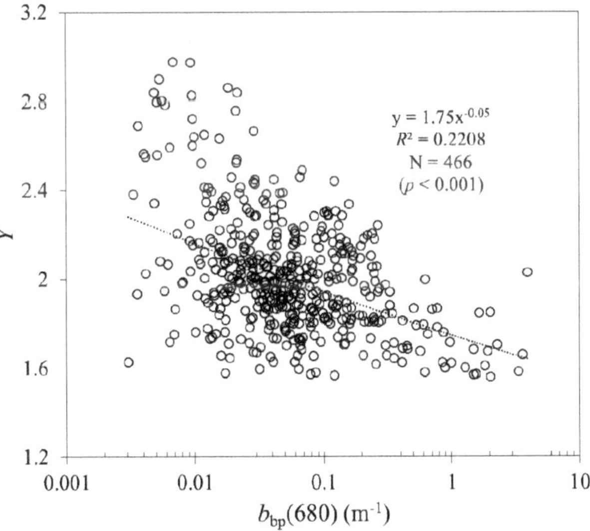

Figure 4. Scatter plot of the Y versus $b_{bp}(680)$ in situ values.

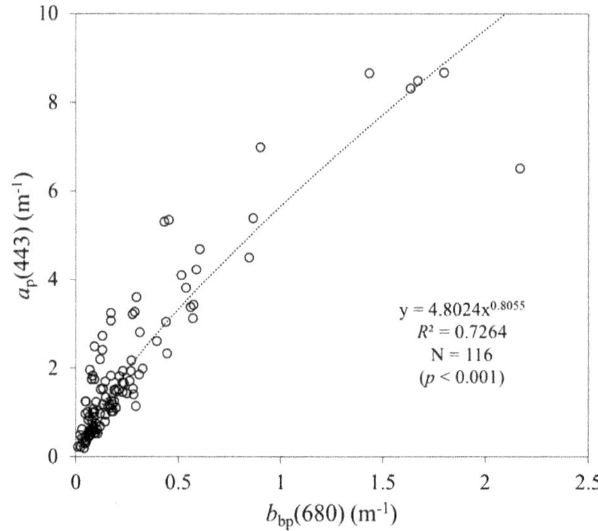

Figure 5. Scatter plot of in situ $a_{\mathrm{p}}(443)$ values plotted versus $b_{\mathrm{bp}}(680)$ values.

Table 3. A QAA_cj algorithm's calibration for the Changjiang estuarine and coastal waters.

Steps	Property	Derivation	Approach
Step 0	$r_{\mathrm{rs}}(\lambda)$	$= \dfrac{R_{\mathrm{rs}}(\lambda)}{\alpha(\lambda)+\beta(\lambda)R_{\mathrm{rs}}(\lambda)}$ $\alpha(\lambda) = 0.3638 + 8.776 \times 10^{-4}\lambda - 9.193 \times 10^{-7}\lambda^2 + 3.17 \times 10^{-10}\lambda^3;$ $\beta(\lambda) = 1.357 + 8.608 \times 10^{-4}\lambda - 6.347 \times 10^{-7}\lambda^2$	Semi-analytical
Step 1	$u(\lambda)$	$= \dfrac{-g_0+\sqrt{g_0^2+4g_1 r_{\mathrm{rs}}(\lambda)}}{2g_1}$, $g_0 = 0.089$, $g_1 = 0.1245$	Semi-analytical
Step 2	$a(680)$	$= a_{\mathrm{w}}(680) + 0.9398x^2 + 0.865x - 0.0852$ $x = \dfrac{R_{\mathrm{rs}}(680)}{R_{\mathrm{rs}}(490)}$	Empirical
Step 3	$b_{\mathrm{bp}}(680)$	$= \dfrac{u(680)a(680)}{1-u(680)} - b_{\mathrm{bw}}(680)$	Analytical
Step 4	Y	$= 1.75 b_{\mathrm{bp}}(680)^{-0.05}$	Empirical
Step 5	$b_{\mathrm{bp}}(\lambda)$	$= b_{\mathrm{bp}}(680)\left(\dfrac{680}{\lambda}\right)^Y$	Semi-analytical
Step 6	$a(\lambda)$	$= \dfrac{(1-u(680))b_{\mathrm{b}}(\lambda)}{u(\lambda)}$	Analytical
Step 7	$a_{\mathrm{g}}(443)$	$= a(443) - a_{\mathrm{p}}(443) - a_{\mathrm{w}}(443)$ $a_{\mathrm{p}}(443) = 4.8024 b_{\mathrm{bp}}(680)^{0.8055}$	Empirical
Step 8	$a_{\mathrm{g}}(\lambda)$	$= a_{\mathrm{g}}(443)e^{-S(\lambda-443)}$, where $S = 0.0112\left[\dfrac{R_{\mathrm{rs}}(555)}{R_{\mathrm{rs}}(490)}\right]^{1.0401}$	Semi-analytical

3.2. In Situ Data for QAA_cj Validation

Both QAA_cj and QAA_v6 algorithms are applied separately to test an accuracy of retrieved optical parameters in the Changjiang estuarine and coastal waters. A reference wavelength $\lambda_0 = 670$ nm has been selected for the $R_{\mathrm{rs}}(\lambda)$ dependence in Equation (7) following the QAA_v6 spectral criterion.

Figure 6 shows the comparison of Y values derived from in situ $b_{\mathrm{bp}}(\lambda)$ by the QAA_v6 (triangles) and QAA_cj (circles) algorithms based on the validation database. It is clear to be seen that values of Y

estimated by QAA_cj range from 1.7 to 2.3, which is much closer to the measured values, whereas, the QAA_v6 estimation is largely underestimated (0.3~1).

Figure 7 shows the comparison of in situ and retrieval results of $a(\lambda)$, in which two short wavelengths (e.g., 443 or 555 nm) and two long wavelengths (e.g., 680 or 715 nm) were compared. Compared with in situ data, QAA_cj has good consistency in short wavelength and long wavelength bands; however, there is a slight overestimation in the short wavelength spectral range. In comparison, QAA_v6 has an obvious underestimation, especially in the long wavelength range. The assessment results for retrieved total absorption from 412 to 715 nm are summarized in Table 4. Statistical results show that two algorithms are similar in the short wavelength range (412, 443, 490, 555 nm), but QAA_cj is more accurate than QAA_v6 in the long wavelength domain (660, 680, 715 nm). Specifically, QAA_cj has a better accuracy in estimating $a(\lambda)$ than QAA_v6 at 715 nm, for which R^2 are 0.73 and 0.30, respectively. Poor RMSE and bias results are mainly caused by the bias of Y value estimation. In addition, the QAA_v6 algorithm performed better than QAA_cj at 412 and 443 nm, which could be explained by the inaccuracies caused by the empirical Equation (18).

Since a_g and a_d have similar absorption features, a_g cannot be extracted from the a_{dg} by the QAA_v6 algorithm. Therefore, QAA_cj-derived $a_g(\lambda)$ values were compared only with the QAA_CDOM-derived $a_g(\lambda)$ values (Figure 8), and the assessment results are shown in Table 5. QAA_cj has a better accuracy in estimating $a_g(443)$ with RMSE, MARE and bias of 0.07, 0.42 and 0.018 m^{-1} compared with those of 0.25, 2.39 and 0.22 m^{-1}, respectively, from QAA_CDOM. The retrival accuracy of $a_g(\lambda)$ at wavelengths of 412 and 490 nm is improved as well (Table 5).

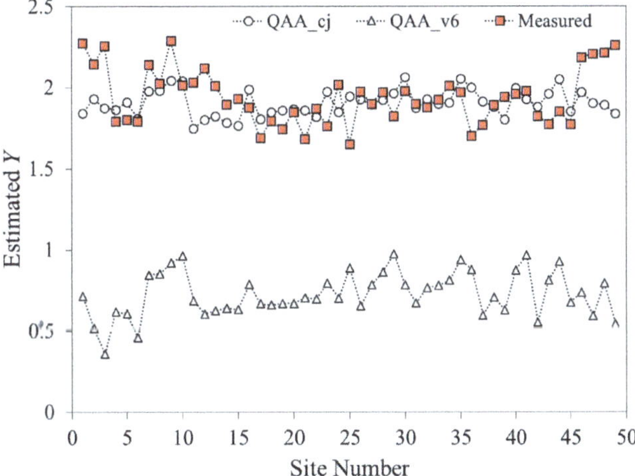

Figure 6. Comparison of in situ and predicted Y values. The filled squares are in situ Y values, empty circles and triangles denote Y values derived from QAA_cj and QAA_v6, respectively.

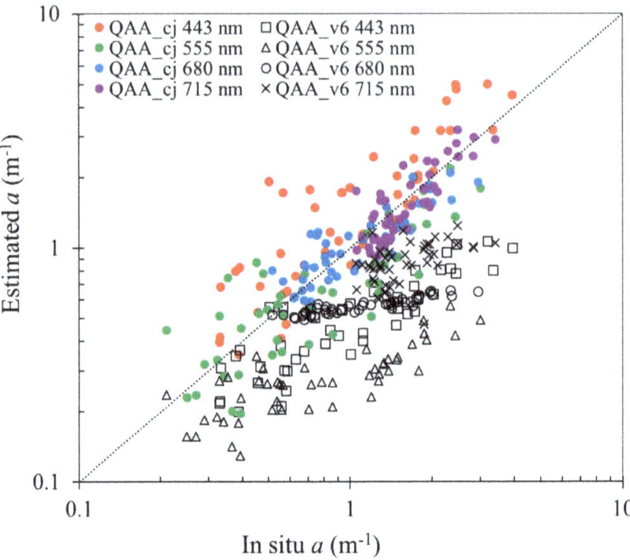

Figure 7. Comparison of in situ and predicted $a(\lambda)$ at 443, 555, 680 and 715 nm based on an in situ data set collected from the Changjiang estuarine and coastal waters. The filled circles and empty symbols denote retrievals from algorithms QAA_cj and QAA_v6, respectively.

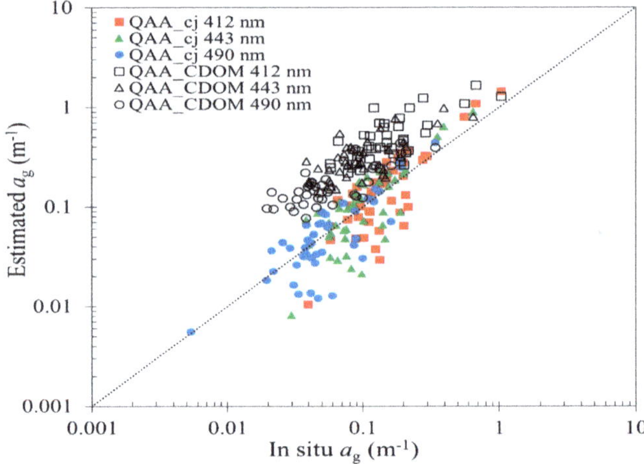

Figure 8. Comparison of in situ and predicted $a_g(\lambda)$ based on an in situ data set collected from the Changjiang estuarine and its adjacent coastal waters. The filled and empty symbols denote retrievals following QAA_cj and QAA_CDOM algorithms, respectively.

Table 4. Comparison statistics of QAA_cj and QAA_v6 based on in situ dataset collected from the Changjiang estuarine and coastal waters (N is the number of validation data).

	Algorithms	N	RMSE (m^{-1})	MARE	Bias (m^{-1})	R^2
$a(412)$	QAA_cj	49	1.09	0.50	0.66	0.82
	QAA_v6	49	1.08	0.48	-0.79	0.71
$a(443)$	QAA_cj	49	0.91	0.52	0.50	0.75
	QAA_v6	49	0.99	0.49	-0.74	0.61
$a(490)$	QAA_cj	49	0.42	0.34	0.027	0.73
	QAA_v6	49	0.93	0.56	-0.71	0.78
$a(555)$	QAA_cj	49	0.64	0.33	-0.23	0.73
	QAA_v6	49	0.93	0.60	-0.68	0.53
$a(660)$	QAA_cj	49	0.71	0.22	-0.085	0.72
	QAA_v6	49	0.80	0.44	-0.62	0.33
$a(680)$	QAA_cj	49	0.54	0.18	-0.084	0.75
	QAA_v6	49	0.88	0.44	-0.62	0.61
$a(715)$	QAA_cj	49	0.56	0.17	-0.057	0.73
	QAA_v6	49	0.95	0.44	-0.78	0.30

Table 5. Comparison statistics between the QAA_cj and QAA_v6 algorithms based on in situ dataset collected from Changjiang estuarine and its adjacent coastal waters (N is the number of validation data).

	Algorithms	N	RMSE (m^{-1})	MARE	Bias (m^{-1})	R^2
$a_g(412)$	QAA_cj	43	0.12	0.41	0.033	0.92
	QAA_CDOM	43	0.41	2.40	0.34	0.61
$a_g(443)$	QAA_cj	43	0.07	0.42	0.018	0.90
	QAA_CDOM	43	0.25	2.39	0.22	0.56
$a_g(490)$	QAA_cj	43	0.035	0.35	0.0023	0.84
	QAA_CDOM	43	0.01	2.48	0.11	0.55

3.3. Satellite Data for QAA_cj Validation

QAA_cj (Table 3, steps 0 to 6) and QAA_v6 (steps 0 to 6) are applied toGOCI to validate the accuracy of $a(\lambda)$ based on the in situ data (Figure 9, Table 6). When applying to GOCI, QAA_cj has a consistency with in situ data at 443, 555, 680 nm, of which R^2 is larger than 0.8 at 680 nm. It shows that the QAA_cj yielded a better accuracy in estimating $a(680)$ with RMSE and bias of -0.025 and 0.10 m^{-1}, compared with those of 0.62 and 0.31 m^{-1} from QAA_v6. However, the inversion result is slightly poor at 745 nm (Figure 9, Table 6).

QAA_cj (Table 3, Steps 7 and 8) and QAA_CDOM algorithms are applied to GOCI to validate the accuracy of $a_g(\lambda)$, compared with in situ data (Figure 10). It is indicated that estimations from the QAA_CDOM have lower agreement with the in situ values. Meanwhile, QAA_cj has a better consistency in estimating $a_g(\lambda)$ at 412, 443 and 490 nm, compared with the QAA_CDOM. Remarkably, the retrieval accuracy is optimal at wavelength of 443 nm, at which MARE is 0.14, R^2 is 0.70, whereas QAA_CDOM yields MARE = 2.34 and R^2 = 0.11 (Table 7).

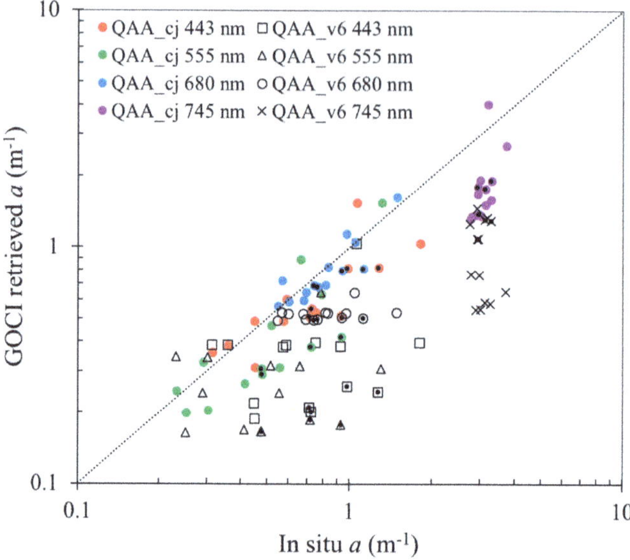

Figure 9. A scattering plot of GOCI retrieved $a(\lambda)$ vs. in situ $a(\lambda)$ data with using the QAA_cj algorithm (filled symbols) and QAA_v6 algorithm (open symbols) at wavelengths of 443, 555, 680, and 745 nm. The symbol with a filled dot inside represents the match point within the time window of ± 3 h.

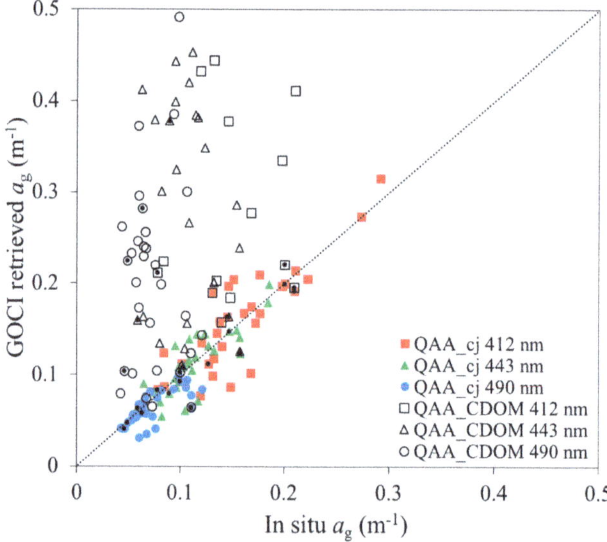

Figure 10. A scattering plot of GOCI retrieved $a_g(\lambda)$ vs. in situ $a_g(\lambda)$ data with using the QAA_cj algorithm (filled symbols) and QAA_CDOM algorithm (open symbols) at wavelengths of 412, 443, and 490 nm. The symbol with a filled dot inside represents the match point within the time window of ± 3 h.

Table 6. Comparison statistics of QAA_cj and QAA_v6 based on in situ dataset and GOCI-derived at $a(443)$, $a(555)$, $a(680)$, $a(745)$ (N is the number of validation data).

	Algorithms	N	RMSE (m^{-1})	MARE	Bias (m^{-1})	R^2
$a(443)$	QAA_cj	14	0.56	0.25	-0.14	0.50
	QAA_v6	14	0.76	0.49	-0.42	0.046
$a(555)$	QAA_cj	14	0.46	0.29	-0.10	0.69
	QAA_v6	14	0.64	0.50	-0.29	0.038
$a(680)$	QAA_cj	14	0.35	0.11	-0.025	0.80
	QAA_v6	14	0.62	0.34	-0.32	0.096
$a(745)$	QAA_cj	14	1.17	0.44	-1.23	0.28
	QAA_v6	14	1.47	0.68	-2.11	0.029

Table 7. Comparison statistics of QAA_cj and QAA_v6 based on in situ dataset and GOCI-derived at $a_g(412)$, $a_g(443)$, $a_g(490)$. (N is the number of validation data).

	Algorithms	N	RMSE (m^{-1})	MARE	Bias (m^{-1})	R^2
$a_g(412)$	QAA_cj	30	0.029	0.16	0.0026	0.72
	QAA_CDOM	30	0.46	2.37	0.36	0.26
$a_g(443)$	QAA_cj	30	0.02	0.14	-0.00025	0.70
	QAA_CDOM	30	0.31	2.34	0.25	0.11
$a_g(490)$	QAA_cj	30	0.017	0.15	-0.0086	0.53
	QAA_CDOM	30	0.17	2.06	0.13	0.0001

4. Discussion

CDOM absorption is a major variable in remote sensing algorithms for deriving concentrations of optically active components of sea water [56]. Based on the spectral absorption characteristic of CDOM, Del Castillo and Miller [14], D'Sa and Miller [18] and Ficek et al. [13] established statistical relationships between a_g and R_{rs} ratio (using different wavelengths) in Mississippi River and the Baltic. Using $R_{rs}(\lambda)$ to calculate a and further to separate out a_{dg} and a_{ph}, semi-analytical models, such as GSM [23] and QAA [28], were developed based on IOCCG dataset and in situ data, while Brando and Dekker [5] and Hoge and Lyon [22] developed models using in situ data in Fitzroy Estuary and U.S. Middle Atlantic Bight.

The QAA_cj algorithm was developed to give an opportunity to obtain more accurate estimates for both total and partial absorption coefficients in highly turbid waters, such as the Changjiang estuarine and coastal waters. For this purpose, we calibrated empirical parameters of the QAA_v6 and QAA_CDOM algorithms. Proofs of improvement made in these algorithms are shown in Tables 4–7 and Figures 6–10. For example, a total absorption coefficient $a(\lambda)$ retrieved by QAA_v6 was underestimated in highly turbid waters, while QAA_cj-based values were closer to in situ data (Table 4, Figure 7). As for CDOM spectral absorption $a_g(\lambda)$, on the one hand it could not be estimated by the QAA algorithm and on the other hand it was overestimated by the QAA_CDOM algorithm (Figure 8), while the QAA_cj retrieved value of $a_g(443)$ has RMSE and R^2 of 0.07 m^{-1} and 0.90, respectively, compared with in situ data (Table 5). Nevertheless, the accuracy of $a_g(\lambda)$ derived from the QAA_cj in highly turbid waters still remains to be a challenge.

We also wish to note that our findings may have great values not only for numerous scientists and for decision makers working for the East China estuarine and coastal waters, but also for many other investigators. As it has been found by Sokoletsky et al. [57], the biogeochemistry-optical relationships for the East China estuarine and coastal waters are very similar to that of the Gironde Estuary [58] and the Southern North Sea [59]. Thus, these findings really allow generalizing all conclusions of our study.

Since the aim of the study was the investigation of underwater IOPs, we used a conversion from the surface remote-sensing reflectance $R_{rs}(\lambda)$ to its underwater analog, $r_{rs}(\lambda)$. Therefore, in this study, we used the formula proposed by Sokoletsky et al. [57] to calculate $r_{rs}(\lambda)$ from $R_{rs}(\lambda)$. Sokoletsky and Shen [47] showed that although relations between $R_{rs}(\lambda)$ and $r_{rs}(\lambda)$ are close for different models, the particular model parameters may play an important role in the inversion results. In addition, we have used only clear and cloudless sky conditions to measure $R_{rs}(\lambda)$ to do calculations more simple and closer to remote-sensing results, Figure 2b.

Equation (4) is an approximation to the exact solution of the radiative transfer equation [60], which may cause a normalized (to the mean value) root-mean-square error of about 20% [61]. Moreover, it is inappropriate to regard parameters of this equation (i.e., g_0 and g_1) as constants. These parameters are associated with the solar zenith angle and water properties, and vary with water composition scattering properties [62]. Consequently, g_0 and g_1 have influence on $b_b(\lambda)$, when using QAA, it will further affect the inversion accuracy of $a(\lambda)$. Lee et al. [62] partitioned and weighted parameter g according to the molecular (water itself) and particulate contributions to the backscattering coefficient. In this study, however, this approach was not exploited.

According to Lee et al. [63], the $a(\lambda_0)$ and Y have an impact on performance of the QAA. Y is a parameter which describes spectral variation of $b_{bp}(\lambda)$ [64], and the variation of Y depends on water composition and size of particles according to the Mie theory [65]. Yang et al. [38] found that Y values have a great impact on the retrieval results, particularly in the shorter spectral bands. Figure 6 shows that the QAA_v6 algorithm has a low accuracy in estimating Y, which perhaps caused by the insufficient capability of this algorithm considering the complex optical features of the Changjiang estuarine and coastal waters. The ranges of Y values derived from the QAA_cj and the QAA_v6 are from 1.5 to 2.5 and from 0.3 to 1, respectively (Figure 6). Even though our algorithm for Y (Equation 19) does not yield a reasonable correlation with the measured values (Figure 6), we have chosen to keep it for generalization purposes, and we are planning to improve the Y model in the following study.

In this study, we have changed the reference wavelength from 670 nm to 680 nm. Although it is an insignificant change of spectral, the retrieval accuracy of calibrated formula was improved effectively when applied to Changjiang estuarine and coastal waters (Tables 4 and 5). We reproduced $a(\lambda)$ and $a_g(\lambda)$ from the GOCI images using the QAA_v6, QAA_CDOM and QAA_cj algorithms, and presented the comparison results in Figures 9 and 10. As shown in Figure 9, differences between our algorithm and in situ data is smaller than those between the QAA_v6 and in situ data for the whole blue to near infrared spectral range. Similar findings were found for the $a_g(\lambda)$ retrieving in the blue spectral domain (Figure 10).

Some contradicting results derived from QAA algorithm were discovered from existing literature as well. For example, investigations by Qin et al. [66] and Shanmugam et al. [67] have shown that there is a lower accuracy for retrieving absorption components in some regions of the ocean by QAA. Zheng et al. [68] have also shown that the accuracy of QAA algorithm (v5) varied greatly in deriving $a(\lambda)$ (from 2% to 28%) and $b_b(\lambda)$ (from 8% to 14%) depending on wavelength and ocean site. In addition, Zhu et al. [69] compared and verified 15 CDOM retrieval algorithms (empirical, semi-analytical, optimization, and matrix inversion algorithms), and pointed out that the QAA_CDOM algorithm was optimal. The reason of this is that CDOM has negligible backscattering, whereas inorganic particles have strong backscattering, even in longer wavelength [27]. Several researches find that QAA_CDOM has a good accuracy in obtaining the $a_p(\lambda)$ from $b_{bp}(\lambda)$ [26,27,69]. However, QAA_CDOM has some empirical parameters, which are required to be calibrated according to specific study area. Figure 10 shows that after calibration, $a_g(\lambda)$ has a significant improvement at 412, 443, 490 nm. It seems obvious that better accuracy in $a_g(\lambda)$ leads to improvement of a retrieval accuracy for a_{ph} and a_d.

5. Conclusions

Through calibration and validation, we improved empirical parameters of QAA_v6 and QAA_CDOM IOPs algorithms, and thus developed a new algorithm, namely, QAA_cj, which is suitable for the highly turbid Changjiang estuary and adjacent areas. Results of validation prove that QAA_cj has a better accuracy in retrieving $a(\lambda)$ and $a_g(\lambda)$, compared with the QAA_v6 and QAA_CDOM. $a(\lambda)$ derived from QAA_cj is in a good agreement with in situ and GOCI data, where RMSE ranges from 0.35 to 1.17 m^{-1}, MARE from 0.11 to 0.52, bias from -1.23 to 0.66 m^{-1} and R^2 is from 0.28 to 0.82. As for $a_g(\lambda)$, RMSE ranges from 0.035 to 0.12 m^{-1}, MARE from 0.14 to 0.42, bias from -0.0086 to 0.033 m^{-1} and R^2 is from 0.53 to 0.92.

The improvement of $a(\lambda)$ and $a_g(\lambda)$ retrieval accuracy will help to provide theoretical basis for the release of satellite product and further study on optical properties. Reliable CDOM products can provide information on the internal movements and nutrients structure of Changjiang diluted water and mechanisms of hydrodynamics in the Changjiang estuarine and coastal waters. The trace of CDOM can advance our understanding of the land-ocean interaction processes through monitoring spatial-temporal distribution of the river plume into sea. In the future, we will focus on exploring the relationships between environmental factors and optical parameters, in combination with satellite data and physical models.

Acknowledgments: Massive field cruise campaigns were funded by the National Natural Science Foundation of China (NSFC) joint cruise programmes. This study is supported by the NSFC projects (no. 41771378 and no. 41271375), Ministry of Science and Technology of China (no. 2016YFE0103200) and the research foundation of SKLEC (Grant no. 2015KYYW04). The authors would like to thank the crew members of the "Runjiang" ship and all young scientists participating in the collection of samples. We also thank the KORDI/KOSC for providing GOCI data. We acknowledge contributors to data accumulation and data processing including Xiaodao Wei, Pei Shang and Yanqun Pan. The insightful and constructive comments from the editor and three anonymous reviewers are also greatly appreciated.

Author Contributions: Yongchao Wang and Fang Shen conceived and designed the study. Fang Shen provided part of the in situ data. Yongchao Wang wrote the programs and performed the experiments, the results were examined by Fang Shen. All authors contributed to the writing and approved the final manuscript.

Conflicts of Interest: The authors declare no conflict of interest.

References

1. Mobley, C.D. *Light and Water: Radiative Transfer in Natural Waters*; Academic Press: Cambridge, MA, USA, 1994.
2. Lee, Z.P.; Carder, K.L.; Peacock, T.G.; Davis, C.O.; Mueller, J.L. Method to derive ocean absorption coefficients from remote-sensing reflectance. *Appl. Opt.* **1996**, *35*, 453–462. [CrossRef] [PubMed]
3. Stramski, D.; Boss, E.; Bogucki, D.; Voss, K.J. The role of seawater constituents in light backscattering in the ocean. *Prog. Oceanogr.* **2004**, *61*, 27–56. [CrossRef]
4. Le, C.F.; Li, Y.M.; Zha, Y.; Sun, D.; Yin, B. Validation of a quasi-analytical algorithm for highly turbid eutrophic water of Meiliang Bay in Taihu Lake, China. *IEEE Trans. Geosci. Remote Sens.* **2009**, *47*, 2492–2500.
5. Brando, V.E.; Dekker, A.G. Satellite hyperspectral remote sensing for estimating estuarine and coastal water quality. *IEEE Trans. Geosci. Remote Sens.* **2003**, *41*, 1378–1387. [CrossRef]
6. Giardino, C.; Brando, V.E.; Dekker, A.G.; Strömbeck, N.; Candiani, G. Assessment of water quality in Lake Garda (Italy) using Hyperion. *Remote Sens. Environ.* **2007**, *109*, 183–195. [CrossRef]
7. Yang, W.; Matsushita, B.; Chen, J.; Fukushima, T. A relaxed matrix inversion method for retrieving water constituent concentrations in case II waters: The case of Lake Kasumigaura, Japan. *IEEE Trans. Geosci. Remote Sens.* **2011**, *49*, 3381–3392. [CrossRef]
8. Bricaud, A.; Morel, A.; Prieur, L. Absorption by dissolved organic matter of the sea (yellow substance) in the UV and visible domains. *Limnol. Oceanogr.* **1981**, *26*, 43–53. [CrossRef]
9. Nieke, B.; Reuter, R.; Heuermann, R.; Wang, H.; Babin, M.; Therriault, J.C. Light absorption and fluorescence properties of chromophoric dissolved organic matter (CDOM), in the St. Lawrence Estuary (Case 2 waters). *Cont. Shelf Res.* **1997**, *17*, 235–252. [CrossRef]
10. Darecki, M.; Weeks, A.; Sagan, S.; Kowalczuk, P.; Kaczmarek, S. Optical characteristics of two contrasting Case 2 waters and their influence on remote sensing algorithms. *Cont. Shelf Res.* **2003**, *23*, 237–250. [CrossRef]
11. Hu, C.; Muller-Karger, F.E.; Taylor, C.J.; Carder, K.L.; Kelble, C.; Johns, E.; Heil, C.A. Red tide detection and tracing using MODIS fluorescence data: A regional example in SW Florida coastal waters. *Remote Sens. Environ.* **2005**, *97*, 311–321. [CrossRef]

12. Bowers, D.G.; Brett, H.L. The relationship between CDOM and salinity in estuaries: An analytical and graphical solution. *J. Mar. Syst.* **2008**, *73*, 1–7. [CrossRef]

13. Ficek, D.; Zapadka, T.; Dera, J. Remote sensing reflectance of Pomeranian lakes and the Baltic. *Oceanologia* **2011**, *53*, 959–970. [CrossRef]

14. Del Castillo, C.E.; Miller, R.L. On the use of ocean color remote sensing to measure the transport of dissolved organic carbon by the Mississippi River Plume. *Remote Sens. Environ.* **2008**, *112*, 836–844. [CrossRef]

15. Griffin, C.G.; Frey, K.E.; Rogan, J.; Holmes, R.M. Spatial and interannual variability of dissolved organic matter in the Kolyma River, East Siberia, observed using satellite imagery. *J. Geophys. Res. Biogeosci.* **2011**, *116*. [CrossRef]

16. Kutser, T.; Pierson, D.C.; Tranvik, L.; Reinart, A.; Sobek, S.; Kallio, K. Using satellite remote sensing to estimate the colored dissolved organic matter absorption coefficient in lakes. *Ecosystems* **2005**, *8*, 709–720. [CrossRef]

17. Mannino, A.; Russ, M.E.; Hooker, S.B. Algorithm development and validation for satellite-derived distributions of DOC and CDOM in the US Middle Atlantic Bight. *J. Geophys. Res. Oceans* **2008**, *113*. [CrossRef]

18. D'Sa, E.J.; Miller, R.L. Bio-optical properties in waters influenced by the Mississippi River during low flow conditions. *Remote Sens. Environ.* **2003**, *84*, 538–549. [CrossRef]

19. Carder, K.L.; Chen, F.R.; Lee, Z.P.; Hawes, S.K.; Kamykowski, D. Semianalytic Moderate-Resolution Imaging Spectrometer algorithms for chlorophyll a and absorption with bio-optical domains based on nitrate-depletion temperatures. *J. Geophys. Res. Oceans* **1999**, *104*, 5403–5421. [CrossRef]

20. Wang, P.; Boss, E.S.; Roesler, C. Uncertainties of inherent optical properties obtained from semianalytical inversions of ocean color. *Appl. Opt.* **2005**, *44*, 4074–4085. [CrossRef] [PubMed]

21. Babin, M.; Stramski, D.; Ferrari, G.M.; Claustre, H.; Bricaud, A.; Obolensky, G.; Hoepffner, N. Variations in the light absorption coefficients of phytoplankton, nonalgal particles, and dissolved organic matter in coastal waters around Europe. *J. Geophys. Res. Oceans* **2003**, *108*. [CrossRef]

22. Hoge, F.E.; Lyon, P.E. Satellite retrieval of inherent optical properties by linear matrix inversion of oceanic radiance models: An analysis of model and radiance measurement errors. *J. Geophys. Res. Oceans* **1996**, *101*, 16631–16648. [CrossRef]

23. Maritorena, S.; Siegel, D.A.; Peterson, A.R. Optimization of a semianalytical ocean color model for global-scale applications. *Appl. Opt.* **2002**, *41*, 2705–2714. [CrossRef] [PubMed]

24. Roesler, C.S.; Perry, M.J. In situ phytoplankton absorption, fluorescence emission, and particulate backscattering spectra determined from reflectance. *J. Geophys. Res. Oceans* **1995**, *100*, 13279–13294. [CrossRef]

25. Lee, Z.; Weidemann, A.; Kindle, J.; Arnone, R.; Carder, K.L.; Davis, C. Euphotic zone depth: Its derivation and implication to ocean-color remote sensing. *J. Geophys. Res. Oceans* **2007**, *112*. [CrossRef]

26. Zhu, W.; Yu, Q. Inversion of Chromophoric Dissolved Organic Matter from EO-1 Hyperion Imagery for Turbid Estuarine and Coastal Waters. *Geosci. Remote Sens. IEEE Trans.* **2013**, *51*, 3286–3298. [CrossRef]

27. Zhu, W.; Yu, Q.; Tian, Y.Q.; Chen, R.F.; Gardner, G.B. Estimation of chromophoric dissolved organic matter in the Mississippi and Atchafalaya river plume regions using above-surface hyperspectral remote sensing. *J. Geophys. Res. Oceans* **2011**, *116*. [CrossRef]

28. Lee, Z.; Carder, K.L.; Arnone, R.A. Deriving inherent optical properties from water color: A multiband quasi-analytical algorithm for optically deep waters. *Appl. Opt.* **2002**, *41*, 5755–5772. [CrossRef] [PubMed]

29. Lee, Z. Diffuse attenuation coefficient of downwelling irradiance: An evaluation of remote sensing methods. *J. Geophys. Res.* **2005**, *110*. [CrossRef]

30. Lee, Z. Remote Sensing of Inherent Optical Properties: Fundamentals, Tests of Algorithms, and Applications. In *Reports of the International Ocean-Colour Coordinating Group*; Lee, Z.-P., Ed.; IOCCG: Dartmouth, NS, Canada, 2006.

31. Lee, Z. Update of the Quasi-Analytical Algorithm (QAA_v6). 2014. Available online: http://www.ioccg.org/groups/Software_OCA/QAA_v6_2014209.pdf (accessed on 19 November 2017).

32. Lee, Z.; Carder, K.L. Absorption spectrum of phytoplankton pigments derived from hyperspectral remote-sensing reflectance. *Remote Sens. Environ.* **2004**, *89*, 361–368. [CrossRef]

33. Lee, Z.; Shang, S.; Hu, C.; Lewis, M.; Arnone, R.; Li, Y.; Lubac, B. Time series of bio-optical properties in a subtropical gyre: Implications for the evaluation of interannual trends of biogeochemical properties. *J. Geophys. Res. Oceans* **2010**, *115*. [CrossRef]
34. Lee, Z.; Lance, V.P.; Shang, S.; Vaillancourt, R.; Freeman, S.; Lubac, B.; Hargreaves, B.R.; del Castillo, C.; Miller, R.; Twardowski, M. An assessment of optical properties and primary production derived from remote sensing in the Southern Ocean (SO GasEx). *J. Geophys. Res. Oceans* **2011**, *116*. [CrossRef]
35. Shang, S.; Lee, Z.; Wei, G. Characterization of MODIS-derived euphotic zone depth: Results for the China Sea. *Remote Sens. Environ.* **2011**, *115*, 180–186. [CrossRef]
36. Wang, W.; Dong, Q.; Shang, S.; Wu, J.; Lee, Z. An evaluation of two semi-analytical ocean color algorithms for waters of the South China Sea. *J. Trop. Oceanogr.* **2009**, *28*, 35–42.
37. Li, S.; Song, K.; Mu, G.; Zhao, Y.; Ma, J.; Ren, J. Evaluation of the Quasi-Analytical Algorithm (QAA) for Estimating Total Absorption Coefficient of Turbid Inland Waters in Northeast China. *IEEE J. Sel. Top. Appl. Earth Obs. Remote Sens.* **2016**, *9*, 4022–4036. [CrossRef]
38. Yang, W.; Matsushita, B.; Chen, J.; Yoshimura, K.; Fukushima, T. Retrieval of Inherent Optical Properties for Turbid Inland Waters From Remote-Sensing Reflectance. *IEEE Trans. Geosci. Remote Sens.* **2013**, *51*, 3761–3773. [CrossRef]
39. Yu, X.; Shen, F.; Liu, Y. Light absorption properties of CDOM in the Changjiang (Yangtze) estuarine and coastal waters: An alternative approach for DOC estimation. *Estuar. Coast. Shelf Sci.* **2016**, *181*, 302–311. [CrossRef]
40. Shen, F.; Zhou, Y.; Hong, G. Absorption Property of Non-algal Particles and Contribution to Total Light Absorption in Optically Complex Waters, a Case Study in Yangtze Estuary and Adjacent Coast. In *Advances in Computational Environment Science*; Springer: Berlin/Heidelberg, Germany, 2012; pp. 61–66.
41. Shen, F.; Salama, M.S.; Zhou, Y.; Li, J.; Su, Z.; Kuang, D. Remote-sensing reflectance characteristics of highly turbid estuarine waters—A comparative experiment of the Yangtze River and the Yellow River. *Int. J. Remote Sens.* **2010**, *31*, 2639–2654. [CrossRef]
42. Yuan, R.; Wu, H.; Zhu, J.; Li, L. The response time of the Changjiang plume to river discharge in summer. *J. Mar. Syst.* **2016**, *154*, 82–92. [CrossRef]
43. Wu, H.; Shen, J.; Zhu, J.; Zhang, J.; Li, L. Characteristics of the Changjiang plume and its extension along the Jiangsu Coast. *Cont. Shelf Res.* **2014**, *76*, 108–123. [CrossRef]
44. Chang, P.H.; Isobe, A. A numerical study on the Changjiang diluted water in the Yellow and East China Seas. *J. Geophys. Res. Oceans* **2003**, *108*. [CrossRef]
45. Zhu, J.; Ding, P.; Hu, D. Observation of the diluted water and plume front off the changjiang river estuary during august 2000. *Oceanol. Limnol. Sin.* **2003**, *34*, 249–255.
46. Mobley, C.D. Estimation of the remote-sensing reflectance from above-surface measurements. *Appl. Opt.* **1999**, *38*, 7442–7455. [CrossRef] [PubMed]
47. Sokoletsky, L.G.; Shen, F. Optical closure for remote-sensing reflectance based on accurate radiative transfer approximations: The case of the Changjiang (Yangtze) River Estuary and its adjacent coastal area, China. *Int. J. Remote Sens.* **2014**, *35*, 4193–4224. [CrossRef]
48. Moore, C.; Barnard, A.; Hankins, D. *Scattering Meter (BB9) User's Guide, Revision A*; WET Labs Inc.: Philomath, OR, USA, 2005; pp. 2–13.
49. Pegau, W.S.; Gray, D.; Zaneveld, J.R.V. Absorption and attenuation of visible and near-infrared light in water: Dependence on temperature and salinity. *Appl. Opt.* **1997**, *36*, 6035–6046. [CrossRef] [PubMed]
50. Choi, J.K.; Park, Y.J.; Ahn, J.H.; Lim, H.S.; Eom, J.; Ryu, J.H. GOCI, the world's first geostationary ocean color observation satellite, for the monitoring of temporal variability in coastal water turbidity. *J. Geophys. Res. Oceans* **2012**, *117*. [CrossRef]
51. Pan, Y.; Shen, F.; Verhoef, W. An improved spectral optimization algorithm for atmospheric correction over turbid coastal waters: A case study from the Changjiang (Yangtze) estuary and the adjacent coast. *Remote Sens. Environ.* **2017**, *191*, 197–214. [CrossRef]
52. Gordon, H.R.; Brown, O.B.; Evans, R.H.; Brown, J.W.; Smith, R.C.; Baker, K.S.; Clark, D.K. A semianalytic radiance model of ocean color. *J. Geophys. Res. Atmos.* **1988**, *93*, 10909–10924. [CrossRef]
53. Yang, X.; Sokoletsky, L.; Wei, X.; Shen, F. Suspended sediment concentration mapping based on the MODIS satellite imagery in the East China inland, estuarine, and coastal waters. *Chin. J. Oceanol. Limnol.* **2017**, *35*, 39–60. [CrossRef]

54. Højerslev, N.K. Analytic remote-sensing optical algorithms requiring simple and practical field parameter inputs. *Appl. Opt.* **2001**, *40*, 4870–4874. [CrossRef] [PubMed]

55. Aas, E. Estimates of radiance reflected towards the zenith at the surface of the sea. *Ocean Sci.* **2010**, *6*, 861. [CrossRef]

56. Kowalczuk, P.; Olszewski, J.; Darecki, M.; Kaczmarek, S. Empirical relationships between coloured dissolved organic matter (CDOM) absorption and apparent optical properties in Baltic Sea waters. *Int. J. Remote Sens.* **2005**, *26*, 345–370. [CrossRef]

57. Sokoletsky, L.; Fang, S.; Yang, X.; Wei, X. Evaluation of empirical and semianalytical spectral reflectance models for surface suspended sediment concentration in the highly variable estuarine and coastal waters of East China. *IEEE J. Sel. Top. Appl. Earth Obs. Remote Sens.* **2016**, *9*, 5182–5192. [CrossRef]

58. Doxaran, D.; Froidefond, J.M.; Castaing, P. A reflectance band ratio used to estimate suspended matter concentrations in sediment-dominated coastal waters. *Int. J. Remote Sens.* **2002**, *23*, 5079–5085. [CrossRef]

59. Nechad, B.; Ruddick, K.G.; Park, Y. Calibration and validation of a generic multisensor algorithm for mapping of total suspended matter in turbid waters. *Remote Sens. Environ.* **2010**, *114*, 854–866. [CrossRef]

60. Lee, Z.; Ahn, Y.; Mobley, C.; Arnone, R. Removal of surface-reflected light for the measurement of remote-sensing reflectance from an above-surface platform. *Opt. Express* **2010**, *18*, 26313–26324. [CrossRef] [PubMed]

61. Sokoletsky, L.G.; Lunetta, R.S.; Wetz, M.S.; Paerl, H.W. Assessment of the water quality components in turbid estuarine waters based on radiative transfer approximations. *Isr. J. Plant Sci.* **2012**, *60*, 209–229. [CrossRef]

62. Lee, Z.; Carder, K.L.; Du, K. Effects of molecular and particle scatterings on the model parameter for remote-sensing reflectance. *Appl. Opt.* **2004**, *43*, 4957–4964. [CrossRef] [PubMed]

63. Lee, Z.; Arnone, R.; Hu, C.; Werdell, P.J.; Lubac, B. Uncertainties of optical parameters and their propagations in an analytical ocean color inversion algorithm. *Appl. Opt.* **2010**, *49*, 369–381. [CrossRef] [PubMed]

64. Gordon, H.R.; Morel, A.Y. *Remote Assessment of Ocean Color for Interpretation of Satellite Visible Imagery: A Review*; Springer Science & Business Media: New York, NY, USA, 1983; p. 114.

65. Aurin, D.A.; Dierssen, H.M. Advantages and limitations of ocean color remote sensing in CDOM-dominated, mineral-rich coastal and estuarine waters. *Remote Sens. Environ.* **2012**, *125*, 181–197. [CrossRef]

66. Qin, Y.; Brando, V.E.; Dekker, A.G.; Patissier, D.B. Validity of SeaDAS water constituents retrieval algorithms in Australian tropical coastal waters. *Geophys. Res. Lett.* **2007**, *34*. [CrossRef]

67. Shanmugam, P.; Ahn, Y.; Ryu, J.; Sundarabalan, B. An evaluation of inversion models for retrieval of inherent optical properties from ocean color in coastal and open sea waters around Korea. *J. Oceanogr.* **2010**, *66*, 815–830. [CrossRef]

68. Zheng, G.; Stramski, D.; Reynolds, R.A. Evaluation of the Quasi-Analytical Algorithm for estimating the inherent optical properties of seawater from ocean color: Comparison of Arctic and lower-latitude waters. *Remote Sens. Environ.* **2014**, *155*, 194–209. [CrossRef]

69. Zhu, W.; Yu, Q.; Tian, Y.Q.; Becker, B.L.; Zheng, T.; Carrick, H.J. An assessment of remote sensing algorithms for colored dissolved organic matter in complex freshwater environments. *Remote Sens. Environ.* **2014**, *140*, 766–778. [CrossRef]

Article

The VIS/NIR Land and Snow BRDF Atlas for RTTOV: Comparison between MODIS MCD43C1 C5 and C6

Jérôme Vidot [1,*], Pascal Brunel [1], Marie Dumont [2], Carlo Carmagnola [2] and James Hocking [3]

1   Météo-France, CMS, F-22300 Lannion, France; pascal.brunel@meteo.fr
2   Météo-France—CNRS, CNRM, UMR 3589, CEN, F-38400 Saint Martin d'Hères, France;
    marie.dumont@meteo.fr (M.D.); carlo.carmagnola@meteo.fr (C.C.)
3   MetOffice, Exeter EX1 3PB, UK; james.hocking@metoffice.gov.uk
*   Correspondence: jerome.vidot@meteo.fr; Tel.: +33-2-96-05-67-66

Received: 27 November 2017; Accepted: 21 December 2017; Published: 23 December 2017

Abstract: A monthly mean land and snow Bidirectional Reflectance Distribution Function (BRDF) atlas for visible and near infrared parts of the spectrum has been developed for Radiative Transfer for Television Infrared Observation Satellite (TIROS) Operational Vertical sounder (TOVS) (RTTOV). The atlas follows the methodology of the RTTOV University of Wisconsin infrared land surface emissivity (UWIREMIS) atlas, i.e., it combines satellite retrievals and a principal component analysis on a dataset of hyper-spectral surface hemispherical reflectance or albedo. The current version of the BRDF atlas is based on the Collection 5 of the Moderate Resolution Imaging (MODIS) MCD43C1 Climate Modeling Grid BRDF kernel-driven model parameters product. The MCD43C1 product combines both Terra and Aqua satellites over a 16-day period of acquisition and is provided globally at 0.05° of spatial resolution. We have improved the RTTOV land surface BRDF atlas by using the last Collection 6 of MODIS product MCD43C1. We firstly found that the MODIS C6 product improved the quality index of the BRDF model as compared with that of C5. When compared with clear-sky top of atmosphere (TOA) reflectance of Spinning Enhanced Visible and InfraRed Imagers (SEVIRI) solar channels over snow-free land surfaces, we showed that the reflectances are simulated with an absolute accuracy of 3% to 5% (i.e., 0.03–0.05 in reflectance units) when either the satellite zenith angle or the solar zenith angle is below 70°, regardless of the MODIS collection. For snow-covered surfaces, we showed that the comparison with in situ snow spectral albedo is improved with C6 with an underestimation of 0.05 in the near infrared.

Keywords: BRDF; RTTOV; MODIS; MCD43C1

---

1. Introduction

The radiative transfer model RTTOV (Radiative Transfer for Television Infrared Observation Satellite (TIROS) Operational Vertical sounder (TOVS) [1,2]) is the fast radiative transfer model (RTM) developed for the assimilation of satellite top of atmosphere (TOA) radiances in many Numerical Weather Prediction (NWP) models. RTTOV is improved and maintained in the frame of the NWP-SAF (Satellite Application Facility) project from the European Organization for the Exploitation of Meteorological Satellites (EUMETSAT). Currently, only TOA radiances in the infrared and the microwave parts of the spectrum are assimilated in NWP models from various polar orbiting and geostationary satellites [3,4]. In version 11 [5], we extended the capability of the RTTOV model to simulate TOA radiances or reflectances for visible to near infrared (VIS/NIR) channels such as the ones of the Moderate Resolution Imaging Spectroradiometer (MODIS) onboard Aqua and Terra, the Visible Infrared Imaging Radiometer Suite (VIIRS) onboard the Suomi-National Polar-orbiting (NPP) and the Joint Polar Satellite System (JPSS) satellites, and all Spinning Enhanced Visible and InfraRed Imagers (SEVIRIs) onboard the series of Meteosat Second Generation (MSG) satellites. If VIS/NIR

TOA radiances are not currently assimilated in NWP models, many applications can benefit from such capabilities. A very promising application of fast VIS/NIR RTM simulations is already developed for weather forecasters. Forecasters are experts at using and interpreting satellite imagery. Having a way of directly comparing cloud and moisture from the model with observations is a valuable tool for verifying the NWP model in near real time. Simulations of VIS/NIR channels offer new insight for cases where thermal infrared or microwave parts of the spectrum are limited, such as for low clouds, fog, or aerosols.

In order to simulate VIS/NIR satellite observations at all locations, we developed a dedicated land surface Bidirectional Reflectance Distribution Function (BRDF) model for RTTOV in order to provide the surface reflectance in both solar and viewing directions [6]. The land surface BRDF model of RTTOV was inspired from the work of Seemann et al. [7] and is similar to the University of Wisconsin infrared land surface emissivity UWIREMIS model [8] implemented in RTTOV. It combines MODIS land surface properties retrievals in the VIS/NIR with a principal component analysis (PCA) over many vegetation and soil reflectances spectra in order to provide a full spectral, temporal, and spatial description of surface reflectance. The model is currently working from 0.4 to 2.5 µm with a spectral resolution of 0.1 nm and provides a global monthly mean land surface BRDF atlas at a spatial resolution of 0.1° for the year 2007. The model also provides a quality index and surface types (land or snow) of the BRDF based on the MODIS flags. The MODIS land surface properties retrieval comes from the operational and global MODIS 16-days BRDF kernel-driven product MCD43C1 Collection 5 (C5). In Vidot and Borbás [6], the comparison of the RTTOV BRDF atlas against the SEVIRI surface black-sky albedo products EUMETSAT Land Satellite Application Facility on Land Surface Analysis (Land-SAF) showed a good spatial and temporal consistency of the RTTOV BRDF atlas when applied on the three SEVIRI VIS/NIR channels. It was found that the RTTOV narrowband black-sky albedo was retrieved with an absolute accuracy of ± 0.01 at 0.6 µm (channel 1) and 1.6 µm (channel 2). It was also found that the RTTOV narrowband black-sky albedo at 0.8 µm (channel 3) was overestimated between 0.01 and 0.03. It was also found over a period of 9 months that the temporal variation of the RTTOV broadband black-sky albedo is consistent with the EUMETSAT Land-SAF SEVIRI products but overestimated by somewhere between 0.01 and 0.02 when considering the best quality index of the RTTOV BRDF atlas. Finally, less agreement is seen in two particular cases: (i) for extreme geometrical conditions when the satellite zenith angle (SZA) > 65° and (ii) for lower quality indices of the RTTOV BRDF atlas. A similar methodology has been used by Zoogman et al. [9] to extend the hyperspectral BRDF model in the ultraviolet (UV). This model was applied to multispectral UV + VIS ozone retrieval on the Global Ozone Monitoring Experiment-2 (GOME-2) instrument with significant improvement in fitting residuals over vegetated scenes.

In this work, we present an improvement of the RTTOV land and snow-covered surface BRDF model by using the last collection 6 (C6) of the MODIS land product. The first part of the paper describes the methodology of the BRDF model and provides the main differences between the two MODIS MCD43C1 collections. The second part is devoted to the difference of the RTTOV BRDF model quality index and surface types mask. The third part of the paper presents the difference of the two collections on TOA simulated reflectance compared to clear-sky SEVIRI observations. The fourth part presents the methodology we used to extend the snow-free land surface BRDF model to snow-covered surface and a dedicated validation exercise with in situ snow spectral albedo measurements. The fifth part gives the conclusion and future developments of this work.

2. Materials and Methods

The RTTOV monthly means land BRDF atlas is currently based on the MODIS C5 MCD43C1 product (see https://lpdaac.usgs.gov/dataset_discovery/modis/modis_products_table/mcd43c1). The MCD43C1 is a Level 3 BRDF kernel-driven product from the operational MODIS land surface products [10]. It combines Terra and Aqua satellites over a 16-day period of acquisition and is provided globally at spatial resolution of 0.05°. This product was derived from the original 1-km spatial resolution of MODIS data and was developed for the global modeling community [11]. The MODIS BRDF is modeled by using the semi empirical linear model of Ross-Li [12] that is given by:

$$R(\theta_{sat}, \theta_{sol}, \Delta\varphi, \lambda) = f_{iso}(\lambda) + f_{vol}(\lambda)K_{vol}(\theta_{sat}, \theta_{sol}, \Delta\varphi) + f_{geo}(\lambda)K_{geo}(\theta_{sat}, \theta_{sol}, \Delta\varphi) \tag{1}$$

where θ_{sat}, θ_{sol}, and $\Delta\varphi$ are the satellite zenith angle, the solar zenith angle, and the azimuth difference between satellite and solar directions, respectively; λ is the wavelength. f_{iso} is due to isotropic scattering, f_{vol} is due to volumetric scattering as from horizontally homogeneous leaf canopies, and f_{geo} is due to geometric-optical surface scattering as from scenes containing 3-D objects that cast shadows and are mutually obscured from view at off-nadir angles. The formulation of the BRDF model kernels K_{vol} and K_{geo} can be found in Lucht et al. [12].

The MCD43C1 product contains the three retrieved BRDF model parameters f_{iso}, f_{vol}, and f_{geo} for each MODIS VIS/NIR channel and three associated pieces of information on quality and inputs (a retrieval quality flag, the percentage of inputs from the original 1-km spatial resolution of MODIS between 0 and 100%, and the percentage of snow coverage between 0 and 100%). The MODIS products are globally able to capture the seasonal cycle of snow-free surface albedo with a typical accuracy better than 0.05 but are also found to be less accurate for higher solar zenith angles [13]. The study of Wang and Zender [14] found that the accuracy over snow deteriorates for SZA > 55°, but Schaaf and Strahler [15] contradicted this finding by arguing an inappropriate use of the extensive quality flags and showed that the MODIS product performs quite well out to the recommended limit for product use of 70° SZA.

We have improved the RTTOV BRDF atlas by using the last collection 6 (C6) of the MODIS MCD43C1 products [16]. The main differences of the two versions are the following. Firstly, the MCD43C1 products are generated daily instead of every 8 days, but still using the 16-day retrieval algorithm. Secondly, all available observations are used at high latitudes in order to improve the number and the quality of the retrieval. Lastly, the current day snow status is used instead of the majority snow/no-snow status from the 16-day period. The generation of the new RTTOV BRDF atlas follows the one that is described in Vidot and Borbás [6] and will be described here. For all original MODIS MCD43C1 data within a month, we extracted the information for a 2 × 2 pixel box (from an original pixel resolution at 0.05° to the final grid at 0.1°). The retrieved BRDF model parameters are averaged over the 4 pixels multiplied by the number of days within each month. As compared with C5, we then have many more pixels to average.

3. Results

3.1. Mask and Quality Index Improvement

Additionally, to the monthly means BRDF atlas, a mask is provided for each of the BRDF datum providing information on the surface type (water, snow-free land with three levels of retrieval quality, snow with two levels of retrieval quality, and no data). On the panel of Figure 1 are shown global maps of the BRDF mask between the one derived from C5 and the one derived from C6 in July 2007 (Figure 1a,b) and in December 2007 (Figure 1c,d). The mask is based on the quality flag of the MODIS MCD43C1 product using an iterative process and is fully described in Vidot and Borbás [6]. There are seven flags in the mask which are defined as:

- Water surfaces: Based on the land/water mask from the UWIREMIS atlas.
- Good quality retrieval: No snow, best, and good MCD43C1 quality for 80% inputs or more.
- Medium quality retrieval: No snow, medium MCD43C1 quality for 80% inputs or more.
- Low quality retrieval: Remaining no snow pixels.
- Snow surfaces: Full snow.
- Bad quality retrieval: Remaining pixels containing snow.
- No data: Remaining pixel from land/water mask with no BRDF retrieval.

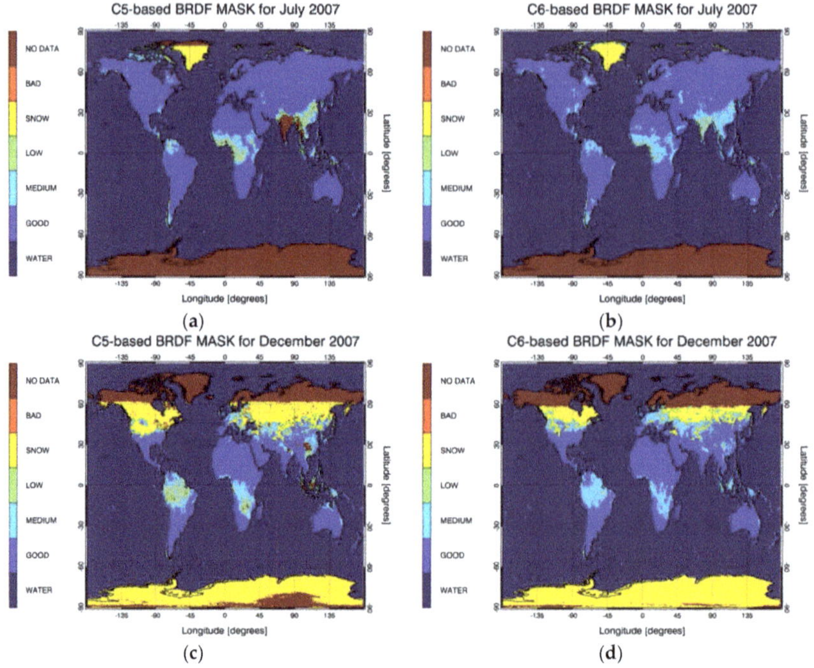

Figure 1. RTTOV Bidirectional Reflectance Distribution Function (BRDF) mask and quality index:
(**a**) For July 2007 based on C5; (**b**) For July 2007 based on C6; (**c**) For December 2007 based on C5;
(**d**) For December 2007 based on C6.

For the two months represented in Figure 1, the number of "Bad" retrieval quality and "No data"
flags (in orange and red) over India (Figure 1a), China, or Indonesia (Figure 1c) is greatly reduced with
C6 as well as the number of "Low" retrieval quality flags (in green over the Tropics). The "Bad" and
"Low" retrieval quality flags from C5 are mainly due to the low retrieval performance of the MODIS
product in areas with strong aerosol and residual cloud contamination. By using the higher frequency
of C6 products, which are provided on a daily basis instead of an 8-day basis, we obtained more cases
with higher MCD43C1 quality leading to a better RTTOV BRDF quality index. The effect of using all
available observations at high latitudes in C6 compared to C5 can be also seen over Greenland where
the number of snow pixels has increased (Figure 1a,b). However, as we wanted to have as much as
possible snow-free surface information in the BRDF model, we put more weight on snow-free retrieval.
The consequence of using the daily basis of C6 is that we have more probability to get snow-free
information within each month. This leads to a reduction of RTTOV BRDF snow flag as seen over high
northern latitudes (Figure 1c,d).

Overall, we compared each flag for three latitudinal areas in both hemispheres for the year
2007. For that, we separated latitudes in polar (latitudes above or below 60°), mid-latitudes (latitudes
between 23.5° and 60° or between −23.5° and −60°), and tropical (latitudes between ±23.5°). Figure 2
shows the temporal evolution of the percentage of each flag compared to the total land pixels for the
three areas (rows) and for the two hemispheres (columns). For almost all areas (except Polar Southern
Hemisphere (SH)), the percentage of "Low" retrieval quality flags has decreased (in green dashed
lines for C5 and green full lines for C6) and the percentage of "Medium" retrieval quality flags has
increased (in cyan dashed lines for C5 and cyan full lines for C6).

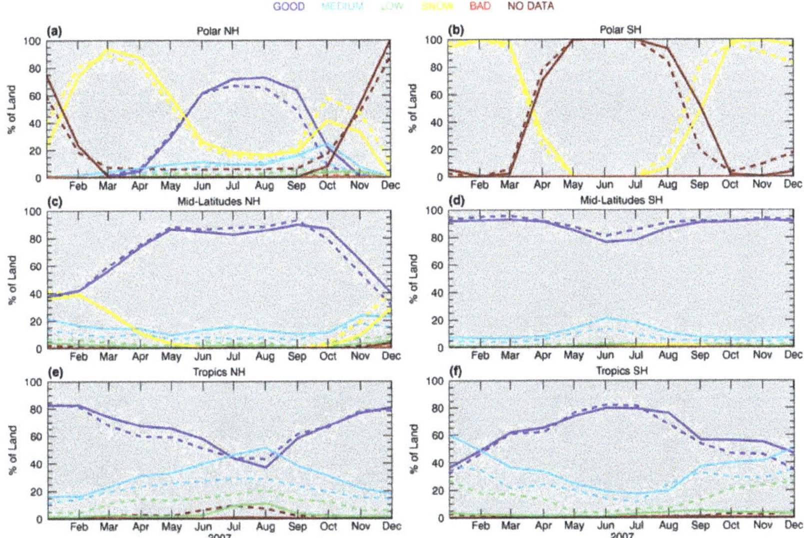

Figure 2. Temporal evolution of the percentage of each flag in 2007. The dashed lines represent C5 and the full lines the C6: (**a**) For Polar regions of the Northern Hemisphere; (**b**) For Polar regions of the Southern Hemisphere; (**c**) For Mid-Latitudes of the Northern Hemisphere; (**d**) For Mid-Latitudes of the Southern Hemisphere; (**e**) For the Tropics of the Northern Hemisphere; (**f**) For the Tropics of the Southern Hemisphere. Colors are similar to those of Figure 1.

For Tropical regions, the percentage of "No data" has been reduced to almost zero (represented in full red lines in Figure 2e,f). The percentages of "Good" retrieval quality flags or "Snow" flags have changed slightly between C5 and C6, but we can conclude that the percentage of "Good" quality retrieval from C6 (full blue lines) is around 70% over Polar Northern Hemisphere (NH) during summer time (Figure 2a). It is between 40% and 85% in Mid-Latitudes NH (Figure 2c) and Tropics (Figure 2e,f) depends on the season and between 80% and 90% in Mid-Latitudes SH (Figure 2d).

3.2. Comparison for SEVIRI TOA Reflectances

The evaluation of the monthly BRDF atlas has been realized by comparing SEVIRI clear-sky top of atmosphere (TOA) observations to simulations on 3 July 2017 in the three visible and near infrared SEVIRI channels (channel 1 centered at 0.6 µm, channel 2 centered at 0.8 µm, and channel 3 centered at 1.6 µm). To simulate clear-sky TOA reflectances, we used the atmospheric profiles (pressure, temperature, and water vapor) and surface pressure from the Global UK MetOffice NWP model [17] and RTTOV version 12. In order to check the ability of representing the BRDF for different geometrical situations, we compared the observed and simulated TOA reflectances every hour from 8:00 UTC to 18:00 UTC. This temporal window was chosen to ensure observations during daytime with various solar illuminations including large zenith angles. It is worth mentioning here that aerosols were not included in the simulations. On the panel of Figure 3 is shown the clear-sky SEVIRI TOA reflectance at 12:00 UTC (left column) and simulated TOA reflectance with C5 (middle column) and with C6 (left column) for the three channels (channels 1 to 3 from top to bottom rows, respectively).

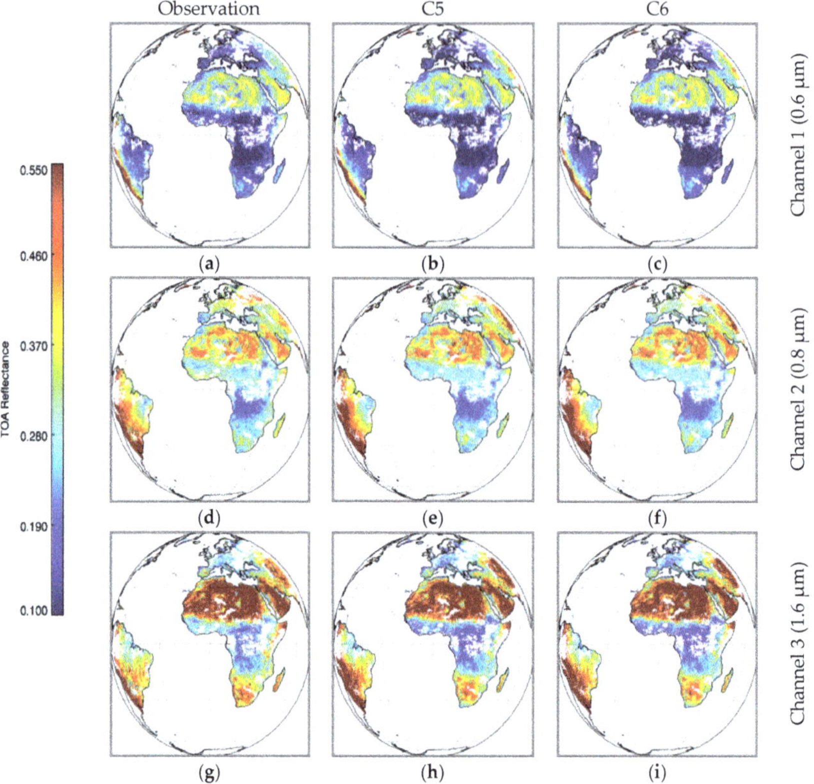

Figure 3. Observed and simulated top of atmosphere (TOA) clear-sky reflectance in three Spinning Enhanced Visible and InfraRed Imagers (SEVIRI) solar channels: (**a–c**) Observed, simulated with C5, and simulated with C6 for channel 1 at 0.6 microns, respectively; (**d-f**) Observed, simulated with C5, and simulated with C6 for channel 2 at 0.8 microns, respectively; (**g–i**) Observed, simulated with C5, and simulated with C6 for channel 2 at 1.6 microns.

The panel of Figure 3 shows that the BRDF model of RTTOV is well suited to represent the spatial distribution of TOA reflectances. However, it can be also noticed that the TOA reflectances are often overestimated at the edge of the disk. The panel of Figure 4 shows the angular effect in the performance of the RTTOV BRDF model to simulate SEVIRI as well as the difference between the two collections. The effect of the viewing zenith angle (VZA), the solar zenith angle (SZA), and the scattering angle (SCA) is shown on Figure 4a–i, respectively. For each angle, the top panel of Figure 4 represents the mean bias of the TOA reflectance observed minus the simulated in function of the angle whereas the bottom panel of Figure 4 represents the standard deviation of the differences between observations and simulations in function of the angle. For VZA and SZA, the data were combined in 5° bins between 0 and 90° whereas for SCA the data were combined in 10° bins between 0 and 180°. The colors refer to the channels whereas the line styles refer to the MODIS collections (plus for C5 and dots for C6). For SEVIRI, there are no observations over land for VZA between 0 and 5°, and the scattering angle range were found to be between 70° and 170°.The results show that the mean bias is less than 0.03–0.05 and the standard deviation is less than 0.05 for most of situations. However, the results show also that the mean bias strongly decreases at higher VZA (above 70°) leading to an overestimation from RTTOV (Figure 4a). The effect is more important in the near infrared (in green) than in the visible (in blue). The standard deviation also greatly increases for VZA above 70° (Figure 4d) and can reach

0.5 for the channel 3. The standard deviation also increases for SZA between 60 and 70° and for low or high SCA. Additionally, we can see that there is a slight underestimation of RTTOV at low VZA that seems to be at its worst with C6 (Figure 4a). This effect can be also seen in Figure 3h,i when compared to Figure 3g for Central Africa. The values of remaining biases and standard deviations might be explained by different factors. The first one comes from the fact that we did not include aerosols in our simulations. Aerosols may have a strong impact on the simulations depending on the aerosol loading. The positive bias at medium angle may also be explained by the Lambertian approximation that is used in the MODIS retrieval and has been shown to underestimate the retrieved surface reflectance [18]. Others errors coming from atmospheric correction in MODIS retrieval or errors from RTTOV simulations are potential explanations. The overestimation of RTTOV as compared with observations at high VZA might be explained by the inaccuracy of the Ross-Li model [12] at these high angles or the inadequacy of such angles in the MODIS retrieval process. Finally, we can conclude that there are no strong differences between C5 and C6, and both collections provide accurate BRDF description for VZA below 70° with TOA reflectance bias below 3%. In comparison, the SEVIRI operational calibration was found to be stable during the years 2004 to 2009, with an offset of −8% at 0.6 μm, −6% at 0.8 μm, and +3.5% at 1.6 μm compared to MODIS [19].

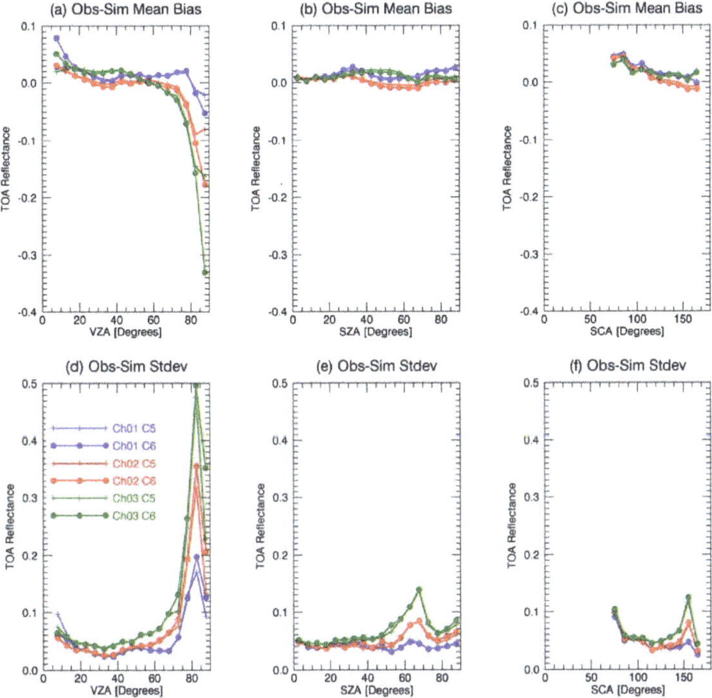

Figure 4. (**a**) Observations minus simulations mean bias of the TOA reflectances as a function of the viewing zenith angle (VZA); (**b**) Same as (**a**) as a function of solar zenith angle (SZA); (**c**) Same as (**a**) as a function of the scattering angle (SCA); (**d–f**) Standard deviation of the observations minus simulations TOA reflectances as a function of VZA, SZA, and SCA, respectively. The different colored lines and styles refer to the Moderate Resolution Imaging Spectroradiometer (MODIS) collection and SEVIRI channels as explained in the legend.

3.3. Comparison with In Situ Snow Spectral Albedo

Since RTTOV version 11.2, we extended the BRDF model to snow surfaces following the methodology developed for the land surfaces. For that we applied the principal component analysis on snow spectral albedo (defined as directional-hemispherical reflectance for direct beam at a certain solar zenith angle or also named black-sky albedo). To cover the whole possible range of snow reflectance variations, we simulated snow spectral albedo with different snow optical properties and for different solar zenith angles. The snow spectral albedo is influenced by many factors, such as the refractive index of ice [20], the snow microphysical properties, and the presence of light absorbing impurities within the snow pack [21]. To simulate the snow spectral albedo, we used the DIScrete Ordinate Radiative Transfer (DISORT) model [22]. The DISORT model was applied to a plane-parallel multi-layer snowpack where the single scattering properties of snow were computed with the Mie theory. The spherical assumption is acceptable to model only the bi-hemispherical reflectance of snow [23]. We refer to the work of Carmagnola et al. [24] (parts 2.2.1) for the full description of the simulation of the snow spectral albedo. In our simulations we used the snow Specific Surface Area (SSA) to characterize the optically relevant size of the snow grain. The snow SSA is impacted by the snow metamorphism. A high value of SSA is often related to fresh snow whereas a small value can be related to melt-frozen snow. Additionally, the presence of impurities also impacts the snow spectral albedo. The DISORT simulations have been successfully compared with in situ measurements [24]. The panel of Figure 5 shows different simulated snow spectral albedos between 0.4 and 2.5 μm for SZA = 0° and SZA = 60° (Figure 5a,b, respectively). The different green to purple lines show the effect of SSA on the simulated snow spectral albedo, whereas the orange to red lines show the effect of Black Carbon Content (BCC). All simulated spectra show typical features of snow albedo (high value in the VIS that decrease with wavelength with peaks near 1.1, 1.3, 1.85, and 2.25 which correspond to the minima of the absorption coefficient of pure snow [21]).

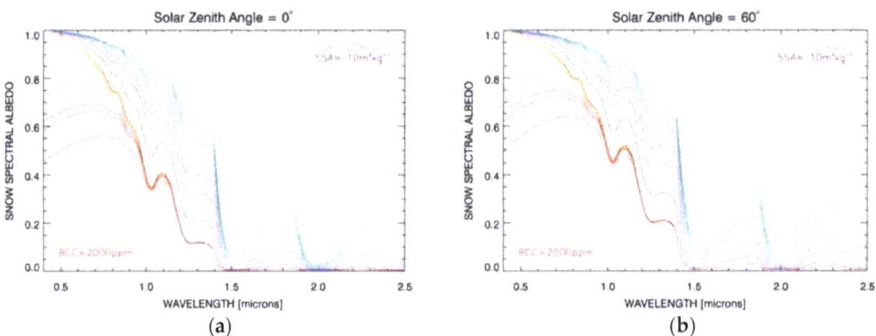

Figure 5. Snow spectral albedo versus wavelength simulated for different values of the snow Specific Surface Area (SSA) (green to purple lines) and for different black carbon content (orange to red lines) for SZA = 0° (**a**) and for SZA = 60° (**b**).

The Figure 5a,b shows also that when the snow microstructure is fine, i.e., for high SSA values (in green), the albedo is high, and when the snow microstructure is coarser, i.e., when the SSA decreases, the snow spectral albedo decreases in the near infrared part. In the visible part, when the content of black carbon increases, the snow spectral albedo decreases. In addition, higher SZA lead to higher albedo. The full database of snow spectral albedos was simulated using SZA between 0 and 85° by steps of 5°, the snow SSA between 5 and 100 $m^2 \cdot kg^{-1}$ by steps of 5 $m^2 \cdot kg^{-1}$, and BCC of 0.1, 0.5, 1, 5, 10, to 50 (by steps of 10), 100 to 1000 (by step of 100), 1000, and 2000 ppm. This database was used in the same way as for land surface to reconstruct hyperspectral BRDF spectra by constraining/fitting the satellite–derived BRDF retrievals at specific channels with the principal components of this representative set of snow spectral albedo.

To validate the snow BRDF RTTOV model, we used in situ snow spectral albedo measurements made at Summit Camp in Greenland (72°36′N, 38°25′W, 3210 m a.s.l.) during two months in 2011 [22]. However, there may be a directional dependence of snow BRDF that is not properly evaluated by the albedo comparison as the snow reflectance is not fully isotropic. A complete validation would require snow BRDF measurements instead of albedos. The instrument is the ASD FieldSpec Pro spectroradiometer with a spectral range of 0.35 to 2.2 μm and a spectral resolution ranging from 3 nm in the UV to 12 nm in the NIR. The measurements accuracy is below 0.3% of the albedo over the full spectrum except between 1.8 and 1.9 μm due to low signal/noise ratio. The measurements were all acquired around 11:00 LT, checked for quality and cloud screening, and corrected from measurements artifacts that are fully described in Carmagnola et al. [24]. We used 19 spectra (8 in May 2011 and 11 in June 2011) with SZA ranging between 49.1° and 58.8° (see Table 1). Figure 6 shows the measured spectra used in this study (in blue for May and in red for June). The variability seen in the measured spectra can be mostly explained by the change in snow properties since the SZA did not change much during the measurements. The interpretation of this variability is beyond the scope of this study and is described in Carmagnola et al. [24].

Table 1. Date and solar zenith angles of the in situ measurements.

Date	SZA
16 May 2011	53.2°
17 May 2011	58.8°
19 May 2011	52.6°
22 May 2011	52.1°
25 May 2011	51.5°
28 May 2011	51.0°
30 May 2011	50.7°
31 May 2011	50.6°
1 June 2011	50.4°
2 June 2011	50.3°
3 June 2011	50.2°
7 June 2011	50.6°
10 June 2011	49.5°
11 June 2011	49.4°
15 June 2011	50.5°
20 June 2011	50.4°
22 June 2011	54.5°
24 June 2011	49.8°
25 June 2011	49.1°

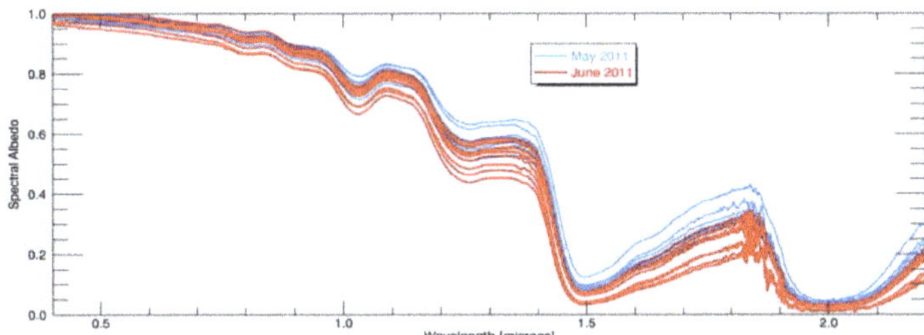

Figure 6. Snow spectral albedo measurement by the ASD spectroradiometer in May (blue lines) and in June (red lines).

Since the RTTOV BRDF model is provided on a monthly basis, we calculated the mean snow spectral albedo for the two months of the in situ measurements. The snow spectral albedo is calculated using Equation (4) of Vidot and Borbás [6] (i.e., as black-sky albedo). It is worth mentioning that in situ snow spectral albedo contains both black-sky albedo and white-sky albedo (bihemispherical reflectance under isotropic radiation) weighted by the proportion of diffuse irradiance [13], which is the function of the aerosol optical depth under clear-sky conditions. In this comparison, we considered that the diffuse component is negligible as compared to the direct component. This is supported by the fact that typical aerosol optical depth is low at Summit (0.05 at 500 nm) [25] and that the snow spectral albedo is similar to the black-sky albedo for moderate SZA (see Figure 1 of Liu et al. [13]). The top panel of Figure 7a displays the mean snow spectral albedo from in situ measurements made in May 2011 (in blue), and the blue shaded area between black lines represents the standard deviation. The red curve represents the snow spectral albedo calculated from the RTTOV BRDF model with C5 whereas the green line represents the snow spectral albedo calculated from the RTTOV BRDF model with C6. The bottom panel of Figure 7a represents the difference between the mean spectral albedo from measurements and the monthly mean calculated snow spectral albedo. The Figure 7b represents the same results for June 2011. MODIS C5 and C6 simulate both the spectral dependence of snow but underestimate the albedo compared to in situ measurements. However, the C6-based BRDF model clearly improves the simulations everywhere in the spectrum as compared those of with C5. This result has been also shown by Wright et al. [25] for MODIS channels. Almost no biases are found in the visible part of the spectrum with C6, whereas it was 0.05 for May and 0.1 for June with C5. In the near infrared part, an underestimation of the albedo still remains in May of a maximum of 0.05.

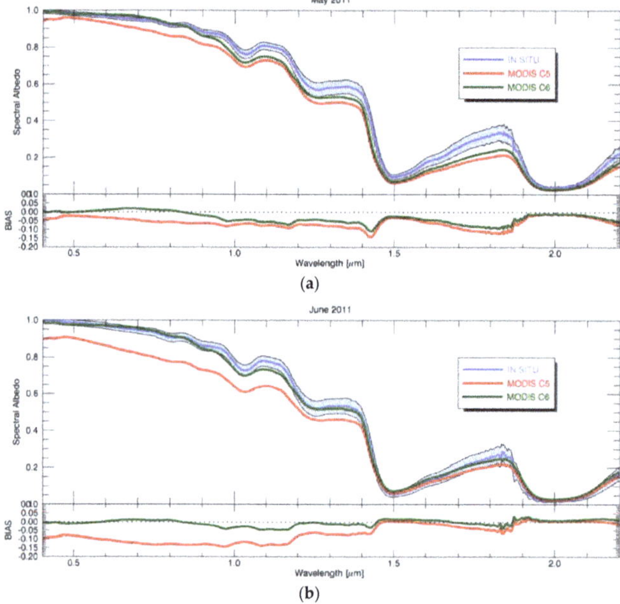

Figure 7. (**a**) Snow spectral albedo (top) and bias (bottom) between then mean spectral albedo measured by the ASD spectroradiometer in May (blue line), and black lines represents the +/− standard deviation. The spectral albedo from C5 and C6 are represented in red and green, respectively; (**b**) Same as (**a**) for June.

4. Conclusions

An improvement of the land and snow surface BRDF model for RTTOV has been studied here. It is based on the last collection C6 of the MODIS MCD43C1 product. The evaluation of this new BRDF model has been proposed for snow-free land surfaces and snow-covered surfaces with two types of observations. For snow-free land surfaces, we compared SEVIRI TOA reflectances at 0.6, 0.8, and 1.6 µm with simulations made using RTTOV version 12 and the Global UK Met-Office NWP model. We complemented our evaluation by using the temporal frequency of observations every hour during daytime to estimate the performance of the BRDF model as a function of the viewing and solar zenith angles as well as the scattering angles. For snow-covered surfaces, we used in situ snow spectral albedo measurements in the VIS/NIR over the Summit Camp in Greenland during two months to check the spectral consistency between the simulations and the observations.

Since the C6 MODIS product is provided on a daily basis (as compared to the 8-days of C5), we firstly showed that the quality index of the RTTOV land and snow surface BRDF model is improved leading to a strong reduction of the lower BRDF retrieval quality index. We showed secondly that both C5 and C6 performed almost equally in simulating SEVIRI clear-sky observations. We concluded that the land surface BRDF model is accurate when VZA and SZA are below 70° with an estimated error below 3% to 5% in TOA reflectance in the three VIS/NIR channels. Thirdly, we showed that the MODIS C6 product improved the simulations for snow surfaces, especially in the visible part of the spectrum where for the two months studied here almost no bias was found. In the near infrared part of the spectrum, an underestimation of the simulated snow spectral albedo was found with C6 but this underestimation was lower than that of the simulations with C5.

In the future, we plan to extend the database to other years since only the year 2007 is implemented in RTTOV. The MODIS C6 product is now provided for 2000 and 2017. We also plan in the framework of the SAF-NWP project to extend the capability of RTTOV into the ultraviolet. This objective is motivated by the use of the high spectral resolution of the RTTOV BRDF model to simulate UV/VIS/NIR hyperspectral spectrometers that will fly onboard the Sentinel-5 and the Meteosat Third Generation (MTG) satellites.

Acknowledgments: This work was funded by the EUMETSAT NWP-SAF project. The MCD43C1 product was retrieved from the online data pool, courtesy of the NASA EOSDIS Land Processes Distributed Active Archive Center (LP DAAC), USGS/Earth Resources Observation and Science (EROS) Center, Sioux Falls, South Dakota. We thank anonymous reviewers for making suggestion to improve the manuscript.

Author Contributions: Jerome Vidot conceived, designed, and performed the experiments; Jerome Vidot and Pascal Brunel analyzed the results; James Hocking, Marie Dumont, and Carlo Carmagnola provided the data for validation; Jerome Vidot wrote the paper.

Conflicts of Interest: The authors declare no conflict of interest.

References

1. Saunders, R.; Matricardi, M.; Brunel, P. An improved fast radiative transfer model for assimilation of satellite radiance observations. *Q. J. R. Meteorol. Soc.* **1999**, *125*, 1407–1425. [CrossRef]

2. Matricardi, M.; Chevallier, F.; Kelly, G.; Thépaut, J.-N. An improved general fast radiative transfer model for the assimilation of radiance observations. *Q. J. R. Meteorol. Soc.* **2004**, *130*, 153–173. [CrossRef]

3. Errico, R.M.; Ohring, G.; Weng, F.; Bauer, P.; Ferrier, B.; Mahfouf, J.; Turk, J. Assimilation of Satellite Cloud and Precipitation Observations in Numerical Weather Prediction Models: Introduction to the JAS Special Collection. *J. Atmos. Sci.* **2007**, *64*, 3737–3741. [CrossRef]

4. Dee, D.P.; Uppala, S.M.; Simmons, A.J.; Berrisford, P.; Poli, P.; Kobayashi, S.; Andrae, U.; Balmaseda, M.A.; Balsamo, G.; Bauer, P.; et al. The ERA-Interim reanalysis: Configuration and performance of the data assimilation system. *Q. J. R. Meteorol. Soc.* **2011**, *137*, 553–597. [CrossRef]

5. Saunders, R.; Hocking, J.; Rundle, D.; Rayer, P.; Matricardi, M.; Geer, A.; Lupu, C.; Brunel, P.; Vidot, J. RTTOV-11 Science and Validation Report. NWP-SAF Report. 2013. Available online: https://www.nwpsaf. eu/site/download/documentation/rtm/docs_rttov11/rttov11_svr.pdf (accessed on 17 November 2017).

6. Vidot, J.; Borbás, É. Land surface VIS/NIR BRDF atlas for RTTOV-11: Model and validation against SEVIRI land SAF albedo product. *Q. J. R. Meteorol. Soc.* **2014**, *140*, 2186–2196. [CrossRef]

7. Seemann, S.W.; Borbas, E.A.; Knuteson, R.O.; Stephenson, G.R.; Huang, H.-L. Development of a Global Infrared Land Surface Emissivity Database for Application to Clear Sky Sounding Retrievals from Multispectral Satellite Radiance Measurements. *J. Appl. Meteorol. Climatol.* **2008**, *47*, 108–123. [CrossRef]

8. Borbas, E.A.; Ruston, B.C. *The RTTOV UWiremis IR Land Surface Emissivity Module*; NWP-SAF Report No NWPSAF-MO-VS-042; EUMETSAT: Darmstadt, Germany, 2010; p. 24.

9. Zoogman, P.; Liu, X.; Chance, K.; Sun, Q.; Schaaf, C.B.; Mahr, T.; Wagner, T. A climatology of visible surface reflectance spectra. *J. Quant. Spectrosc. Radiat. Transf.* **2016**, *80*, 39–46. [CrossRef]

10. Schaaf, C.B.; Gao, F.; Strahler, A.H.; Lucht, W.; Li, X.; Tsang, T.; Strugnell, N.C.; Zhang, X.; Jin, Y.; Muller, J.-P.; et al. First operational BRDF, albedo nadir reflectance products from MODIS. *Remote Sens. Environ.* **2002**, *83*, 135–148. [CrossRef]

11. Gao, F.; Schaaf, C.B.; Strahler, A.H.; Roesch, A.; Lucht, W.; Dickinson, R. MODIS bidirectional reflectance distribution function and albedo Climate Modeling Grid products and the variability of albedo for major global vegetation types. *J. Geophys. Res.* **2005**, *110*. [CrossRef]

12. Lucht, W.; Schaaf, C.B.; Strahler, A.H. An Algorithm for the retrieval of albedo from space using semiempirical BRDF models. *IEEE Trans. Geosci. Remote Sens.* **2000**, *38*, 977–998. [CrossRef]

13. Liu, J.; Schaaf, C.B.; Strahler, A.H.; Jiao, Z.; Shuai, Y.; Zhang, Q.; Roman, M.; Augustine, J.A.; Dutton, E.G. Validation of Moderate Resolution Imaging Spectroradiometer (MODIS) albedo retrieval algorithm: Dependence of albedo on solar zenith angle. *J. Geophys. Res.* **2009**, *114*. [CrossRef]

14. Wang, Z.; Zender, C.S. MODIS snow albedo bias at high solar zenith angles relative to theory and to in situ observations in Greenland. *Remote Sens. Environ.* **2010**, *114*, 563–575. [CrossRef]

15. Schaaf, C.B.; Strahler, A.H. Commentary on Wang and Zender—MODIS snow albedo bias at high solar zenith angles relative to theory and to in situ observations in Greenland. *Remote Sens. Environ.* **2011**, *115*, 1296–1300. [CrossRef]

16. Schaaf, C.B.; Wang, Z. MCD43C1 MODIS Terra + Aqua BRDF Albedo Model Parameters Daily L3 Global 0.05Deg CMG V006; NASA EOSDIS Land Processes DAAC. 2015. Available online: https://e4ftl01.cr.usgs.gov/MOTA/ (accessed on 17 November 2017).

17. Walters, D.; Boutle, I.; Brooks, M.; Melvin, T.; Stratton, R.; Vosper, S.; Wells, H.; Williams, K.; Wood, N.; Allen, T.; et al. The Met Office Unified Model Global Atmosphere 6.0/6.1 and JULES Global Land 6.0/6.1 configurations. *Geosci. Model Dev.* **2017**, *10*, 1487–1520. [CrossRef]

18. Wang, Y.; Lyapustin, A.I.; Privette, J.L.; Cook, R.B.; SanthanaVannan, S.K.; Vermote, E.F.; Schaaf, C.L. Assessment of biases in MODIS surface reflectance due to Lambertian approximation. *Remote Sens. Environ.* **2010**, *114*, 2791–2801. [CrossRef]

19. Meirink, J.F.; Roebeling, R.A.; Stammes, P. Inter-calibration of polar imager solar channels using SEVIRI. *Atmos. Meas. Tech.* **2013**, *6*, 2495–2508. [CrossRef]

20. Warren, S.G.; Wiscombe, W.J. A Model for the Spectral Albedo of Snow. II: Snow Containing Atmospheric Aerosols. *J. Atmos. Sci.* **1980**, *37*, 2734–2745. [CrossRef]

21. Wiscombe, W.J.; Warren, S.G. A Model for the Spectral Albedo of Snow. I: Pure Snow. *J. Atmos. Sci.* **1980**, *37*, 2712–2733. [CrossRef]

22. Stamnes, K.; Tsay, S.; Wiscombe, W.; Jayaweera, K. Numerically stable algorithm for discrete-ordinate-method radiative transfer in multiple scattering and emitting layered media. *Appl. Opt.* **1988**, *27*, 2502–2509. [CrossRef] [PubMed]

23. Grenfell, T.C.; Warren, S.G. Representation of a nonspherical ice particle by a collection of independent spheres for scattering and absorption of radiation. *J. Geophys. Res.* **1999**, *104*, 31697–31709. [CrossRef]

24. Carmagnola, C.M.; Dominé, F.; Dumont, M.; Wright, P.; Strellis, B.; Bergin, M.; Dibb, J.; Picard, G.; Libois, Q.; Arnaud, L.; et al. Snow spectral albedo at Summit, Greenland: Measurements and numerical simulations based on physical and chemical properties of the snowpack. *Cryosphere* **2013**, *7*, 1139–1160. [CrossRef]

25. Wright, P.; Bergin, M.; Dibb, J.; Lefer, B.; Domine, F.; Carman, T.; Carmagnola, C.; Dumont, M.; Courville, Z.; Schaaf, C.; et al. Comparing MODIS daily snow albedo to spectral albedo field measurements in Central Greenland. *Remote Sens. Environ.* **2014**, *140*, 118–129. [CrossRef]

 *remote sensing*

Article

Estimation of Penetration Depth from Soil Effective Temperature in Microwave Radiometry

Shaoning Lv [1,2,*], **Yijian Zeng** [1], **Jun Wen** [3], **Hong Zhao** [1] and **Zhongbo Su** [1]

1   Department of Water Resources, Faculty of Geo-Information Science and Earth Observation (ITC), University of Twente, P.O. Box 217, 7500 AE Enschede, The Netherlands; y.zeng@utwente.nl (Y.Z.); h.zhao@utwente.nl (H.Z.); z.su@utwente.nl (Z.S.)
2   Key Laboratory of Land Surface Process and Climate Change in Cold and Arid Region, Northwest Institute of Eco-Environment and Resources, Chinese Academy of Sciences, Lanzhou 730000, Gansu, China
3   College of Atmospheric Sciences, the Plateau Atmosphere and Environment Key Laboratory of Sichuan Province, Chengdu University of Information Technology, Chengdu 610225, China; jwen@cuit.edu.cn
*   Correspondence: s.lv@utwente.nl

Received: 27 February 2018; Accepted: 21 March 2018; Published: 26 March 2018

Abstract: Soil moisture is an essential variable in Earth surface modeling. Two dedicated satellite missions, the Soil Moisture and Ocean Salinity (SMOS) and the Soil Moisture Active Passive (SMAP), are currently in operation to map the global distribution of soil moisture. However, at the longer L-band wavelength of these satellites, the emitting behavior of the land becomes very complex due to the unknown deeper penetration depth. This complexity leads to more uncertainty in calibration and validation of satellite soil moisture product and their applications. In the framework of zeroth-order incoherent microwave radiative transfer model, the soil effective temperature is the only component that contains depth information and thus provides the necessary link to quantify the penetration depth. By means of the multi-layer soil effective temperature (Lv's T_{eff}) scheme, we have determined the relationship between the penetration depth and soil effective temperature and verified it against field observations at the Maqu Network. The key findings are that the penetration depth can be estimated according to Lv's T_{eff} scheme with the assumption of linear soil temperature gradient along the optical depth; and conversely, the soil temperature at the penetration depth should be equal to the soil effective temperature with the same linear assumption. The accuracy of this inference depends on to what extent the assumption of linear soil temperature gradient is satisfied. The result of this study is expected to advance understanding of the soil moisture products retrieved by SMOS and SMAP and improve the techniques in data assimilation and climate research.

Keywords: microwave remote sensing; soil moisture; Maqu network; penetration depth; soil effective temperature

---

1. Introduction

Soil moisture is a key variable in weather forecast and climate research because it plays a role in both the energy and water cycles [1–4]. It controls how much water returns to the atmosphere via land-atmosphere interactions and it also carries the energy in terms of the latent heat flux when evaporated to reshape the atmospheric circulation. Therefore, availability of accurate and near real-time global soil moisture is critical for the improvement of weather forecast and climate projection skills [5–7].

Since the 1970s, satellite remote sensing has been used to estimate global soil moisture with microwave frequencies and more recently focus has been on L-band (1400–1427 MHz), which is sensitive to the dielectric constant as well as is a protected radio astronomy band with minimum radio frequency interference (RFI). Both the Soil Moisture and Ocean Salinity (SMOS) [8] and the

Soil Moisture Active Passive (SMAP) [9] missions operate at L-band for providing the brightness temperature and soil moisture data products. With the efforts from SMOS and SMAP missions, abundant data have been produced and applied in various studies [10–14].

However L-band radiometry for monitoring soil moisture is strongly affected by the soil temperature and soil moisture [15,16], which usually leads to questions on where the satellite is exactly sensing [17–19]. Corresponding to the satellite missions, plenty of in situ soil moisture monitoring networks have been established to calibrate and validate (Cal/Val) L-band brightness temperature (T_b) or soil moisture data [20]. Usually, the soil moisture and soil temperature sensors are installed at certain depths (e.g., 2.5 cm, 5 cm, 10 cm or deeper) based on experiences or to match numerical simulations of soil moisture and soil temperature. However, such in situ defined depths do not precisely match the satellite sensing depths, at which the T_b or soil moisture data are retrieved. Hence errors may arise because the Cal/Val data are not correspondingly sampled, in other words, are not comparable to the satellite observations. Additionally, different satellite soil moisture products may have different sensing depths, as different frequencies are used. As such, the various satellite soil moisture products may lack consistency and generate ambiguity in Cal/Val and their applications [21].

In regard to the dielectric constant and the soil effective temperature, the SMOS/SMAP sensors may measure soil moisture deeper over the dry soil than over the wet soil. Even for the same region, the sensing depths may vary in a certain range, depending on the soil moisture and soil temperature profiles [19,22]. For that reason, different satellite soil moisture products can be only made inter-comparable, after defining exactly the sensing depths. Furthermore, data assimilation approach has been deemed as the most feasible method for estimating the soil moisture profile [22,23], by extrapolating the remotely sensed surface information to lower depths in the soil via a coupled heat and moisture flow model. It is, therefore, critical to understand which depth is sensed by a satellite sensor for a sound retrieval and use of soil moisture and soil temperature profile information. The exact sensing depth strongly affects the accuracy of the soil moisture and soil temperature profile retrieval, as the soil moisture and soil temperature near the land surface has strong gradients and can be varying dramatically. One way to infer soil moisture sensing depth is by correlating brightness temperature and in-site soil moisture time-series so that the soil moisture layer that corresponds best with brightness temperature is considered to be the sensing layer [24,25]. To get soil moisture at different depth, this method always requires precise and vertically dense soil profile measurement or simulation. Another way is to use models to compute the sensing depth according to its definition [26,27] or an empirical model [26,28]. However, these methods need either vertically dense profile information or relay on prior knowledge, which is hard to acquire in practice. Besides soil, it should be noted that vegetation also has an impact on the penetration depth. The attenuation by vegetation is mainly due to the vegetation water content. Usually, the penetration depth only focuses on soil because comparatively, the influence of atmospheric attenuation at L-band is almost negligible.

As the penetration depth (henceforth we use penetration depth synonymously with sensing depth and emission depth) is defined by energy attenuation, it is possible to infer it from T_{eff}. In general, all current two-layer T_{eff} schemes use a weighting function for the soil temperature between upper layer and deeper layer. Such weighting function can be a constant [29], a fitting function [30,31], or an exponential function [15,16]. The weighting function is supposed to reflect the impact of soil moisture on the soil effective temperature. However, there is no variable of depth contained in Choudhury's [29], Wigneron's [30] or Holmes' [31] T_{eff} schemes. As indicated by the integral scheme, the weighting function would be more representative if it considers the influence of both soil moisture and soil temperature. However, it is difficult to quantify its effect on soil effective temperature, because soil temperature also affects soil moisture (e.g., as in dielectric constant models). In other words, T_{eff} is a weighted mean of the soil temperature along the vertical profile. Therefore, it must be $T_{\min} < T_{\mathit{eff}} < T_{\max}$ (if $T_{\min} \neq T_{\max}$, e.g., non-uniform profile which is always the case for a land surface subject to radiative heating and cooling). Considering the diurnal variation and a semi unbounded soil column, $T_{\max}$ and $T_{\min}$ usually appear at the surface skin or the deep layer where the soil temperature is almost constant. When the above condition is satisfied the sample layer covers the variation of T_{eff}. This also means as the soil temperature profile is continuous, there must be a layer where its soil temperature equals to T_{eff}.

In order to investigate the relationship between the penetration depth and T_{eff}, in terms of T_{eff} calculation, we first review the hypotheses of the coherent/incoherent microwave radiative transfer models and the definition of satellite sensing depth (Section 2). We then analytically quantify the relationship between the penetration depth and T_{eff} (Section 3). Next, we use the field soil moisture and soil temperature observation at Maqu Network to verify the developed approach and demonstrate its application to SMAP (Section 4). Finally, we discuss the uncertainties of the developed method (Section 5) and conclude with some final thoughts for future work (Section 6).

2. Theoretical Background

In this section, we first review the default assumption in zeroth-order incoherent model in which there is only one emissivity for all layers and then reformulate the Lv's T_{eff} scheme in terms of the optical depth τ and clarify the definition of penetration depth.

2.1. Microwave Radiative Transfer Model

The SMOS and SMAP soil moisture retrieval algorithms are based on the following equation

$$T_B = \varepsilon T_{eff} \tag{1}$$

where T_B is the brightness temperature detected by the radiometer, ε is the unique emissivity in the zeroth-order incoherent microwave radiative transfer model (e.g., based on Fresnel reflectivity equation) and T_{eff} is the soil effective temperature which utilizes the net radiative energy affected by the soil moisture/temperature gradient in the profile [32]. Equation (1) implies that ε does not contain depth information. Meanwhile T_{eff} is expressed in terms of soil physical temperature of different layers, usually of two layers as

$$T_{eff} = w_1 T_1 + w_2 T_2 \tag{2}$$

where T_1 represents the soil temperature at 0–5 cm and T_2 is for 40–80 cm or even deeper depending on the soil texture. The w_1 and w_2 are weighting function which are mainly affected by soil moisture, wavelength and slightly by soil temperature [16,33] (Note: All specific variables in this study are listed in Table 1) The sum of weighting function should satisfy:

$$w_1 + w_2 = 1 \tag{3}$$

In the above equations, the unique emissivity is a variable to simplify the coherent microwave radiative transfer model, assuming that the dielectric and temperature properties of the soil are uniform throughout the emitting layer.

Combining Equations (2) and (3) leads to following expressions:

$$\begin{cases} 1 = w_1 + w_2 \\ T_{eff} = w_1 T_1 + w_2 T_2 \end{cases} \tag{4}$$

As we know, $T_{eff} \in (T_1, T_2)$ if $T_1 < T_2$. The opposite case is also possible as $T_{eff} \in (T_2, T_1)$ if $T_1 > T_2$. For a special case, when $T_1 = T_{eff}$ (or $T_2 = T_{eff}$), the only possible solution is $w_1 = 1$ (or $w_2 = 1$). $w_1 = 1$ is the necessary and sufficient condition for $T_1 = T_{eff}$.

In the following, we will prove that the soil temperature at one time of the optical depth equals to T_{eff} with linear soil temperature gradient assumption. The accuracy of this inference depends on whether the linear assumption is satisfied which is basically the case if more layers are observed (e.g., via the use of Lv's multilayer T_{eff} scheme).

Table 1. Variables used in this study.

Abbreviation	Definition	Unit	Expression
T_{eff}	soil effective temperature	K	Equations (5), (6), (12) and (13)
T_b	brightness temperature	K	
θ	soil moisture	Vol/Vol	
T_{max}	maximum soil temperature along soil temperature profile	K	
T_{min}	minimum soil temperature along soil temperature profile	K	
T_i	soil temperature at *i*th layer	K	
w_i	weighting function for T_{eff}	-	Defined in [29,31,34]
Δx_i	soil thickness at *i*th layer	m	
$x_{(i)}$	soil depth (at *i*th layer)	m	$x_i = \sum_{j=1}^{i} \Delta x_j$
$\Delta\tau_i$	optical thickness at *i*th layer	m	$\Delta\tau_i = \Delta x_i \frac{2\pi}{\lambda} \frac{\varepsilon''}{\sqrt{\varepsilon'}} = \Delta x_i \cdot \alpha(x)$
$\tau_{i(x)}$	optical depth at *i*th layer (or corresponding to soil depth *x*)	m	$\tau_i = \sum_{j=1}^{i} \Delta\tau_j$
T_{nor}/T_{inor}	normalized soil temperature (at *i*th layer)	-	$T_{(i)nor} = \frac{T_{(i)} - T_{surf}}{T_{deep} - T_{surf}}$
T_{surf}	skin temperature	K	
T_{deep}	soil temperature at deep layer that the soil temperature could be considered as constant	K	
a	Soil temperature gradient	K/τ	$a = dT/d\tau$
$\alpha(x)$	attenuation parameter	-	$\alpha(x) = \frac{4\pi}{\lambda}\varepsilon''(x)/2[\varepsilon'(x)]^{\frac{1}{2}}$
τ_{deep}	τ deep enough that the soil temperature could be considered as constant	-	$\tau_{deep} \approx 5$

2.2. Soil Effective Temperature

The concept of soil effective temperature T_{eff} is developed to describe the emissive capacity of a soil column. According to the Rayleigh-Jeans approximation, in the microwave domain the emitted energy from the soil is proportional to the thermodynamic temperature [35] as shown in Equation (1), where the ε is the emissivity that is strongly related to soil moisture, while T_{eff} is the effective temperature and is formulated by [36] as:

$$T_{eff} = \int_0^{\infty} T(x)\alpha(x)\exp\left[-\int_0^x \alpha(x')dx'\right]dx \tag{5}$$

where $\alpha(x) = \frac{4\pi}{\lambda}\varepsilon''(x)/2[\varepsilon'(x)]^{\frac{1}{2}}$. Equation (5) states that T_{eff} at the soil surface is a superposition of the intensities emitted at various depths within the soil.

An accurate computation of T_{eff} is thus critical for obtaining relevant values of soil emissivity from brightness temperature measurements. It follows that soil moisture can be retrieved from the estimate of soil emissivity [35]. However, the soil moisture and soil temperature profile information is usually limited in a field experiment, because discrete observation sensors are usually installed empirically at limited vertical intervals. Recently, a new scheme (hereafter, Lv's scheme, [16,33]) has been derived directly from Equation (5) as

$$T_{eff} = T_1\left(1 - e^{-\Delta\tau_1}\right) + \sum_{i=2}^{n-1} T_i\left(1 - e^{-\Delta\tau_i}\right)\prod_{j=1}^{i-1} e^{-\Delta\tau_j} + T_n\prod_{j=1}^{n-1} e^{-\Delta\tau_j} \tag{6}$$

in which $\Delta\tau_i = \Delta x_i \frac{4\pi}{\lambda}\frac{\varepsilon''}{2\sqrt{\varepsilon'}}$, a parameter related to wavelength λ, to soil moisture through the dielectric constant (ε'—real part, ε''—imaginary part), and to sampling depth Δx_i for each layer. Comparing to other two-layer schemes, Lv's scheme uses an exponential function to distribute the weight among different layers. In Equation (6), T_i depicts the mean soil temperature of the ith layer, with $1 - e^{-\tau_i}$ indicating its weight in the calculation of T_{eff}. Parameter τ_i is the key variable for Lv's scheme, and is a function of Δx_i. While λ is fixed for any specified sensor and the dielectric constant is varying with soil moisture and temperature, Δx_i is the only remaining variable which needs to be determined.

As stated in Lv et al. (2016b), Δx_1 could be determined by considering τ_1 as a function of Δx_1 and the integral exponential function $\int e^{-\tau} d\tau = 1 - e^{-\tau}$,

$$ e^{-\tau_{1s}} = \frac{1}{\tau_1} \int_0^{\tau_1} e^{-\tau} d\tau = \frac{1}{\tau_1} \left(1 - e^{-\tau_1} \right) \tag{7} $$

where τ_{1s} is calculated using the depth where the first layer sensors are installed (Δx_{1s}). With Equation (7), τ_1 can be determined as well as the Δx_1 used in Equation (6). The physical meaning of Δx_1 could be inferred from Equation (6) that T_1 matches the layer-averaged soil temperature integrated from the surface to the sampling depth Δx_1, which is used for calculating $1 - e^{-\tau_1}$. It is to note that Δx_1 (i.e., the bulk sampling layer thickness) is different from Δx_{1s} (i.e., the exact installation depth). Therefore, the soil moisture and soil temperature detected at Δx_{1s} represents average values from surface to Δx_1, so that Δx_{1s} will be called the representative depth for the first layer. The representative depth is computed from the known installation depth for soil moisture and soil temperature sensors and has no relation to the deeper layers below. Let $\tau_1 = \Delta x_1 \cdot \frac{4\pi}{\lambda} \cdot \frac{\varepsilon''}{2\sqrt{\varepsilon'}}$ (noting $\Delta \tau_i = \tau_i - \tau_{i-1}$, and $\tau_{i-1} = 0$ for the first layer). Since soil depth at ith layer can be expressed as $x_i = \sum_{j=1}^{i} \Delta x_j$, it follows $\tau_i = \sum_{j=1}^{i} \Delta \tau_j$. Hence, τ monotonically increases with soil column depth x. With $[\tau, T]$ instead of $[x, T]$ we can compute the correlation coefficient (cc) along the profile.

2.3. Penetration Depth

For the non-isothermal case, Njoku and Entekhabi (1996) defined the penetration depth (e.g., the temperature sensing depth, hereafter as PD1) as the depth, which satisfies the following condition:

$$ \frac{\int_0^{\Delta x_T} T(x)\alpha(x) \exp\left[-\int_0^x \alpha(x')dx'\right] dx}{\int_0^{\infty} T(x)\alpha(x) \exp\left[-\int_0^x \alpha(x')dx'\right] dx} = \frac{1}{e} \tag{8} $$

The equals to one time of the optical thickness (or optical depth, Napierian absorbance) with the linear assumption and is the natural logarithm of the ratio of incident to transmitted radiant intensity through soil at L-band. The optical thickness gives a measurement about the attenuation of radiation through the medium (e.g., soil in this study). According to Equation (8), the penetration depth will be affected by those factors influencing the soil effective temperature, including soil temperature, soil moisture, and wavelength λ [36]. For SMOS and SMAP missions, the wavelength is a given constant, $\lambda = 21$ cm. The impact of soil moisture and soil temperature on the soil effective temperature is functional through the dielectric models, while the soil moisture's influences dominate over the soil temperature one. When soil temperature is neglected and wavelength is fixed, a monotonic relationship between the soil moisture and the penetration depth ($\Delta x_T = f(\theta)$) could be founded. However, it is not clearly stated how the change in penetration depth is related to the in situ observation. Δx_T is a characteristic length, the value of which is somewhat arbitrary and could not be computed without knowing detailed soil moisture and soil temperature profiles [37].

Meanwhile, there is another definition of the penetration depth (PD2) [32]. The radiative transfer theory has shown that while the thickness of the emitting layer may actually exceed 1 m for low-frequency radiation, the magnitude of its contribution becomes infinitesimal after a comparatively shallow depth (e.g., compared to 1 m layer). This comparatively shallow depth, which provides most of the measurable energy contribution, is also called the penetration depth (PD2) [38]. Nevertheless, the PD2 is somewhat less quantitative since the magnitude of its contribution cannot be quantified. Hence, the penetration depth in this study refers to only PD1, e.g., Δx_T.

3. Method and Data

3.1. Predigest of Wilheit's T_{eff} Scheme

Equation (5) is Wilheit's is integral scheme and Equations (6) and (7) have been explained in previous studies ([16,33]). This subsection shows how Equation (5) is reformulated with the integral optical depth ($d\tau$), instead of depth (dx) with Lv's scheme. In the above sections we stated that at the soil surface there is a superposition of the intensities emitted at various depths within the soil. In Lv's scheme,

$$
\begin{cases}
\Delta\tau_i = \Delta x_i \frac{2\pi}{\lambda} \frac{\varepsilon''}{\sqrt{\varepsilon'}} = \Delta x_i \cdot \alpha(x) \\
\tau_i = \sum_{j=1}^{i} \Delta\tau_j \\
x_i = \sum_{j=1}^{i} \Delta x_j
\end{cases}
\tag{9}
$$

Apparently, the parameter τ is the same as the concept of optical depth. According to $\Delta\tau = \Delta x \cdot \frac{4\pi}{\lambda} \cdot \frac{\varepsilon''}{2\sqrt{\varepsilon'}}$ and $\tau = \Delta\tau_1 = 1$ (i.e., here τ equals to one time of the optical depth), the penetration depth Δx_s can be expressed as,

$$
\Delta x_s = \frac{\lambda}{2\pi} \cdot \frac{\sqrt{\varepsilon'}}{\varepsilon''}
\tag{10}
$$

Equation (10) is equivalent to the penetration depth as identified by previous studies [18,28,32] although it is inferred from Lv's scheme. It is worth noting that the concept of penetration depth does not only indicate the depth where radiation is reduced to $1/e$ of its original value, but also indicates the depth where the physical temperature represents the average temperature of that emitting layer. Considering the pre-mentioned assumption that the dielectric and temperature properties of the soil are uniform throughout the emitting layer, this means $\exp\left[-\int_0^x \alpha(x')dx'\right] = e^{-\tau}$ and $\alpha(x)dx = d\tau$. Therefore Equation (5) could be simplified as

$$
T_{eff} = \int_0^\infty T(x)e^{-\tau}d\tau
\tag{11}
$$

As such, Equation (11) can be rewritten by replacing physical depth $x \in [0, \infty)$ with the optical thickness $\tau \in [0, \infty)$ as follows:

$$
T_{eff} = \int_0^\infty T(\tau)e^{-\tau}d\tau
\tag{12}
$$

With Equation (12), we replace integral item dx with $d\tau$. τ is also used in the description of radiometry in atmosphere and vegetation and it should also work with soil column. With $d\tau$, T_{eff} becomes concise and convenient for following analysis.

3.2. Characteristic Expression of T_{eff}

The essence of T_{eff} calculation is a series of weighting values which reflect the soil temperature gradient (for example, the Choudhury's scheme) [29] and further the impact of soil moisture, for example, Wigneron's [34] and Holmes' scheme [31]. Keeping this in mind, we can simplify the soil temperature gradient by normalizing it as follows:

$$
T_{inor} = \frac{T_i - T_{surf}}{T_{deep} - T_{surf}}
\tag{13}
$$

where T_{surf} is the soil temperature at the soil surface. T_{deep} is the soil temperature at the soil bottom where soil temperature could be considered as constant at inter-annual scale. T_i is the physical soil temperature at ith layer. Besides, we already know that in Lv's scheme, Wilheit's integral scheme could be simplified as Equation (12), where τ_i is the optical depth at ith layer. One τ_i value corresponds to only one physical soil depth for a certain soil temperature/moisture combination at any moment ($\tau_i = \tau_{i-1} + \Delta x_i \frac{2\pi}{\lambda} \frac{\varepsilon''}{\sqrt{\varepsilon'}}$). As such, we can deem soil temperature a function of τ. Furthermore, since the soil depth is between $[0, +\infty)$ as is τ, we can use $1 - e^{-\tau}|_0^\infty$ to represent the variation between $[0, 1]$.

To indicate how the normalized soil temperature was applied to calculate T_{eff}, the simplest case (i.e., the linear case) was demonstrated as follows. The linear case could be expressed as

$$T_i = T_{surf} + a\tau_i \left(\tau < \tau_{deep}\right) \tag{14}$$

and for the layers where $\tau_i \geq \tau_{deep}$:

$$T_i = T_{surf} + a\tau_{deep} = T_{deep} \tag{15}$$

where a is the soil temperature gradient with optical depth (unit: K/τ). τ_{deep} is where it is deep enough that the soil temperature could be considered as constant. For example, when $\tau_{deep} = 5$, the contribution from deeper layer $\tau > 5$ is $e^{-5} \approx 0.0067$. Therefore, the soil temperature below τ_{deep} has negligible impact on T_{eff}. In other words, it does not matter where exactly τ_{deep} is as long as it is deep enough while Equation (14) is valid. We suggest $\tau_{deep} \geq 5$. According to Equation (12), we can calculate the normalized soil temperature:

$$\begin{aligned} T_{inor} \quad &= \frac{T_i - T_{surf}}{T_{deep} - T_{surf}} \\ &= \frac{T_{surf} + a\tau_i - T_{surf}}{T_{surf} + a\tau_{ideep} - T_{surf}} \\ &= \frac{\tau_i}{\tau_{deep}} \quad \left(\tau < \tau_{deep}\right) \end{aligned} \tag{16}$$

Then, $0 \leq \Delta T_{nor} \leq 1$. Put Equation (14) in Lv's scheme as

$$\begin{aligned} T_{eff} \quad &= \int_0^\infty T(\tau)e^{-\tau}d\tau \\ &= \int_0^{\tau_{deep}} \left(T_{surf} + a\tau\right)e^{-\tau}d\tau + \int_{\tau_{deep}}^\infty \left(T_{surf} + a\tau_{deep}\right)e^{-\tau}d\tau \\ &= \int_0^{\tau_{deep}} T_{surf}e^{-\tau}d\tau + \int_0^{\tau_{deep}} a\tau e^{-\tau}d\tau + \left(T_{surf} + a\tau_{deep}\right)\int_{\tau_{deep}}^\infty e^{-\tau}d\tau \\ &= T_{surf}\left(1 - e^{-\tau_{deep}}\right) + a\left[1 - e^{-\tau_{deep}} \cdot \left(\tau_{deep} + 1\right)\right] + \left(T_{surf} + a\tau_{deep}\right) \cdot e^{-\tau_{deep}} \\ &= T_{surf} + a \qquad \left(if\, e^{-\tau_{deep}} \approx 0\right) \end{aligned} \tag{17}$$

Hence, $T_{eff} = T_{surf} + a$ (i.e., considering $e^{-\tau_{deep}} \approx 0$). Comparing to Equation (14), this means T_{eff} equals to the soil temperature at $\tau = 1$, with the linear assumption of soil moisture and soil temperature profile. The same normalization as in Equation (16) with T_{eff} in Equation (12), we have:

$$\frac{T_{eff} - T_{surf}}{T_{deep} - T_{surf}} = \frac{T_{eff} - T_{surf}}{T_{surf} + a\tau_{deep} - T_{surf}} = \frac{T_{surf} + a - T_{surf}}{a\tau_{deep}} = \frac{1}{\tau_{deep}} \tag{18}$$

Using $\frac{T_i - T_{surf}}{T_{deep} - T_{surf}}$ to normalize Equation (12) on both sides, we get,

$$
\begin{aligned}
\frac{T_{eff} - T_{surf}}{T_{deep} - T_{surf}} &= \int_0^\infty \frac{T - T_{surf}}{T_{deep} - T_{surf}} e^{-\tau} d\tau \\
&= \int_0^\infty T_{nor} e^{-\tau} d\tau \\
&= \int_0^{\tau_x} \frac{\tau}{\tau_{deep}} e^{-\tau} d\tau \qquad when \, \tau_x \to \infty \\
&= \frac{1}{\tau_{deep}} \left[1 - e^{-\tau_x} \cdot (\tau_x + 1) \right] \qquad when \, \tau_x \to \infty \\
&= \frac{1}{\tau_{deep}}
\end{aligned}
\tag{19}
$$

where τ_x is a mark for τ at x. $1 - e^{-\tau_x} \cdot (\tau_x + 1)$ is the characteristic expression for T_{eff} calculation in linear case because it is not related to the gradient a and τ_{deep}. The term $1 - e^{-\tau_x} \cdot (\tau_x + 1)$ describes the distribution of radiation along τ. With Equation (14), Equation (19) is an analytic solution for Equations (5) and (12) after normalization. It reflects the cumulative energy starting from $1 - e^{-\tau_x} \cdot (\tau_x + 1)|_0 = 0$ to deep layer where $1 - e^{-\tau_x} \cdot (\tau_x + 1)|_{+\infty} = 1$. Since this distribution is not related to a and τ_{deep}, $1 - e^{-\tau_x} \cdot (\tau_x + 1)$ is universal for all linear cases. Therefore, $T_{eff} \sim \tau \sim x$ relationship (Equations (9) and (19)) is quantified by $1 - e^{-\tau_x} \cdot (\tau_x + 1)$ which is fundamental if we want to determine $T_b \sim T_{eff} \sim \tau \sim x$ (Equations (1), (9) and (19)) in future. It is to note that $0 < \tau_x < \tau_{deep}$. Equation (19) is based on the $\frac{dT}{d\tau}$ linear assumption. Here we use correlation coefficient (cc) between T profile and τ profile to measure this linear assumption. Usually, T profile and τ profile are hard to acquire from either field observation or reanalysis data because only a few layers of soil are measured or modelled.

After normalizing soil temperature in Equation (16) and Equation (12), it appears that the soil temperature at the penetration depth equals to T_{eff} while $dT/d\tau$ is linear.

3.3. In-Situ Data, MERRA-2 and SMAP

Maqu network is located in the northeast margin of the Tibetan Plateau (Figure 1). The average elevation is about 3300 m above the sea level. The network was built in 2008 and continuously provides soil moisture and soil temperature profile information at 20 sites since then [39]. Since its establishment, the Maqu Network has provided accurate soil moisture and soil temperature measurements for evaluating soil moisture data from satellites [40,41]. The vegetation of the Maqu network consists of meadow and grass less than 1 m in height with roots extending tens of centimeters in depth. An accumulated humus layer of around 10 cm is mixed with the soil. Bushes and trees are scarce in this region, while desert dunes appear along the river off and on. Besides these 20 sites of profile monitoring, there is a complete land-atmosphere interaction observation site which consists of a boundary layer meteorology tower, an eddy covariance system and two dense soil moisture and soil temperature profile measurements.

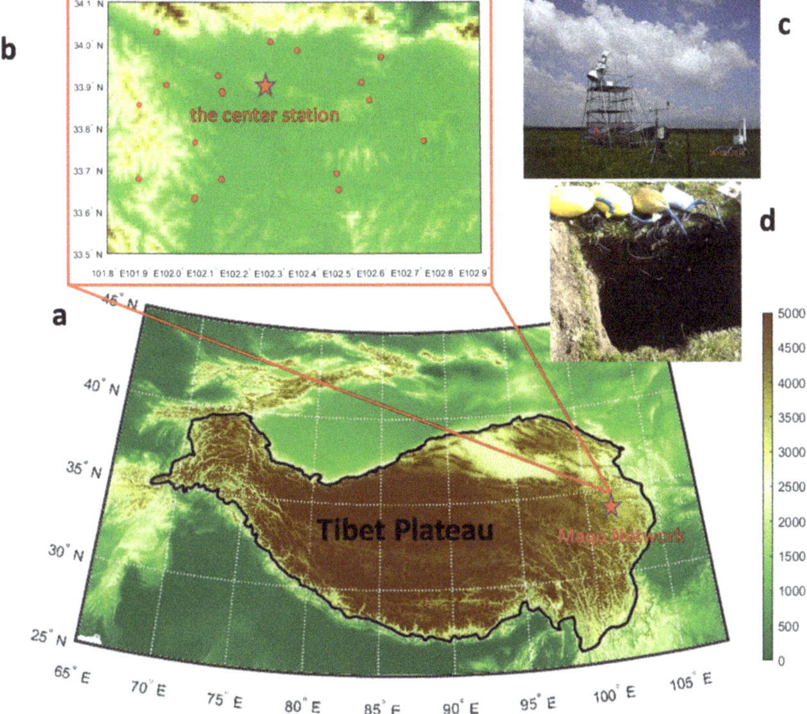

Figure 1. (**a**) Geographical location of the Maqu network on the Tibetan Plateau. The background indicates the elevation from USGS 1 km topography and the border in black is where elevation >2500 m; (**b**) The distribution of all sites at the Maqu network and the center site (ELBARA) located in the center; (**c**) ELABRA; (**d**) the detailed soil moisture and soil temperature profile.

A vertically dense soil moisture and soil temperature profile observation at the center station of Maqu network is used in this study (Figures 1d and 2). To facilitate the comparison with SMAP soil moisture products, the profile data used in this study is provided by ECH20 5TM soil moisture/soil temperature sensors, covering the period from 6 August to 27 November 2016 (Figure 2). It has 20 sensors installed at 19 layers, with two duplicate sensors at 2.5 cm. The depth configuration is illustrated in Figure 2d. Soil moisture data is calibrated with soil texture, bulk density and organic matter content. Furthermore, soil samples are collected near the micro-meteorological observing system, indicating that the soil consists of sand fraction of 26.95% and clay of 9.86% at 0.05 m, respectively, while 29.2% and 10.15% at 0.2 m, 31.6% and 10.43% at 0.4 m [42]. The layer settings for the other sites are 5 cm, 10 cm, 20 cm, 40 cm, and 80 cm or simply 5 cm and 10 cm with an additional infrared sensor for the skin temperature. With such a vertically dense profile observation, we can compute cc for $[\tau, T]$ profile at any moment. Mironov's dielectric constant model was for calculating the real and complex parts of dielectric constants in this study [43].

The in situ data above is used to compute the temporal variation of penetration depth. To extend the understanding of the relationship between soil effective temperature and penetration depth, we also use MERRA-2 (The Modern-Era Retrospective analysis for Research and Applications, Version 2) [44] and SMAP Level 3 radiometer global daily 36 km EASE-Grid soil moisture, Version 4 product in this study to illustrate a spatial distribution of penetration depth. MERRA-2 is supposed to replace former MERRA dataset with the advances made in the assimilation system that enable assimilation of modern hyperspectral radiance and microwave observations, along with GPS-Radio Occultation datasets. The spatial resolution used in this study is 0.625 × 0.5 degree. While soil properties are retrieved from MERRA-2 constant fields, only soil temperature, soil moisture and surface temperature at SMAP over passing time are collected. Since March 2015, SMAP is providing promising global soil moisture distribution with its passive radiometer every 2–3 days [45]. The spatial resolution for passive sensor is 36 km that is double the resolution compared with MERRA-2. In this case, the SMAP L3 product is downscaled with nearest neighbor interpolation method to rebuild the soil moisture map matching with MERRA-2. Since there is just one frequency operated by SMAP, the soil moisture detected should reflect the average emission capacity within a certain depth of soil column, but not a particularly fixed depth. This average soil moisture synthesizes the strong penetration due to longer wavelength at L-band as another virtual concept of T_{eff}. As noted in the introduction part, vegetation also affects the penetration depth. In this study, the in situ soil moisture and soil temperature data at Maqu Center Station were used to directly compute the T_{eff} and penetration depth within the soil column, sparing the need to consider the vegetation effect. For SMAP soil moisture product, the impact of vegetation is already removed during the retrieval procedure. It should also be noted that there are other sites at Maqu Network but none of them has vertically dense profile measurement. Without such measurement, the calculated penetration depth strongly depends on the fixed soil moisture layer. In contrast to the SMAP soil moisture product which is derived from average emissivity, the soil moisture at fixed depth is not the same as the average soil moisture. Therefore, it is not appropriate to compute penetration depth for the rest sites at Maqu network.

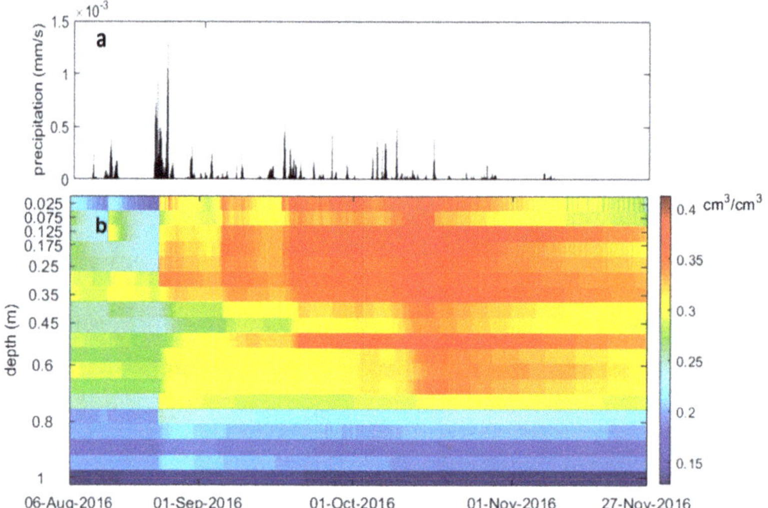

Figure 2. *Cont.*

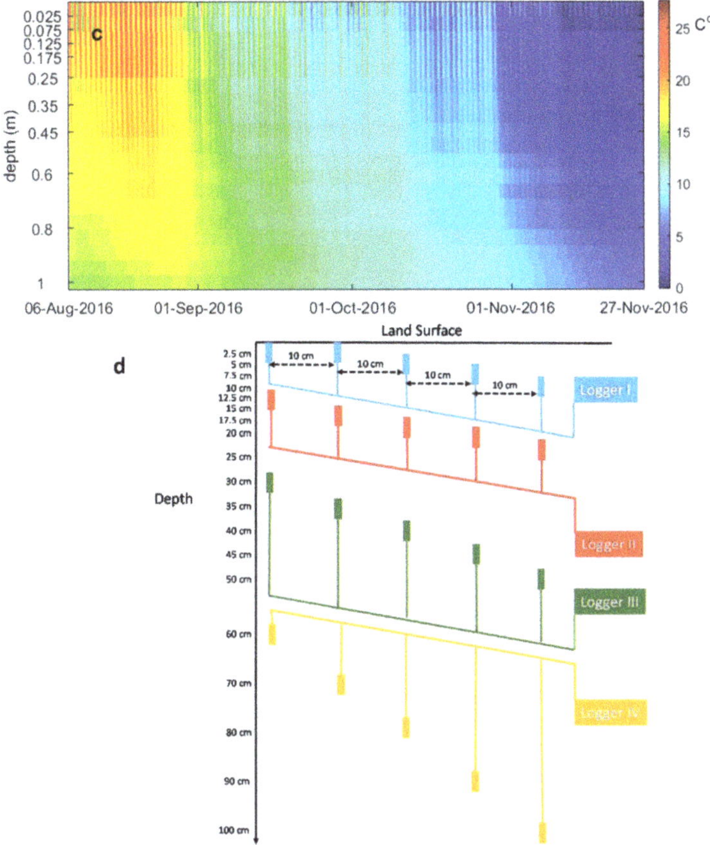

Figure 2. (**a**) precipitation; (**b**) the time series of soil moisture and (**c**) soil temperature profiles at Maqu Network Center Station; (**d**) the installation configuration of 20 sensors.

4. Results

In this section, the penetration depth is first calculated according to Equation (10) for SMOS/SMAP at 1.4 GHz. The result is intended to give a broad view on how penetration depth is affected by soil moisture and soil temperature theoretically. After that, Equation (8) and Equation (11) are applied at Maqu Center Station, where a dedicated penetration depth time series is generated for further analysis. With the penetration depth acquired at Maqu Center Station, we then can select accordingly the soil moisture and soil temperature observation to compare with the integral soil effective temperature by Equation (12).

Figure 3 uses the operational channel of SMOS/SMAP (L-band, 1.4 GHz) as examples to show how the soil moisture and soil temperature affect the penetration depth. It is clear that soil moisture is the dominant factor in affecting the penetration depth when the soil is dry, while soil temperature has more impact on the penetration depth for the wet soil. With the range of soil moisture of 0.01–0.6 cm^3 cm^{-3} and soil temperature of 0–60 °C, the penetration depth ranges from 3–70 cm for L-band. When the soil is very dry (i.e., soil moisture is less than 0.01 cm^3 cm^{-3}), the penetration depth is the greatest. Generally, the penetration depth would be 12 cm for L-band at 0.3 cm^3 cm^{-3} and 30 °C. If the soil is not so dry, the effect of soil temperature needs to be considered. For instance, in Figure 3, the penetration depth could be 11 cm when soil moisture is 0.55 cm^3 cm^{-3} and soil temperature is 50 °C. Nevertheless, the same penetration depth is also associated with soil moisture of 0.2 cm^3 cm^{-3} and soil temperature of 10 °C. However, an error of soil temperature both in measurement and model simulation larger than 10 °C is rare and soil moisture dominate the penetration depth especially for dry soil. In this case, although the calculation of the penetration depth strongly depends on the dielectric constant model but the difference can be ignored. If the layers configuration were too sparse, the estimation would not be so precise in practice as in Figure 3. This is partly the reason why a vertically dense soil moisture and soil temperature profile was mounted at the Maqu Center Site with dense layers (19 layers within the top one meter). Such intensive layering would greatly minimize the uncertainty introduced by the dielectric models.

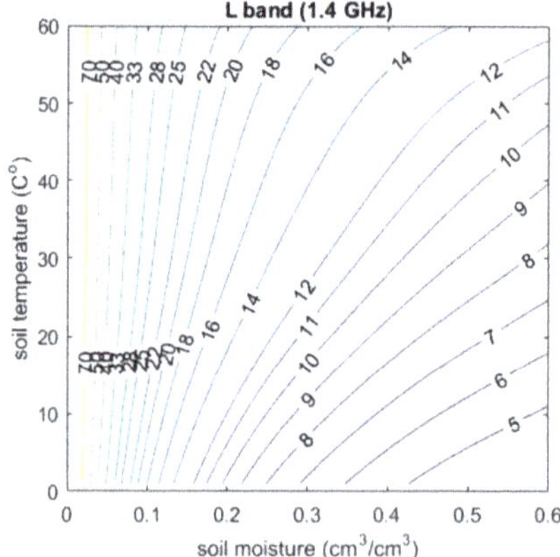

Figure 3. Penetration depth at L band (1.4 GHz). The ranges of penetration depth (in centimeters) were shown as contour lines, depending on the soil moisture and soil temperature. Mironov's dielectric constant model was used here for calculating the real and complex parts of dielectric constants.

Figure 4 shows the time series of the penetration depth (Blue) and correlation coefficient (Red) between the soil temperature at the penetration depth and the corresponding soil effective temperature. With vertically dense soil moisture/temperature profile measurement at Maqu Center Station, the penetration depth in Figure 4 is computed by Equations (8) and (12). While soil moisture ranges from 0.15 to 0.45 cm^3 cm^{-3}, the penetration depth varies from 6 to 10 cm at the center site. The average penetration depth is about 9 cm for the time before August 25 and 7 cm for the rest. As can be seen the penetration depth is strongly correlated with soil moisture which explains the variation in the period. Meanwhile, the penetration depth has its diurnal changes (around 1 cm) and is affected dominantly by soil temperature.

From the foregoing we highlighted the essence of T_{eff} as the mid-level of soil temperature profile in terms of Equations (1), (5) and (12) which could represent the average soil temperature of the soil column. Similarly, the soil moisture detected by satellites are also supposed to be the average soil moisture of the soil column in the view of emission depth. With T_{eff} computed from MERRA-2 and global soil moisture map acquired from SMAP, Figure 5 illustrates a global distribution of the penetration depth by Equation (10).

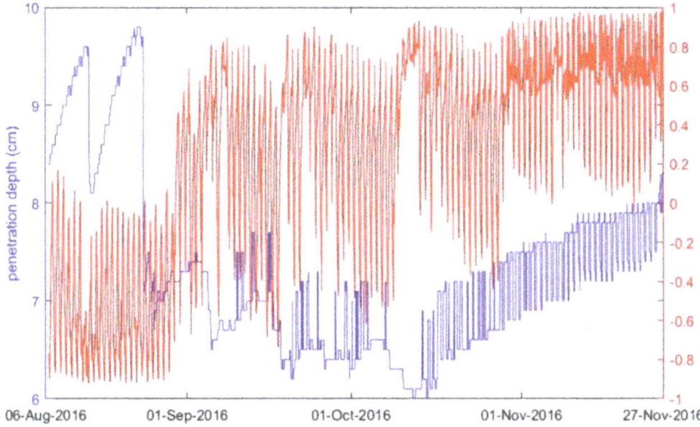

Figure 4. The time series of the penetration depth (Blue) and correlation coefficient (Red) between the soil temperature at the penetration depth and the corresponding soil effective temperature at Maqu Center Station as computed from the soil temperature/moisture profiles between 6 August and 27 November 2016.

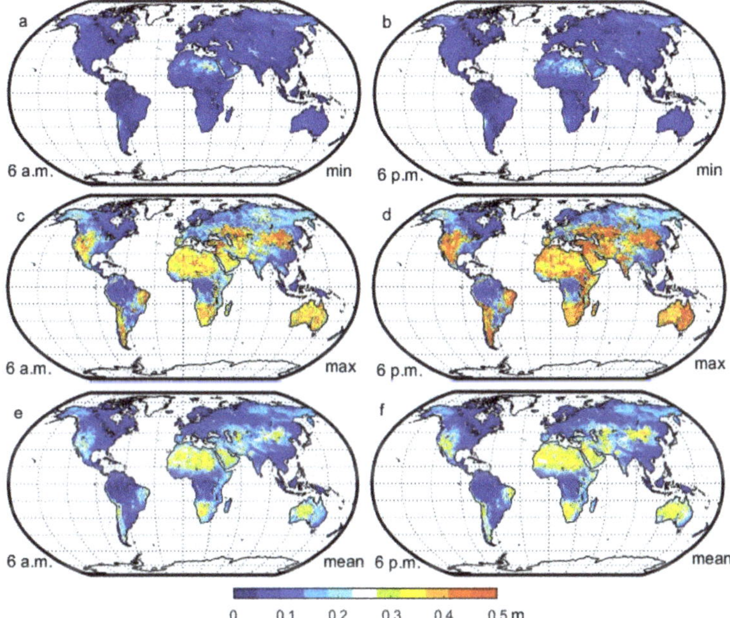

Figure 5. Global map of the penetration depth (PD) for SMAP with (**a**) minimum at 6 a.m.; (**b**) minimum at 6 p.m.; (**c**) maximum at 6 a.m.; (**d**) maximum at 6 p.m.; (**e**) mean at 6 a.m.; (**f**) mean at 6 p.m. Data used are SMAP soil moisture passive L3 product and the corresponding soil effective temperature calculated from MERRA-2 for 2016. The SMAP soil moisture and soil effective temperature are considered as the mid-level values for each pixel vertically.

The minimum cases (Figure 5a,b) reflect the penetration depth especially after the rainfall when the soil moisture is then higher. If there is a sufficient rain event, the surface soil layer would be fully moist even the soil column would be dry. Therefore, the penetration depth would be less than

0.1 m or even 0.05 m around the globe. In comparison, the cases for maximum penetration depth (Figure 5c,d) occur after a long drying period when the soil moisture has drawn down from the surface to deeper layers. This situation depends on how dry the soil could be so such regions coincide with the arid regions like central Asia, Australia and Sahara where the penetration depth is over 0.3 m. The annual mean value of penetration depth (Figure 5e,f) is from 0.05 to 0.2 m except the extremely dry regions. According to Equation (7) in Lv's scheme, the soil moisture/temperature sampled at 0.05 m represents the radiative contribution from 0 m to more than 0.1 m (depending on the specific profile) so 0.05 m samples may match with the satellite signal in most regions. In general, there is not too much difference between 6 a.m. and 6 p.m. while the latter may have deeper penetration depth of a few centimeters because T_{eff} is higher.

5. Discussion

Because the penetration depth is defined as the depth at which the intensity of the radiation inside the medium reduces to $1/e$ (about 37%) of its original value at the surface (or from its source origin to the surface from a sensing point of view), it means there are about 37% signal comes beneath the penetration depth. To characterize the penetration depth, several factors (see below) need to be considered and their uncertainties quantified, among which the validity of the assumption of linear temperature gradient is of most significance which we will briefly discuss as follows.

Several factors influence the characteristics of penetration depth which means that for Cal/Val the satellite soil moisture product, it is very important to know the overpassing time, because the penetration depth can vary diurnally. On the other hand, Equations (12) and (19) are derived based on the linear assumption (i.e., $\frac{dT}{d\tau} = const$). If the soil moisture and soil temperature profiles are complex or have great gradients, the remotely sensed soil moisture may not be at the penetration depth as calculated by Equation (12).

To which extent the linear assumption could be satisfied is quantified by the absolute value of correlation coefficient $|cc|$ between the soil temperature profile and the optical depth profile at Maqu Center Station. Figure 6 gives a glance about the validity of the linear assumption. It could be seen that the best accuracy was achieved when the assumption is valid if the data satisfy ($|cc| > 0.8$), which accounts for 10.89% during the experiment period. The $|cc| > 0.8$ appears mainly around 10:00 o'clock (local time) and about 40% of the observation period. The second occurrence peak is around 18:00 o'clock (local time) which coincides with SMOS/SMAP descending/ascending overpassing time. Other than the validity of the liner assumption, the assumption of $e^{-\tau_{deep}} \approx 0$ in Equation (18) can affect the correlation coefficient as well.

Equation (19) indicates that the soil temperature value at one certain depth could represent the soil temperature profile in terms of soil effective temperature (e.g., Lv's two-layer scheme vs. Wilheit's integral scheme) and that depth is the penetration depth. The soil temperature observed at 5 cm, 10 cm, 40 cm and the penetration depth in Figure 4 are compared with the soil effective temperature calculated by Wilheit's scheme in Figure 7. It is seen that the soil temperature variation at 2.5 cm (Figure 7a) has a slight underestimation bias and the distribution of points is much scattered than at the penetration depth (Figure 7d). For the soil temperature approximately between 15 and 20 °C, the difference can reach about 5 °C. When the soil temperature is out of that temperature range, the difference become relatively smaller. The variation range of soil temperature at 10 cm (Figure 7b) matches better but is still worse than that at the penetration depth. From Figure 4, it is known that the average penetration depth is about 7 cm after 20 August. Therefore the relatively similar accuracy is reasonable between Figure 7b,d. The soil temperature at 40 cm (Figure 7c) has a positive/negative bias after/before 20 August. RMSE reaches 3.32 °C. Obviously 40 cm cannot represent the soil profile (Figure 7c) to calculate T_{eff}.

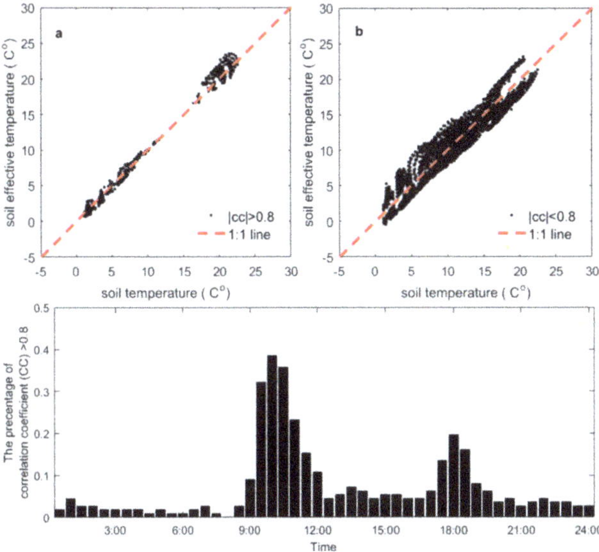

Figure 6. Comparison of soil temperature at the penetration depth vs. soil effective temperature at Maqu Center Station. The absolute correlation coefficient ($|cc|$) divided the time series into two groups where $|cc| > 0.8$ (**a**) and $|cc| < 0.8$ (**b**). The bottom figure shows the daily distribution of the moment when correlation coefficient $|cc| > 0.8$.

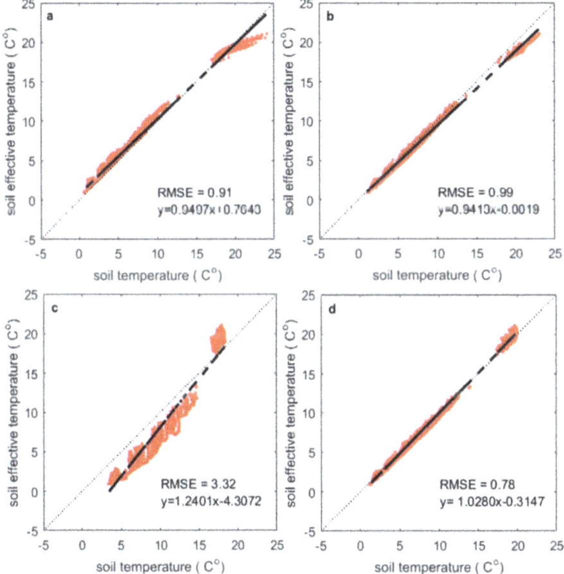

Figure 7. Comparison of soil effective temperature calculated by Wilheit's integral scheme against soil temperature observed at Maqu Center Station: (**a**) 2.5 cm; (**b**) 10 cm; (**c**) 40 cm observation and (**d**) the penetration depth. Data are shown only when $|cc| > 0.8$ and the dashed line is the regression line. The period is from 6 August to 27 November 2016.

For those data with $|cc| > 0.8$ (i.e., $\frac{dT}{d\tau} = const$) (see Figure 6), the soil temperature at the penetration depth is very close to as expected by Equation (19). The penetration depth contains all the factors which affect the soil effective temperature, such that the soil effective temperature varies with the penetration depth.

6. Conclusions

The concept of penetration depth in microwave radiometry was published more than four decades ago and its importance has long been realized by the microwave remote sensing community. The penetration depth is a characteristic length in the soil column which should be considered as a dynamic whole instead of just a few centimeters at the surface layer and becomes especially important with the increased wavelength for the dedicated soil moisture mission SMOS/SMAP. Because the penetration depth is defined based on the integral effective soil temperature, it leaves a gap between the simple two-layer schemes (e.g., commonly used in the operational soil moisture retrieval) and the integral one. Nevertheless, the implication behind the penetration depth has so far not been fully investigated and explained. The casually referred $1/e$ residual is just a "qualitative" number and how it is linked with satellite soil moisture sensing depth has not been analytically determined.

In this study, with rigorous mathematical derivations based on Lv's T_{eff} scheme, we have proved that the penetration depth is not only a "qualitative" number but a characteristic depth which synthesizes emitting behavior of a soil column in microwave radiometry. In Lv's scheme, the optical depth τ appears in T_{eff} and thus unifies the radiative transfer processes in atmosphere, vegetation and soil. More specifically, τ quantifies the attenuation of radiation transfer in a medium, being the air dielectric properties in the atmosphere, vegetation optical thickness (mainly the water content in leaves) in vegetation, as well as soil dielectric properties (e.g., mainly soil moisture) in the soil. The use of the integral τ instead of soil depth x is proposed in this study for the first time and with τ appearing in the T_{eff} formula, we can determine essential characteristics in T_{eff}. By means of τ, it is proved that the penetration is not just the depth where the energy is reduced to $1/e$ of its original value but it is also the median value of soil temperature in the soil column. The penetration depth is strongly related to soil moisture but also has diurnal variation which may have an amplitude of several centimeters at the center station of Maqu network.

The question of at which depth L-band soil moisture monitoring satellites such as SMOS/SMAP measure has confused the soil moisture community at large. As stated in introduction, the sensitive layer is supposed to be the depth where these satellites are sensing in previous studies. In SMAP retrieval [46], vegetation and soil surface roughness are accounted in terms of emissivity calibration. The final soil moisture product is derived from a smooth emission model. A precise estimation of vegetation and roughness is critical before determining the penetration depth. Particularly, the global map of penetration depth in Figure 5 depends on a correct vegetation calibration. Therefore, the penetration depth over dense vegetation zone, for instance the tropics may be even smaller. In contrast, in the dessert area with few or no vegetation, the penetration depth is usually large and vegetation calibration is not so important except after rainfall events. The conclusion in this study would be most useful to the transition zone where soil moisture variation is larger and affects climate/hydro-process more intensively. Different to the sensitive layer view, we proposed in this study a median value view and found the soil temperature median value under the linear assumption. From the hypothesis of zeroth-order incoherent microwave transfer frame, the median soil temperature layer represents not only T_{eff} but also provides the depth information contained in this frame. The soil moisture retrieved from microwave (e.g., L-band) observation should be the average radiative emission capacity of the soil column and there should be a median soil moisture depth as well. This study has successfully developed such a new method to find this median soil moisture depth by relating the penetrating depth in terms of temperature to radiative energy attenuation. This is done by building up a $T_b \sim T_{eff} \sim \tau \sim depth$ relationship with median theory in which a median value of T could be found at the penetration depth with certain condition. The method is verified with in situ data from the Maqu observation site and the conclusion is valid whenever the field condition satisfies the assumption. This is critical to the application of SMOS/SMAP soil moisture product because a difference of several centimeters between the depth of in situ measurement and the satellite sensing depth will lead to systematic bias in evaluating the satellite products. Based on an application of the developed method to SMAP passive L3 soil moisture product and the corresponding soil effective temperature calculated from MERRA-2 for 2016, it may be concluded that it is appropriate to use 5 cm depth of soil moisture measurement as a ground reference to calibrate and validate satellites soil moisture product because 5 cm captures the main signal source on average. However, for some extreme cases like arid region or the region after a long drought event, 5 cm may not represent the dominant emission layer. In other words, it means that even though the satellite product is precise, we may still get biased conclusion, if the ground measurement is inappropriately organized, and the comparability between satellite and in situ measurement is not established [47]. The developed method should also be beneficial to the Earth surface modelling in improving the consistency in the dynamics of the soil moisture processes and satellite observations.

Acknowledgments: We acknowledge the financial support received from the National Science Foundation of China (Grant No. 41530529 & 41575013) and the Chinese Scholarship Council (CSC) for Shaoning Lv. This work was also funded in part by the Netherlands Organization for Scientific Research (Project No. ALW-GO/14-29), the ESA MOST Dragon IV programme (Monitoring Water and Energy Cycles at Climate Scale in the Third Pole Environment (CLIMATE-TPE)).

Author Contributions: Shaoning Lv, Yijian Zeng and Zhongbo Su conceived and designed the research; Jun Wen contributed maintaining the Tibet-obs networks; Hong Zhao contributed soil properties data; Shaoning Lv wrote the paper.

Conflicts of Interest: The authors declare no conflict of interest.

References

1. Yamada, T.J.; Kanae, S.; Oki, T.; Hirabayashi, Y. The onset of the West African monsoon simulated in a high-resolution atmospheric general circulation model with reanalyzed soil moisture fields. *Atmos. Sci. Lett.* **2012**, *13*, 103–107. [CrossRef]

2. Douville, H.; Chauvin, F.; Broqua, H. Influence of soil moisture on the Asian and African monsoons. Part I: Mean monsoon and daily precipitation. *J. Clim.* **2001**, *14*, 2381–2403. [CrossRef]

3. Douville, H.; Conil, S.; Tyteca, S.; Voldoire, A. Soil moisture memory and West African monsoon predictability: Artefact or reality? *Clim. Dyn.* **2007**, *28*, 723–742. [CrossRef]

4. Lim, Y.J.; Hong, J.; Lee, T.Y. Spin-up behavior of soil moisture content over East Asia in a land surface model. *Meteorol. Atmos. Phys.* **2012**, *118*, 151–161. [CrossRef]

5. De Rosnay, P.; Drusch, M.; Boone, A.; Balsamo, G.; Decharme, B.; Harris, P.; Kerr, Y.; Pellarin, T.; Polcher, J.; Wigneron, J.P. AMMA land surface model intercomparison experiment coupled to the Community Microwave Emission Model: ALMIP-MEM. *J. Geophys. Res. Atmos.* **2009**, *114*. [CrossRef]

6. Kerr, Y.H. Soil moisture from space: Where are we? *Hydrogeol. J.* **2006**, *15*, 117–120. [CrossRef]

7. Dedieu, G.; Deschamps, P.Y.; Kerr, Y.H. Satellite estimation of solar irradiance at the surface of the Earth and of surface albedo using a physical model applied to meteosat data. *J. Clim. Appl. Meteorol.* **1987**, *26*, 79–87. [CrossRef]

8. Kerr, Y.H.; Waldteufel, P.; Richaume, P.; Wigneron, J.P.; Ferrazzoli, P.; Mahmoodi, A.; Al Bitar, A.; Cabot, F.; Gruhier, C.; Juglea, S.E.; et al. The SMOS soil moisture retrieval algorithm. *IEEE Trans. Geosci. Remote Sens.* **2012**, *50*, 1384–1403. [CrossRef]

9. Entekhabi, D.; Njoku, E.G.; O'Neill, P.E.; Kellogg, K.H.; Crow, W.T.; Edelstein, W.N.; Entin, J.K.; Goodman, S.D.; Jackson, T.J.; Johnson, J.; et al. The Soil Moisture Active Passive (SMAP) Mission. *Proc. IEEE* **2010**, *98*, 704–716. [CrossRef]

10. Dumedah, G.; Walker, J.P.; Rudiger, C. Can SMOS data be used directly on the 15-km discrete global grid? *IEEE Trans. Geosci. Remote Sens.* **2014**, *52*, 2538–2544. [CrossRef]

11. Reul, N.; Tenerelli, J.; Chapron, B.; Vandemark, D.; Quilfen, Y.; Kerr, Y. SMOS satellite L-band radiometer: A new capability for ocean surface remote sensing in hurricanes. *J. Geophys. Res. Oceans* **2012**, *117*. [CrossRef]

12. Dente, L.; Su, Z.B.; Wen, J. Validation of SMOS soil moisture products over the Maqu and Twente regions. *Sensors* **2012**, *12*, 9965–9986. [CrossRef] [PubMed]

13. Tranchant, B.; Testut, C.E.; Renault, L.; Ferry, N.; Birol, F.; Brasseur, P. Expected impact of the future SMOS and Aquarius Ocean surface salinity missions in the Mercator Ocean operational systems: New perspectives to monitor ocean circulation. *Remote Sens. Environ.* **2008**, *112*, 1476–1487. [CrossRef]

14. Tranchant, B.; Testut, C.E.; Ferry, N.; Renault, L.; Obligis, E.; Boone, C.; Larnicol, G. Data assimilation of simulated SSS SMOS products in an ocean forecasting system. *J. Oper. Oceanogr.* **2008**, *1*, 19–27. [CrossRef]

15. Lv, S.; Zeng, Y.; Wen, J.; Zheng, D.; Su, Z. Determination of the optimal mounting depth for calculating effective soil temperature at l-band: Maqu case. *Remote Sens.* **2016**, *8*, 476. [CrossRef]

16. Lv, S.; Wen, J.; Zeng, Y.; Tian, H.; Su, Z. An improved two-layer algorithm for estimating effective soil temperature in microwave radiometry using in situ temperature and soil moisture measurements. *Remote Sens. Environ.* **2014**, *152*, 356–363. [CrossRef]

17. Burke, W.J.; Schmugge, T.; Paris, J.F. Comparison of 2.8- and 21-cm microwave radiometer observations over soils with emission model calculations. *J. Geophys. Res. Oceans* **1979**, *84*, 287–294. [CrossRef]

18. Rao, K.; Chandra, G.; Rao, P.N. Study on penetration depth and its dependence on frequency, soil moisture, texture and temperature in the context of microwave remote sensing. *J. Indian Soc. Remote Sens.* **1988**, *16*, 7–19. [CrossRef]

19. Njoku, E.G.; Kong, J.-A. Theory for passive microwave remote-sensing of near-surface soil-moisture. *J. Geophys. Res.* **1977**, *82*, 3108–3118. [CrossRef]

20. Colliander, A.; Jackson, T.J.; Bindlish, R.; Chan, S.; Das, N.; Kim, S.B.; Cosh, M.H.; Dunbar, R.S.; Dang, L.; Pashaian, L.; et al. Validation of SMAP surface soil moisture products with core validation sites. *Remote Sens. Environ.* **2017**, *191*, 215–231. [CrossRef]

21. Su, Z.; de Rosnay, P.; Wen, J.; Wang, L.; Zeng, Y. Evaluation of ECMWF's soil moisture analyses using observations on the Tibetan Plateau. *J. Geophys. Res. Atmos.* **2013**, *118*, 5304–5318. [CrossRef]

22. Njoku, E.G.; Entekhabi, D. Passive microwave remote sensing of soil moisture. *J. Hydrol.* **1996**, *184*, 101–129. [CrossRef]

23. Entekhabi, D.; Nakamura, H.; Njoku, E.G. Solving the inverse problems for soil-moisture and temperature profiles by sequential assimilation of multifrequency remotely-sensed observations. *IEEE Trans. Geosci. Remote Sens.* **1994**, *32*, 438–448. [CrossRef]

24. Escorihuela, M.J.; Chanzy, A.; Wigneron, J.P.; Kerr, Y.H. Effective soil moisture sampling depth of L-band radiometry: A case study. *Remote Sens. Environ.* **2010**, *114*, 995–1001. [CrossRef]

25. Scheeler, R.; Kuester, E.F.; Popovic, Z. Sensing depth of microwave radiation for internal body temperature measurement. *IEEE Trans. Antennas Propag.* **2014**, *62*, 1293–1303. [CrossRef]

26. Zhou, F.C.; Song, X.N.; Leng, P.; Li, Z.L. An effective emission depth model for passive microwave remote sensing. *IEEE J. Sel. Top. Appl. Earth Obs. Remote Sens.* **2016**, *9*, 1752–1760. [CrossRef]

27. Nolan, M.; Fatland, D.R. Penetration depth as a DInSAR observable and proxy for soil moisture. *IEEE Trans. Geosci. Remote Sens.* **2003**, *41*, 532–537. [CrossRef]

28. Zhao, S.; Zhang, L.; Zhang, T.; Hao, Z.; Chai, L.; Zhang, Z. An empirical model to estimate the microwave penetration depth of frozen soil. In Proceedings of the 2012 IEEE International Geoscience and Remote Sensing Symposium (IGARSS), Munich, Germany, 22–27 July 2012; pp. 4493–4496.

29. Choudhury, B.J.; Schmugge, T.J.; Mo, T. A Parameterization of Effective Soil-Temperature for Microwave Emission. *J. Geophys. Res. Oceans* **1982**, *87*, 1301–1304. [CrossRef]

30. Wigneron, J.P.; Laguerre, L.; Kerr, Y.H. A simple parameterization of the L-band microwave emission from rough agricultural soils. *IEEE Trans. Geosci. Remote Sens.* **2001**, *39*, 1697–1707. [CrossRef]

31. Holmes, T.R.H.; de Rosnay, P.; de Jeu, R.; Wigneron, R.J.P.; Kerr, Y.; Calvet, J.C.; Escorihuela, M.J.; Saleh, K.; Lemaitre, F. A new parameterization of the effective temperature for L band radiometry. *Geophys. Res. Lett.* **2006**, *33*. [CrossRef]

32. Ulaby, F.T.; Moore, R.K.; Fung, A.K. *Microwave Remote Sensing Active and Passive-Volume III: From Theory to Applications*; Artech House, Inc.: Norwood, MA, USA, 1986.

33. Lv, S.; Zeng, Y.; Wen, J.; Su, Z. A reappraisal of global soil effective temperature schemes. *Remote Sens. Environ.* **2016**, *183*, 144–153. [CrossRef]

34. De Rosnay, P.; Wigneron, J.; Holmes, T.; Calvet, J. Parameterizations of the effective temperature for L-band radiometry. Inter-comparison and long term validation with SMOSREX field experiment. In *Thermal Microwave Radiation-Applications for Remote Sensing*; IET Electromagnetic Waves Series 52; Mtzler, C., Rosenkranz, P.W., Battaglia, A., Wigneron, J.P., Eds.; Christian Matzler: London, UK, 2006; Volume 573, ISBN 0-86341-573-3 and 978-086341-573-9.

35. Wilheit, T.T. Radiative-transfer in a plane stratified dielectric. *IEEE Trans. Geosci. Remote Sens.* **1978**, *16*, 138–143. [CrossRef]

36. Zwieback, S.; Hensley, S.; Hajnsek, I. Assessment of soil moisture effects on L-band radar interferometry. *Remote Sens. Environ.* **2015**, *164*, 77–89. [CrossRef]

37. Owe, M.; Van de Griend, A.A. Comparison of soil moisture penetration depths for several bare soils at two microwave frequencies and implications for remote sensing. *Water Resour. Res.* **1998**, *34*, 2319–2327. [CrossRef]

38. Wigneron, J.P.; Chanzy, A.; de Rosnay, P.; Rudiger, C.; Calvet, J.C. Estimating the effective soil temperature at L-band as a function of soil properties. *IEEE Trans. Geosci. Remote Sens.* **2008**, *46*, 797–807. [CrossRef]

39. Su, Z.; Wen, J.; Dente, L.; van der Velde, R.; Wang, L.; Ma, Y.; Yang, K.; Hu, Z. The Tibetan Plateau observatory of plateau scale soil moisture and soil temperature (Tibet-Obs) for quantifying uncertainties in coarse resolution satellite and model products. *Hydrol. Earth Syst. Sci.* **2011**, *15*, 2303–2316. [CrossRef]

40. Zheng, D.; Wang, X.; van der Velde, R.; Zeng, Y.; Wen, J.; Wang, Z.; Schwank, M.; Ferrazzoli, P.; Su, Z. L-band microwave emission of soil freezesc-thaw process in the Third Pole Environment. *IEEE Trans. Geosci. Remote Sens.* **2017**. [CrossRef]

41. Wang, Q.; van der Velde, R.; Su, Z. Use of a discrete electromagnetic model for simulating Aquarius L-band active/passive observations and soil moisture retrieval. *Remote Sens. Environ.* **2017**, *205*, 434–452. [CrossRef]

42. Zhao, H.; Zeng, Y.; Lv, S.; Su, Z. Analysis of soil hydraulic and thermal properties for land surface modelling over the tibetan plateau. *Earth Syst. Sci. Data Discuss.* **2018**. [CrossRef]

43. Mironov, V.L.; Dobson, M.C.; Kaupp, V.H.; Komarov, S.A.; Kleshchenko, V.N. Generalized refractive mixing dielectric model for moist soils. *IEEE Trans. Geosci. Remote Sens.* **2004**, *42*, 773–785. [CrossRef]

44. Gelaro, R.; McCarty, W.; Suárez, M.J.; Todling, R.; Molod, A.; Takacs, L.; Randles, C.A.; Darmenov, A.; Bosilovich, M.G.; Reichle, R.; et al. The Modern-Era Retrospective Analysis for Research and Applications, Version 2 (MERRA-2). *J. Clim.* **2017**, *30*, 5419–5454. [CrossRef]

45. O'Neill, P.; Chan, S.; Njoku, E.; Jackson, T.; Bindlish, R. *Soil Moisture Active Passive (SMAP), Algorithm Theoretical Basis Document Level 2 & 3 Soil Moisture (Passive) Data Products, Revison B*; Jet Propulsion Laboratory, California Institute of Technology: Pasadena, CA, USA, 2015.

46. Entekhabi, D.; Yueh, S.; O'Neill, P.E.; Kellogg, K.H.; Allen, A.; Bindlish, R.; Brown, M.; Chan, S.; Colliander, A.; Crow, W.T. *SMAP Handbook Soil Moisture Active Passive: Mapping Soil Moisture and Freeze/thaw from Space*; National Aeronautics and Space Administration (NASA): Washington, DC, USA, 2014.

47. Zeng, J.; Li, Z.; Chen, Q.; Bi, H.; Qiu, J.; Zou, P. Evaluation of remotely sensed and reanalysis soil moisture products over the Tibetan Plateau using in-situ observations. *Remote Sens. Environ.* **2015**, *163*, 91–110. [CrossRef]

 *remote sensing* 

Article

The Combined ASTER MODIS Emissivity over Land (CAMEL) Part 1: Methodology and High Spectral Resolution Application

E. Eva Borbas [1,*], Glynn Hulley [2], Michelle Feltz [1], Robert Knuteson [1] and Simon Hook [2]

[1]   Space Science and Engineering Center, University of Wisconsin-Madison, Madison, WI 53706, USA; michelle.feltz@ssec.wisc.edu (M.F.); robert.knuteson@ssec.wisc.edu (R.K.)
[2]   California Institute of Technology Jet Propulsion Laboratory, Pasadena, CA 91109, USA; Glynn.Hulley@jpl.nasa.gov (G.H.); Simon.Hook@jpl.nasa.gov (S.H.)
*   Correspondence: eva.borbas@ssec.wisc.edu; Tel.: +1-608-263-0228

Received: 28 February 2018; Accepted: 12 April 2018; Published: 21 April 2018

Abstract: As part of a National Aeronautics and Space Administration (NASA) MEaSUREs (Making Earth System Data Records for Use in Research Environments) Land Surface Temperature and Emissivity project, the Space Science and Engineering Center (UW-Madison) and the NASA Jet Propulsion Laboratory (JPL) developed a global monthly mean emissivity Earth System Data Record (ESDR). This new Combined ASTER (Advanced Spaceborne Thermal Emission and Reflection Radiometer) and MODIS (Moderate Resolution Imaging Spectroradiometer) Emissivity over Land (CAMEL) ESDR was produced by merging two current state-of-the-art emissivity datasets: the UW-Madison MODIS Infrared emissivity dataset (UW BF) and the JPL ASTER Global Emissivity Dataset Version 4 (GEDv4). The dataset includes monthly global records of emissivity and related uncertainties at 13 hinge points between 3.6–14.3 μm, as well as principal component analysis (PCA) coefficients at 5-km resolution for the years 2000 through 2016. A high spectral resolution (HSR) algorithm is provided for HSR applications. This paper describes the 13 hinge-points combination methodology and the high spectral resolutions algorithm, as well as reports the current status of the dataset.

Keywords: emissivity; infrared; surface; land; hyperspectral; radiation

---

1. Introduction

Land Surface Temperature and Emissivity (LST&E) data are critical variables for studying a variety of Earth surface processes and surface–atmosphere interactions such as evapotranspiration, surface energy balance, and water vapor retrievals. LST&E have been identified as an important Earth System Data Record (ESDR) by National Aeronautics and Space Administration (NASA) and many other international organizations (NASA Strategic Roadmap Committee #9, 2005; Global Climate Observing System (GCOS), 2003; Climate Change Science Program (CCSP), 2006 and the recently established International Surface Temperature Initiative) [1].

Accurate knowledge of LST&E at high spatial (1 km) and temporal (hourly) scales is a key requirement for many energy balance models to estimate important surface biophysical variables such as evapotranspiration and plant-available soil moisture [2,3]. LST&E data are essential for balancing the Earth's surface radiation budget. For example, a surface emissivity error of 0.1 will result in climate models having errors of up to 7 Wm^{-2} in their upward long-wave radiation estimates, which is a much larger term than the surface radiative forcing (~2–3 Wm^{-2}) due to an increase in greenhouse gases [4]. LST&E are also used to monitor land-cover/land-use changes [5], and in atmospheric retrieval schemes [6].

LST&E products are generated with accuracies that vary depending on the input data, including ancillary data such as atmospheric water vapor, as well as algorithmic approaches. For example, certain Moderate Resolution Imaging Spectroradiometer (MODIS) products (MOD11) use an infrared (IR) split window algorithm applied to two or more bands in conjunction with an emissivity estimate based on the land classification to produce the LST. Conversely, other MODIS products (MOD21) [7] use a physics-based approach involving a radiative transfer model to first correct the data to a surface radiance, and then use a model to extract the temperature and emissivities in the spectral bands. This physics-based approach is also adopted for the Advanced Spaceborne Thermal Emission and Reflection Radiometer (ASTER) measurements. Validation of these approaches has shown that they are complementary, with the split-window approach better suited over heavily vegetated regions, and the physics-based approach better suited for semi-arid and arid regions. Figure 1 shows an example of the ASTER Global Emissivity Dataset Version 3 (ASTER GEDv3) [8] and the UW-Madison MODIS Infrared emissivity dataset Baseline Fit (UW BF) [9] mean emissivity at 9.1 μm over Africa for the summer season (July–September) between 2000 and 2008. There is good overall agreement between the two databases; however, each has their own benefits and drawbacks. For example, the ASTER GEDv3 is able to better capture the deeper quartz minimum at 9.1 μm compared to the UW BF, which has only one band available in this region (MODIS band 29, 8.5 μm). For the UW BF database, the emissivity in the mid to long-wave region (8–12 μm) is not well defined, because MODIS only has three bands in this region (bands 8.5 μm, 11 μm, and 12 μm). This results in an imperfect spectral shape in the two quartz doublet regions at 8.5 μm and 12 μm. The advantages of the UW BF include its moderate spatial resolution (5 km), its uniform temporal coverage (monthly), and that its emissivities span the entire IR region (3.6–12 μm). In contrast, although there are more bands in the ASTER GEDv3 available to define the spectral shape in the mid–long IR region (five bands, bands at 8.3 μm, 8.6 μm, 9.3 μm, 10.6 μm, and 11.3 μm), there are no bands in the short-wave infrared (SWIR) region around 3.8–4.1 μm, which limits its use in models and other atmospheric retrieval schemes. On the plus side, ASTER GEDv3 has high spatial resolution (~100 m) and high accuracy over arid regions. In terms of temporal sampling, the MODIS product has been used to create monthly emissivity estimates, whereas constraints on the ASTER data collection limit the derived emissivity dataset to multi-year climatologies. By combining the two measurements, the Combined ASTER and MODIS Emissivity over Land (CAMEL) dataset takes advantage of the strengths of each dataset while mitigating the problems of each.

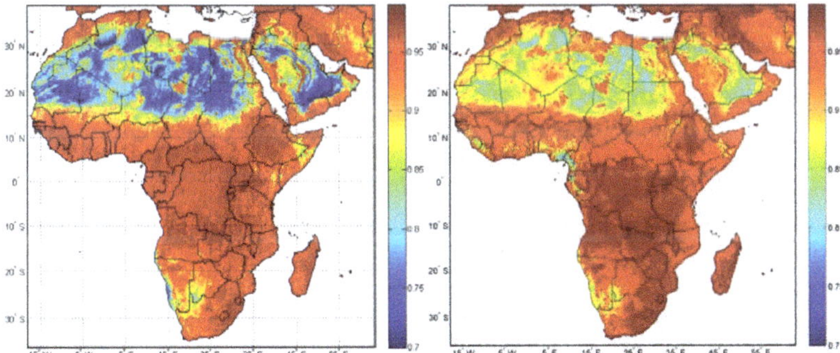

Figure 1. (**Left**) Advanced Spaceborne Thermal Emission and Reflection Radiometer Global Emissivity Dataset (ASTER GEDv3) mean summertime (July–September) emissivity for band 12 (9.1 μm), (**Right**) UW-Madison MODIS Infrared emissivity dataset Baseline Fit emissivity (UW BF) mean summertime (July–September) emissivity at 9.1 μm, which is the same as Moderate Resolution Imaging Spectroradiometer (MODIS) Band 29, 8.5 μm.

NASA has recognized the importance of LST&E, and identified the need to develop long-term, consistent, and calibrated data and products that are valid across multiple missions and satellite sensors. Under the NASA Making Earth Science Data Records for Use in Research Environments (MEaSUREs) program, a monthly mean unified Low Earth Orbit (LEO) based Land Surface Emissivity (LSE) ESDR at 5 km has been produced by merging two current state-of-the-art emissivity databases, the UW-Madison MODIS based UW BF, and the Jet Propulsion Laboratory (JPL) ASTER GED Version4 (GEDv4) [10], which is called the CAMEL. The CAMEL LSE ESDR has been further extended to hyperspectral resolution using a Principal Component (PC) regression approach similar to the UW high spectral resolution (HSR) algorithm [11].

This document is the first part of a two-part series that will describe the NASA MEaSUREs LSE ESDR called CAMEL Version 1.0 [12] in detail, including their methodologies, data products, and technical aspects. Part II discusses the uncertainty determination and current validation efforts of the CAMEL database.

2. Data

In this section, the two input emissivity databases are introduced: the ASTER GEDv4 and the UW BF emissivity, along with the selected laboratory measurements that are needed for HSR application.

2.1. The ASTER Global Emissivity Dataset

In 2009, a level-3 mean, gridded ASTER Global Emissivity Dataset Verison3 (ASTER GEDv3) was generated using all ASTER clear-sky data available since 2000. The emissivity retrieval was based on an improved Temperature Emissivity Separation (TES) algorithm with a water vapor scaling (WVS) atmospheric correction approach [13,14]. The ASTER GEDv3 is output on $1° \times 1°$ grids at 100-m, 1-km, and 5-km spatial resolutions. The product has been validated extensively over a set of pseudo-invariant sites, and results indicate agreement to within 1.5% [15]. Additionally, ASTER GEDv3 has shown good agreement with other coarser sensor LSE products such as Atmospheric Infrared Sounder (AIRS) [16] and MODIS [13,17]. The ASTER GEDv3 is currently being distributed at the NASA Land Processes Distributed Active Archive Center (LP DAAC) at the U.S. Geological Survey (USGS) Earth Resources Observation and Science (EROS) with global coverage, and has been available since the end of 2012.

2.1.1. ASTER Vegetation and Snow Cover Adjustment

Since the ASTER GEDv3 product represents a mean emissivity climatology of ASTER data acquired over an 11-year period (2000–2010), an emissivity adjustment is necessary over heterogeneous land cover types that are subject to annual and inter-annual land cover changes (e.g., due to snow and ice melt, and agricultural practices). The emissivity of vegetation and snow is fairly high and constant (~0.98–1.0). As a result, surfaces with high amounts of vegetation or snow cover reduce the amount of spectral variation. This relationship is used to adjust the ASTER GEDv3 emissivity product by either increasing or decreasing the spectral contrast as a simple function of the amount of snow or vegetation relative to the reference mean state. The methodology to create the vegetation and snow adjusted ASTER GEDv4 emissivity is described in Hulley et al. [14]. This allows the ASTER GEDv4 emissivity product to be produced at the same monthly resolution of the UW products. The snow cover amount is obtained from the standard MODIS snow cover maps (MOD10 product). The vegetation amount is obtained by applying the National Oceanic and Atmospheric Administration (NOAA) National Environmental Satellite, Data and Information Service (NESDIS) "Green Vegetation Fraction" approach to the NASA MOD13A3 monthly gridded normalized difference vegetation index (NDVI) product, whereby the current vegetation influence is estimated as f = (NDVI_current - NDVI_min)/(NDVI_max - NDVI_min) [18]. The NDVI is a well-tested and proven indicator of partial and emerging vegetation growth.

2.1.2. Aggregation of ASTER GED to 5-km Resolution

The ASTER GEDv3 is produced in $1° \times 1°$ grids with a resolution of $0.001°$ (~100 m), and consequently, the spectral emissivities are first aggregated to the UW database resolution of $0.05°$ (5 km) before merging. It has been shown that a simple aggregation from fine to coarse resolution is valid only if the scene is homogeneous in emissivity and surface temperature [19]. In this case, the effective emissivity is simply an average of individual pixels, i. $\bar{\varepsilon}(v) = 1/n \sum \varepsilon(i, v)$ i=1:n, where $\bar{\varepsilon}(v)$ is the effective spectral emissivity for wavelength v at the coarser resolution scale, and $\varepsilon(i, v)$ is the spectral emissivities for each pixel i at the finer resolution scale. This approach was used successfully in validating AIRS emissivities with the ASTER emissivity product over large homogenous sand seas [16]. However, over more heterogeneous cover types, this assumption breaks down due to a higher variability in the surface temperature distribution. A potential solution is to aggregate the surface emitted radiance for each pixel at the finer resolution scale (e.g., ASTER at 100 m), and then normalize with the radiance of an effective surface temperature at the coarser pixel scale (e.g., MODIS at 5 km), as follows:

$$\bar{\varepsilon}_v = \frac{\frac{1}{n} \sum_{i=1}^{n} \varepsilon_{i,v} \cdot B_v(T_{i,s})}{B_v(\overline{T_s})} \tag{1}$$

where $B_v(T_{i,s})$ is the radiance for temperature $T_{i,s}$ for each pixel i at the finer resolution scale, and $B_v(\overline{T_s})$ is the radiance for an effective temperature $\overline{T_s}$ at the coarser resolution scale. This method was used to intercompare emissivities from the ASTER GEDv4 with MODIS emissivities at a coarser scale over the southwestern United States (USA) [17].

The MODIS products known as MOD11C3 and ASTER GEDv4 are both level-3 gridded products. Since both the original pixel resolution of the ASTER GEDv3 (100 m) and MODIS (1 km) are resampled to 0.05 degree, any misregistration and geolocation inconsistencies between ASTER/MODIS are very likely to be negligible, particularly for thermal data. In addition, ASTER's geometric accuracy and pixel geolocation knowledge have exceeded the original goals of the project [20]. For Terra and Aqua MODIS instruments, specific correction approaches are implemented to ensure that the geolocation of individual MODIS observations are at the sub-pixel accuracy level [21]. Additionally, the MODIS Terra and Aqua band-to-band registration accuracy is under 50 m [22,23].

2.2. The UW Baseline Fit and High Spectral Resolution Land Surface Emissivity Database

At the University of Wisconsin-Madison, a monthly MODIS global IR land surface emissivity database (UW BF) was developed based on the standard monthly mean MODIS emissivity product at 10 wavelengths (3.6, 4.3, 5.0, 5.8, 7.6, 8.3, 9.3, 10.8, 12.1, and 14.3 μm) at 5-km spatial resolution. The baseline fit method [24], which was based on a conceptual model developed from laboratory measurements of surface emissivity, is applied to fill in the spectral gaps between the six available MODIS/MYD11 emissivity bands. The 10 wavelengths in the UW BF emissivity database were chosen as hinge points to capture as much of the shape of the higher resolution emissivity spectra as possible. This approach was extended by the method described in Borbas [11] and Masiello et al. [25] to provide 416 spectral points from 3.6 μm to 14.3 μm. The UW HSR emissivity algorithm is based on a principal component analysis (PCA) regression using the eigenfunction representation of high spectral resolution laboratory measurements from the ASTER spectral library [26].

In the next section, the input data for the UW database are discussed. The quality and accuracy of those input data is important for determining the uncertainty of the CAMEL dataset.

2.2.1. Input MODIS MOD11 Products

The operational MODIS MOD11 surface temperature and emissivity products are generated over land in clear-sky conditions for day and night in 5-min granules for both NASA's Earth Observing System (EOS) Terra and Aqua satellites. The 5-min granule products are averaged to daily, eight-day, and monthly time scales, and the Level-3 gridded products are produced at 5-km spatial resolution.

The UW BF emissivity data is comprised of the monthly mean surface emissivity products (MOD11C3) that include IR land surface emissivity at six IR bands (20, 22, 23, 29, 31, and 32) located in the 3.6–4.2 μm and the 8–13 μm atmospheric windows. The approach to derive the monthly mean land surface temperature and emissivity assumes emissivity, which is known as the day–night algorithm [27], does not change between day and night at the same location over a period of a few days. This assumption is one of the sources of uncertainty for the CAMEL product. The Collection 4.0/4.1 (Col 4.0/4.1) MOD11C3 products are used as input to the UW BF database, even though newer Collections (Col 5 and 6) have been released since then. We have found that the Col 4.0/4.1 version has the best quality products. Significant differences were found between the MYD11 Col 4 and 5 data: higher emissivity values at the reststrahlen band over desert areas (see Figure 2); an increase in minimum emissivity for bands 20, 22, 23, and 29 (3.7μm, 3.9 μm, 4.0 μm, and 8.5 μm, respectively) by ~0.1; and a loss of variability for bands 31 and 32 (11 μm and 12 μm, respectively).

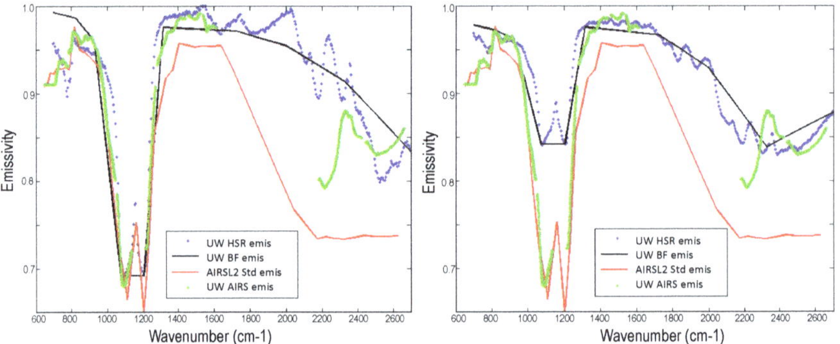

Figure 2. Emissivity comparison on January 2003 over Sahara Desert (Lat = 25.075N, Lon = 26.058E) of UW BF (solid black line), UW high spectral resolution (HSR) (blue dots), the AIRS L2 (V5.0) Standard products (red line), and the UW/AIRS (green dots) emissivity products. (**Left**) The UW BF and UW HSR emissivity products have been derived from the Col 4, and (**Right**) from the Col 5 MODIS emissivity products.

Due to these differences, the NASA LP DAAC decided in the beginning of 2007 to continue to produce Col 4 data beyond December 2006, but using the only available Col 5 MODIS input data such as the cloud mask, L1B data, and atmospheric profiles. This version of the MYD11 data is called Col 4.1. More information about the MYD11 Col 4.1 product may be found in the C4.1 LST Document [28]. The update (mostly due to the changes in the cloud mask) between Col 4 and Col 4.1 in January 2007 caused minimal inconsistencies in the UW BF database (see left panels of Figure 3.).

The processing of the MOD11C3 Col 4.1 and 5 products was discontinued in 2017 and replaced by a new Col 6 product. Figure 3 shows a time series comparison of Col 4/4.1 (left, currently used as input to UW BF) and Col 6 (right) for bands 20, 29, 31, and 32 MO/YD11C3 monthly mean emissivity products for Aqua (red) and Terra (blue) MODIS over a Namib desert location. The Col 6 band 20 (3.76 μm) and band 29 (8.5 μm) emissivity are—we believe—mistakenly identical for both Terra and Aqua MODIS. It appears that band 29 values have been copied into the band 20 variable. Bands 31 and 32 show very little time variation. These facts make the Col 6 MOD11C3 emissivity data unusable for our project. Despite the Col 4/4.1 MOD11 products being the best quality for our purposes, there are some problems in that dataset as well for longer time-scale purposes. The increasing emissivity of the Terra band 29 was probably due to the band 29 cross-talk error [29]. This will likely be mitigated in the future by reprocessing the data on the Col 6.1 L1B cross-talk corrected radiances. Col 4.1 emissivity values also start to decrease significantly, especially for the long-wave region, starting in 2009 January, which is an issue in the UW BF emissivity database. This artifact may be caused by an initialization issue during processing. This defect is eliminated by ASTER GEDv4 data in the CAMEL database, which is presented in Figure 4.

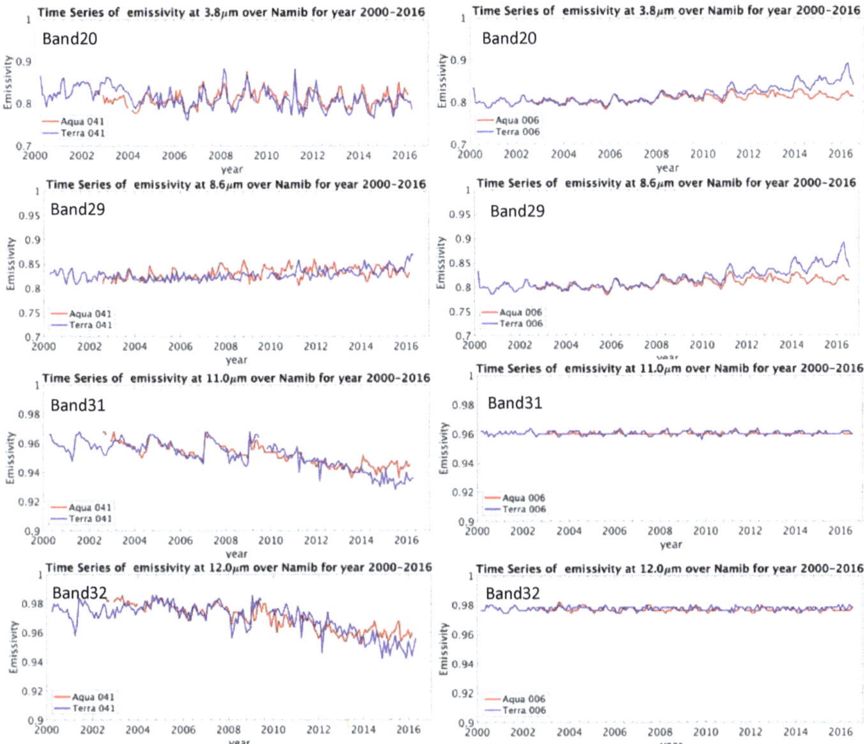

Figure 3. Time series of Col 4/4.1 (**left**) and Col 6 (**right**) band 20, 29, 31, and 32 MxD11C3 monthly mean emissivity for Aqua (red) and Terra (blue) MODIS over a Namib desert location (Lat: 24.25S, Lon:15.25E). Note: In Col 6, band 20 appears to be a copy of band 29 for both Terra and Aqua MODIS.

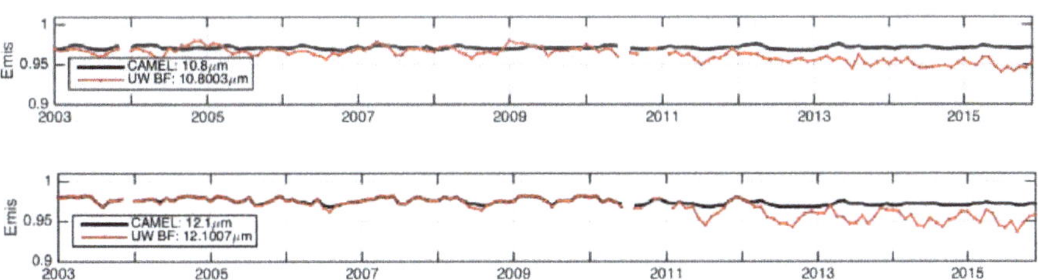

Figure 4. Time series of the Combined ASTER and MODIS Emissivity over Land (CAMEL) database and its input UW BF dataset at 10.8 μm (**top panel**) and 12.1 μm (**bottom panel**) is shown over the Atmospheric Radiation Measurement (ARM) Southern Great Plains (SGP) Cart Site. The CAMEL algorithm effectively eliminates the emissivity degradation observed after 2009 for the long-wave MODIS MOD11 product based UW BF hinge-points.

2.3. The Laboratory Measurements

The CAMEL HSR algorithm takes advantage of a wide variety of laboratory measurements of terrestrial materials (minerals, soils, vegetation, fresh water, salt water, snow, ice, etc.) that have been collected at high spectral resolution for a continuous IR range [30,31]. The laboratory measurements have the advantage of being performed using short path lengths and under purged conditions to minimize the effects of water vapor absorption (and other gases). They also take advantage of laboratory spectrometers that have resolving powers of 1000 or more. The laboratory measurements used to derive the emissivity in this paper were drawn from the MODIS emissivity library (http://www.icess.ucsb.edu/modis/EMIS/html/em.html) at the University of California, Santa Barbara, and the ASTER spectral library [26], including spectra from the John Hopkins University (JHU) Spectral Library, the JPL Spectral Library, and the United States (U.S.) Geological Survey (USGS) Spectral Library.

The MEaSUREs CAMEL database was extended to high spectral resolution using a PC regression analysis similar to the UW HSR algorithm. While the UW HSR algorithm includes 123 selected laboratory measurements, the CAMEL HSR algorithm now includes three sets of laboratory spectra, specifically 55 selected spectra (called version 8) for general use, 82 spectra (called version 10; version 8 + carbonates) for non-vegetated cases, and four snow/ice selected spectra (version 12). The three sets of CAMEL laboratory data and the UW HSR 123 selected laboratory measurements are presented in Figure 5. If the snow fraction is larger than 0.5, only the new snow/ice set of PCs based on laboratory measurements is used. The three new sets of laboratory sets better characterize the emissivity spectra of the non-vegetated surface types and snow-covered areas.

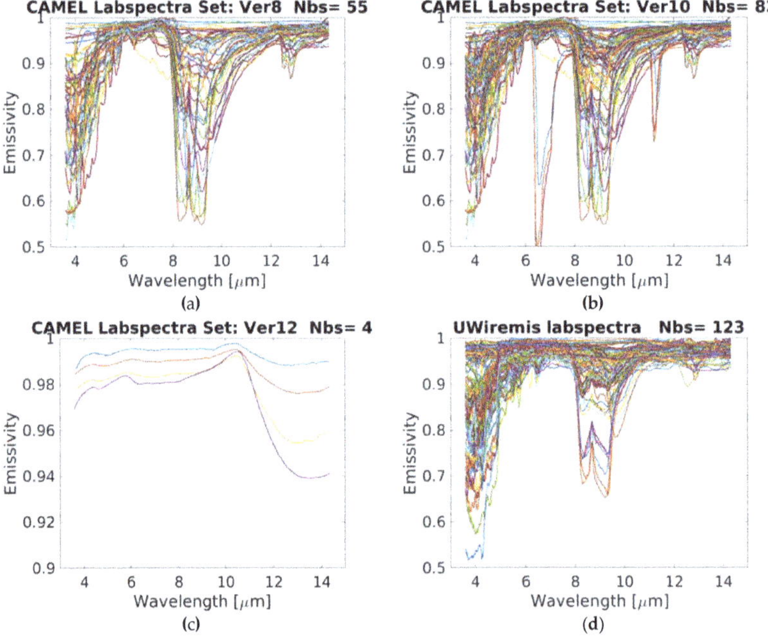

Figure 5. The CAMEL high spectral resolution (HSR) emissivity algorithm now includes three sets of laboratory spectra: (**a**) 55 selected spectra for general use (called version 8), (**b**) 82 spectra for surface types including carbonates (called version 10; version 8 + carbonates), and (**c**) four snow/ice selected spectra (version 12). The UW HSR 123 selected laboratory measurements (**d**) are shown for comparison.

3. Method

As noted earlier, CAMEL is produced by combining the UW-Madison MODIS based UW BF database and the JPL ASTER GEDv4. A limitation of the UW BF database is that emissivity in the thermal IR region (TIR) region (8–12 µm) is not well defined, because MODIS only has three bands in this region (bands 29, 31, 32). This results in an imperfect TIR spectral shape in the two quartz doublet regions at 8.5 µm and 12 µm. The advantages are its moderate spatial resolution (5 km), uniform temporal coverage (monthly), and emissivities, which span the entire IR region (3.6–12 µm). A disadvantage of the ASTER GED is that although there are more bands to define the spectral shape in the TIR region (five bands, 8–12 µm), there are no bands in the mid-wave infrared (MIR) region around 3.8–4.1 µm, which limits its use in models and other atmospheric retrieval schemes. The advantages are its high spatial resolution (~100 m) and high accuracy over arid regions. The two datasets have been integrated together to capitalize on the unique strengths of each product's characteristics. This integration involved two preparatory steps: (1) ASTER GEDv3 emissivities are adjusted for vegetation and snow cover variations over heterogeneous regions to produce ASTER GEDv4, (2) ASTER GEDv3 emissivities are aggregated from 100-m resolution to the UW BF 5-km resolution, and two processing steps: (i) the spectral emissivities are merged together to generate the CAMEL product at 13 hinge points from 3.6 µm to 12 µm, and (ii) the 13 hinge points have been further extended to hyperspectral resolution using a PC-regression approach. The preparation of the ASTER data in steps 1 and 2 has been discussed in Sections 2.1.1 and 2.1.2. The two processing steps are summarized in Figure 6, and are discussed below in more detail.

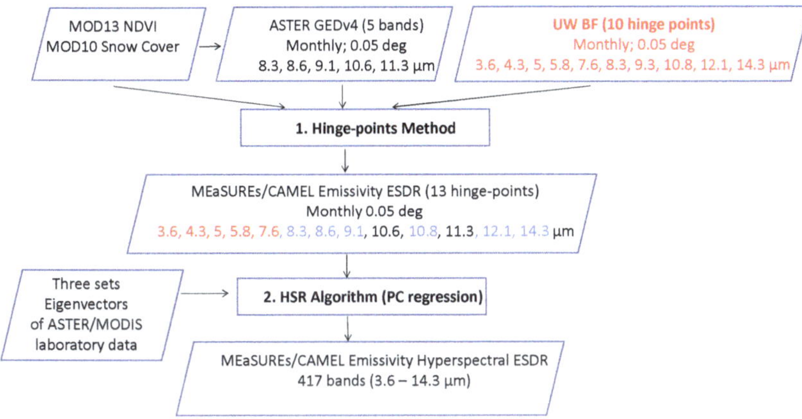

Figure 6. Making Earth System Data Records for Use in Research Environments (MEaSUREs) CAMEL Emissivity Earth System Data Record (ESDR) flowchart.

3.1. Emissivity Hinge-Points Methodology

The merging of the spectral emissivities from the five ASTER bands with the 10 hinge-point bands from the UW BF database and the determination of the CAMEL emissivity hinge points are summarized in Table 1, and described below.

CAMEL hinge points between 3.6–7.6 µm: In the ASTER band gap of the SWIR and MIR region, the CAMEL emissivities between 3.6–7.6 µm are determined by the UW BF values only, and keep the location of the hinge points.

CAMEL hinge points at 10.6 µm and 11.3 µm: The 10.6 µm and 11.3 µm hinge points are added based on the additional observations from ASTER band 10.6 µm and 11.3 µm. CAMEL values at these hinge points are determined from the ASTER GEDv4 observations only.

Table 1. Method to create CAMEL emissivity for the 13 hinge points.

CAMEL Channel Number	CAMEL Wavelength [μm]	UWBF Channels	ASTER Channels	CAMEL Combining Method
1	3.6	Y	-	UWBF1
2	4.3	Y	-	UWBF2
3	5.0	Y	-	UWBF3
4	5.8	Y	-	UWBF4
5	7.6	Y	-	UWBF5
6	8.3	Y	Y	ASTER1 + (CAMEL7(UWBF6, ASTER2) − ASTER2)
7	8.6	Y	Y	Weighted Mean(UWBF6, ASTER2)
8	9.1	-	Y	ASTER 3 + (CAMEL7(UWBF6, ASTER2) − ASTER2)
9	10.6	-	Y	ASTER4
10	10.8	-	-	Linear Interpolation(ASTER4, ASTER5)
11	11.3	-	Y	ASTER5
12	12.1	Y	-	UWBF9 but if ASTER5 > BF9, UWBF9 + diff(UWBF9, ASTER5) * weight if snowfrac > 0.5
12 *	12.1	Y	-	UWBF9 + diff(CAMEL10, UWBF8)
13	14.3	Y	-	UWBF10 but if ASTER5 > BF9, UWBF9+ diff(UWBF9, ASTER5) * weight if snowfrac > 0.5
13 *	14.3	Y	-	CAMEL12

* CAMEL channels 12 and 13 have separate combining methods based on whether the pixel is defined as snow or non-snow using 0.5 snow fraction as a threshold.

CAMEL hinge point at 8.6 μm: This is the only hinge point where the MODIS band 29 (8.55 μm) and ASTER band 11 (8.6 μm) overlap; both have very similar spectral response functions. Since these two bands match closely, we used a weighting rule based on the uncertainties using a "combination of states of information" approach. In this approach, two pieces of information (e.g., two spectral emissivities $\varepsilon(1, \nu)$ and $\varepsilon(2, \nu)$ can be merged in a probabilistic manner by weighting each input based on its relative uncertainty, i.e., $\varepsilon(\nu) = [1/(w1 + w2)] [w1{\cdot}\varepsilon(1, \nu) + w2{\cdot}\varepsilon(2, \nu)]$, where w is a weighting factor based on an uncertainty, σ, as follows: $w = 1/\sigma$. To apply this method, we used 90% and 10% weights as the corresponding uncertainties for ASTER GEDv4 and UW BF on a pixel-by-pixel basis. Given the lack of uncertainty estimates in the MODIS (MOD11) product, the 90/10% weights are determined based on test case studies. In the future, when uncertainty estimates of the MODIS products are available, those weights will be adjusted and objectively defined based on the uncertainties in the input products (MODIS and ASTER). Since the ASTER GEDv4 has more bands that more accurately define the quartz doublets, ASTER band 11 (8.6 μm) gets the 90% weight for arid and semi-arid regions, while for all of the other cases, the UW BF hinge point 6 (8.3 μm) is weighted by the 90%. To determine the arid and semi-arid region, the ASTER NDVI (< 0.2) and ASTER 9.1-μm band ($\leq$0.85) is used. Additionally, over the heavily vegetated tropical rainforests, the MODIS MOD11 emissivity suffers from cloud contamination, resulting in a low emissivity value at the reststrahlen band. To avoid this artifact over this region ($\pm$20 degree latitude band where ASTER NDVI is larger than 0.7 and UW BF emissivity at 8.6 μm is less than 0.96), the ASTER 11 (8.6 μm) is weighted by 90%.

CAMEL hinge points at 8.3 μm and 9.1 μm: The baseline fit procedure that was used in generating the UW BF product extends emissivity from MODIS band 29 (8.6 μm) to inflection points at 8.3 μm and 9.1 μm. The location of these inflection points was maintained, but the UW BF emissivities are improved by replacing the interpolated inflection points with retrieved ASTER emissivities from corresponding bands 10 (8.3 μm) and 12 (9.1 μm), and then adjusting them by the emissivity difference between the new CAMEL 8.6 μm and ASTER 8.6 μm bands. This significantly improves the spectral shape in the Si-O stretching region (8–12 μm).

CAMEL hinge point 10.8 μm: The CAMEL emissivity at the 10.8-μm hinge point is determined as the linear combination of the ASTER band 10.6 μm and 11.3 μm emissivity.

CAMEL hinge points at 12.1 μm and 14.3 μm: The UW BF emissivities at 12.1 μm and 14.3 μm are adjusted by the differences between the UW BF 12.1 μm and ASTER 11.3 μm emissivities to be consistent with the 10.6–11.3 μm region (mostly ASTER-based observations) and improve the spectral shape in this TIR spectral region. A weighting factor is applied based on the difference between the UW BF 12.1 μm and ASTER 11.3 μm emissivities. If the UW BF emissivity value at 12.1 μm is larger than the value at ASTER 11.3 μm, it is likely that the input MOD11 data is not degraded (as is shown on the bottom panel of Figure 4 before 2011), and the weighting factor is 0, with no need for adjustment. If the difference is negative, suggesting that the UW BF 12.1 μm emissivity has likely degraded, so it is smaller than the ASTER 11.3 μm values, then the weighting factor varies based on the ASTER 11.3 μm emissivity value. The weighting factor is 1 or 2, depending on if the ASTER GEDv4 value is larger (likely vegetated or snow/ice covered surface) or smaller than 0.95 (likely mixed or unvegetated surface), respectively. If the weighting factor of 2 produces emissivity values larger than 1, the weighting factor is reduced to 1.5.

Figure 7 shows an example of how much the combined 8.6-μm CAMEL emissivity field differs from the input UW BF and ASTER GEDv4 data for February 2004. The CAMEL emissivity agrees with the UW BF data over vegetated areas (white area) and is higher (yellow-orange) for non-vegetated and snow-covered areas (see Figure 7d). Furthermore, the CAMEL emissivities agree with the ASTER GEDv4 emissivities over the arid, non-vegetated areas such as the Sahara Desert and Siberia (white areas), and is lower (blue) for vegetated scenes (See Figure 7e).

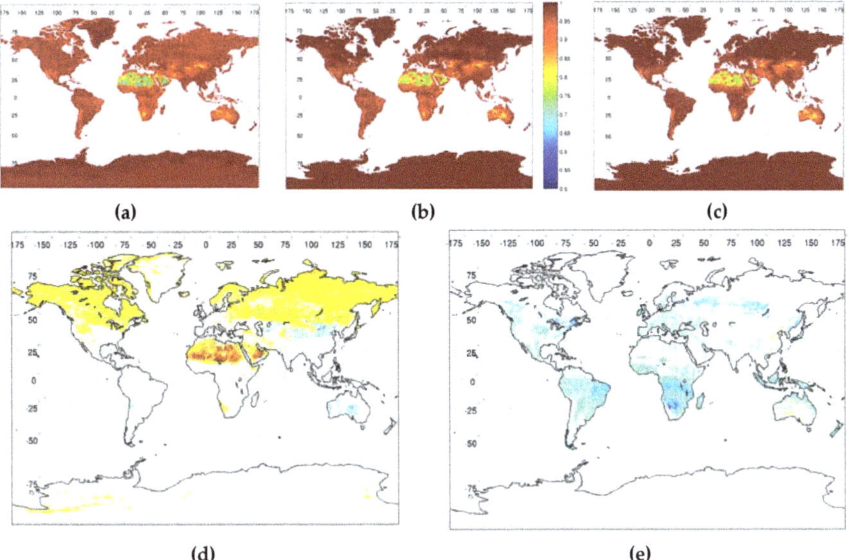

Figure 7. The UW BF (**a**), CAMEL (**b**) and the ASTER GEDv4 (**c**) emissivity at 8.6 μm for February 2004. The difference maps of the emissivity between CAMEL and UW BF database (**d**) and ASTER GED v4 (**e**) are also shown for February 2004.

3.2. High Spectral Resolution Methodology

The MEaSUREs CAMEL 13 hinge-point database was extended to high spectral resolution to capture the small-scale fluctuations in emissivity that were not captured by the CAMEL 13 hinge point dataset.

The CAMEL HSR algorithm uses a PC regression analysis, which is similar to the method developed for the UW HSR Algorithm [11,25]. We assume that emissivity can be derived as a linear combination of the first couple of eigenvectors (or principal components (PCs)) of the laboratory measurements, and that the linear relationship between emissivity spectra and PCs are the same at both moderate and high spectral resolution.

The PCs (eigenvectors) are generated using three sets of selected laboratory measurements (chosen to represent various surface types), and are regressed to the CAMEL 13 hinge points as follows:

$$\vec{e} = \vec{c}\,U \tag{2}$$

where $\vec{e}\,[nh]$ is the CAMEL emissivity on 13 hinge points, $\vec{c}\,[npc]$ is the PCA coefficient vector, and $U[npc,nh]$ is the matrix of the PCs of the lab emissivity spectra on the reduced spectral resolution. The nh stands for the number of hinge points, which is 13 in our case, and npc is the number of eigenvectors. The coefficient can be calculated then:

$$\vec{c} = \vec{e}\,U^T \left(U U^T \right)^{-1} \tag{3}$$

After calculating the coefficients ($\vec{c}$), the high spectral resolution emissivity values are determined using Equation (2) at the same latitude and longitude point by using the high spectral resolution PCs of the laboratory sets. This time, the U matrix sizes are $[npc,nhsr]$, where $nhsr$ stands for the number of high spectral resolution wavenumber points.

In the HSR emissivity algorithm, the ($\vec{e}$) is actually the difference of the emissivity spectra from the mean (E_{lab}) of the selected laboratory dataset (e.g., "general", "general+carbonates", or "snow/ice"), so Equation (3) becomes:

$$\vec{c} = (\vec{e} - E_{lab})U^T \left(U U^T \right)^{-1} \tag{4}$$

Then, the high spectral resolution emissivity $\vec{e_h}$ is calculated:

$$\vec{e_h} = \vec{c}\,U_h + \vec{\lambda_h} \tag{5}$$

There are two primary updates in the CAMEL HSR algorithm from the UW HSR algorithm: one is the number of principal components (PCs) determination (this will be discussed in the next section), and the other is the selection of the laboratory measurements. While the UW HSR algorithm includes only one set of laboratory measurements for all of the surface types, the CAMEL HSR algorithm uses three sets of laboratory measurements based on the surface scene type and coverage: one for general purpose (55 spectra), one for arid areas, which includes more carbonate measurements (82 spectra), and a separate snow/ice set, which includes only snow and ice laboratory measurements (four spectra) (see Section 2.3 for more details). With these three separate categories, false spectral features have been avoided that occasionally occurred in the UW HSR emissivities. For example, the quartz doublet feature sometimes erroneously appeared over fully snow-covered areas.

Table 2 summarizes the methodology used to determine which laboratory set would be used for a given pixel. First, a carbonate test is performed that is based on the laboratory measurements. A pixel falls into the non-vegetated and carbonate category if the ASTER NDVI is less than 0.2, the CAMEL emissivity at 10.6 μm is larger than the 11.3 μm emissivity by more than 0.009, and the CAMEL emissivity in the SWIR region is lower than 0.9. If the carbonate test is true, the Version 10 laboratory dataset is assigned, and if it fails, then the "general" Version 8 laboratory dataset is assigned to the pixel. In the general category, the quartz doublet feature can be present, which requires more PCs to capture it accurately.

Table 2. Determination of the number of PCs and the version number of laboratory datasets for each pixel.

Tests	Version # of Laboratory Dataset	Number of PCs
[1]Carbonate test: yes	10 (general_carbonates)	5
[1]Carbonate test: no, but CAMEL9.1 <= 0.85	8 (general)	9
All the others	8 (general)	7
MOD10 snow fraction >= 0.5	12 (snow/ice)	2

[1]Carbonate test: $(CAMEL_{10.6} - CAMEL_{11.3}) > 0.009$ & ASTER NDVI < 0.2 & $CAMEL_{3.6} < 0.9$.

To determine the scene over a bare and sandy area, the 9.1-μm emissivity is used. If it is lower than 0.85, the surface probably contains quartz. Figure 8 shows an example of applying the PC regression fit to the CAMEL LSE ESDR product at 13 points over the Namib Desert, Namibia. Comparisons of the CAMEL emissivity with lab emissivity spectra from field sand samples show very good agreement, particularly in the quartz doublet regions at 8.5 μm and 12.5 μm when compared with the UW HSR. Biases and root mean square errors were reduced by 3% and 4%, respectively, by using the CAMEL product instead of the UW HSR product. This case failed the carbonate test, but the 9.1-μm CAMEL emissivity was less than 0.85; hence, the Version 8 (general) laboratory set with nine PCs was determined by the CAMEL HSR algorithm. The determination of the number of PCs is explained in more detail in the next section.

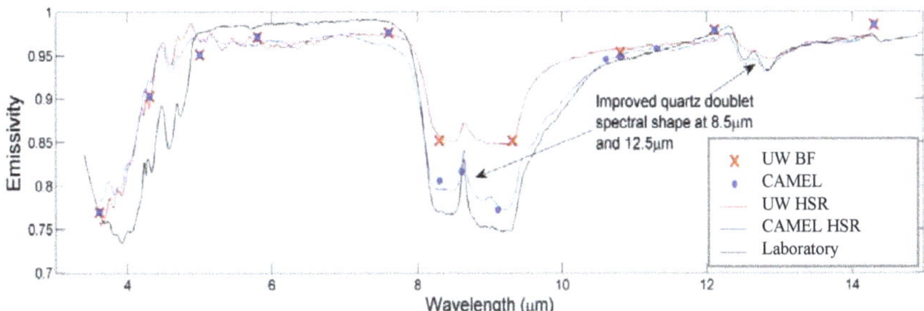

Figure 8. The advantages of combining the ASTER GEDv4 and UW BF databases are evident here, showing the emissivity spectra over the Namib Desert, Namibia. UW BF emissivity for January 2004 (crosses) and hyperspectral fit (red line), the CAMEL 13 hinge-point emissivity (blue dots) and hyperspectral fit (blue line), and lab spectra (black) of sand samples collected over the Namib Desert. Note the improved spectral shape in CAMEL HSR (blue) in the quartz doublet regions between 8–10 μm and 12–13 μm.

The snow fraction based on the MODIS MOD10 product has been added to the ASTER GEDv4, and hence to the CAMEL products, to improve the emissivity determination over snow and ice. In the CAMEL algorithm, if the snow fraction is larger than 0.5, then the snow/ice laboratory dataset (Version 12) is used with two PCs. The emissivity over fully snow and ice-covered areas is improved [32], but the 0.5 threshold assumption may need to be modified for partially snow/ice covered areas. A blended average emissivity between the snowy and not snowy emissivity spectra for a single pixel based on the snow fraction may be a better approach in the future.

3.2.1. Determining of the Number of Principal Components to Use

In the PCA regression method, the first PCs with the highest eigenvalues represent real variations in the data while the last, least significant PCs most often represent random white noise. In this study, the maximum number of PCs allowed is 13, due to the number of spectral points of the input CAMEL hinge-point emissivities. However, use of the maximum number or close to the maximum number of PCs sometimes makes the solution unstable.

To determine the appropriate number of PCs to use, we first determine the percentage cumulative variance (PCV) function and eigenvalues of the three laboratory datasets. This was then followed by spectral reconstruction for the selected case sites, which were selected to cover the major global surface types, such as the sandy desert location over Namib (see also Figure 8), the rocky carbonated surface type over Yemen, seasonal vegetation cover at the Atmospheric Radiation Measurement (ARM) Cart site, a mountainous region at Mt. Massive, which is covered by forest and snow in the winter, and the permanently snow-covered location at Greenland. For these locations, the number of PCs was chosen to reconstruct high spectral emissivity spectra. The reconstructed spectra were subsequently compared to laboratory measurements or other in situ measurements. More details about the validation are provided in Part 2 of this paper [32].

The left panels of Figure 9 show the number of PCs that were chosen for the three lab datasets, as well as the optimal number of PCs that were determined where the PCV value equaled 0.999. The right panels illustrate the eigenvalues of the laboratory datasets. The nature of the eigenvalues becomes less significant after the first eight eigenvectors for Version 8 and Version 10, and after two eigenvectors for Version 12, which indicates that the optimal number of PCs can be as low as nine for Version 8 and Version 10, and two for Version 12. The optimal number of PCs was then finalized based on inspecting the case studies.

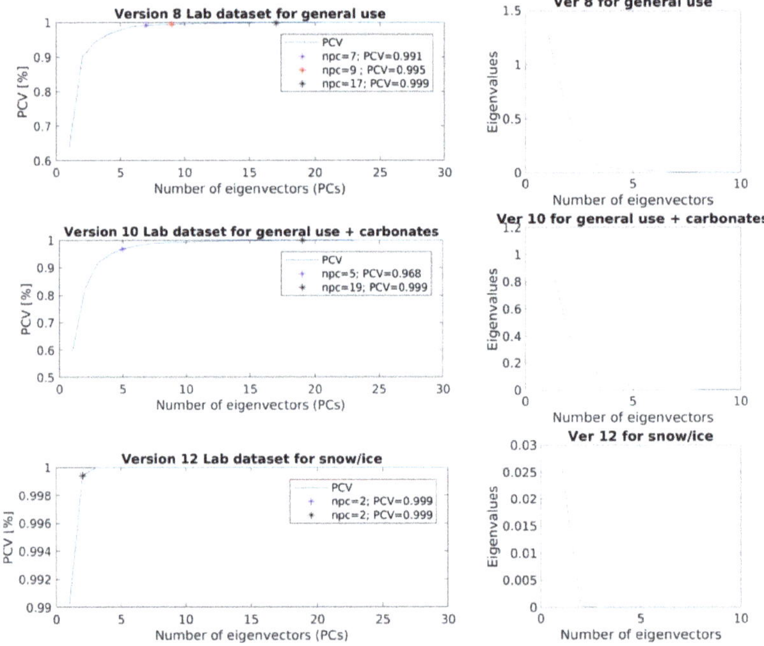

Figure 9. (**Left**) Percentage cumulative variance (PCV) function of the three selected laboratory measurement sets as a function of the number of principal components (PCs). The chosen number of PCs is indicated with blue or red stars. The legend contains the corresponding PCV values. Black stars stand for the number of PCVs, which reached the 0.999 value. (**Right**) The mean eigenvalues of the laboratory datasets for the first 10 eigenvectors.

For example, for the Yemen case (Figure 10a), the optimal number of PCs would be 19 (Figure 9, middle panel), but the emissivity spectra with more than five PCs starts capturing less of the carbonated (dip) feature in the 6–7 µm spectral region. For a general case such as that over the ARM cart site (Figure 10b), the emissivity spectra with seven PCs captures more of the flat shape of the emissivity spectra between the 5–8 µm region than with that nine PCs. However, for the same "general" laboratory set, over the Namib Desert (Figure 10c), nine PCs better captures the quartz doublet feature. For the snow/ice laboratory dataset, the chosen number of PCs is identical with the optimal number of PCs (equal two) determined by the PCV function and eigenvalues (see the bottom panels of Figures 9 and 10d).

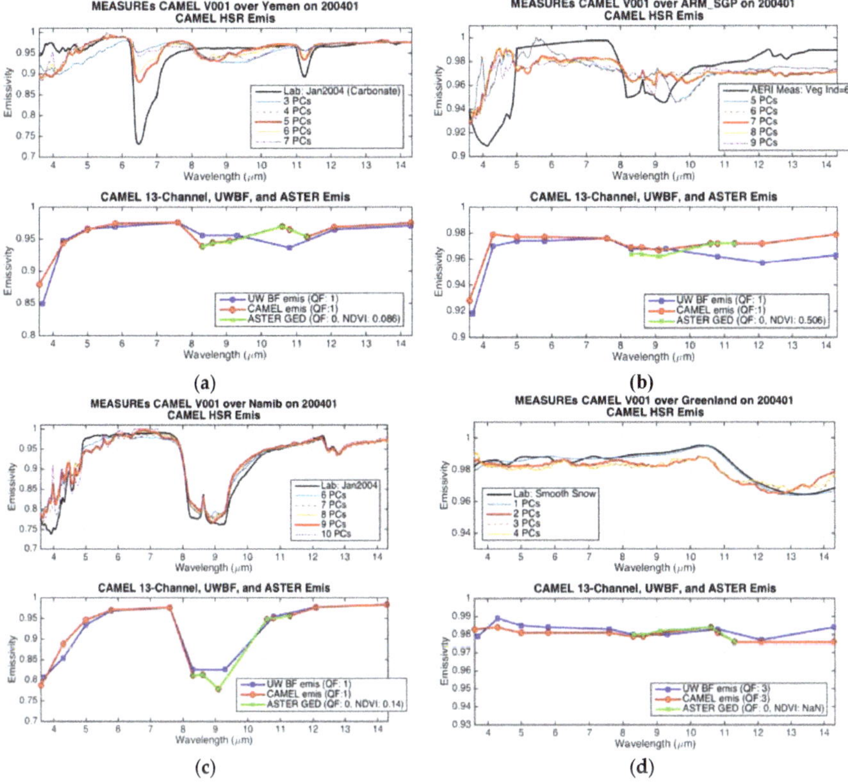

Figure 10. Emissivity on January 2004 at (**a**) Yemen, (**b**) Atmospheric Radiation Measurement (ARM) Southern Great Plains (SGP) Cart site, (**c**) Namib Desert, and (**d**) Greenland case sites. High spectral resolution emissivity from CAMEL with a different number of PCs used (different colored lines) and lab or Atmospheric Emitted Radiance Interferometer (AERI) measurements (black) are shown in the top row panels. The selected number of PCs is solid red for each case.

4. CAMEL Products

As noted earlier, the CAMEL dataset includes monthly global records of emissivity and uncertainty at 13 hinge points between 3.6–14.3 µm, as well as PCA coefficients at 5-km resolution for the years 2000 to 2016. A HSR algorithm is also provided for HSR applications. Detailed information about these file specifics is provided in the CAMEL Users' Guide [33].

The input UW BF emissivity database of the CAMEL dataset is based on the Aqua/MODIS MYD11 monthly mean emissivity product. For the time period between 2000 and 2002, when the

Aqua satellite was not yet launched, the Terra/MODIS MOD11C3 monthly mean emissivities are used to produce the UW BF data, which are subsequently input to the CAMEL database. To check the data consistency and continuity of the Aqua and Terra MODIS, time series of CAMEL, ASTER, Aqua/MODIS MYD11, and Terra/MODIS MOD11 are compared over four case study sites. Figure 11 demonstrates unbiased and consistent agreement between Aqua and Terra MOD11 products over a Rocky Mountain case site for five common wavelengths. The peaks in the CAMEL products during the winter months reflect the snow cover over the area.

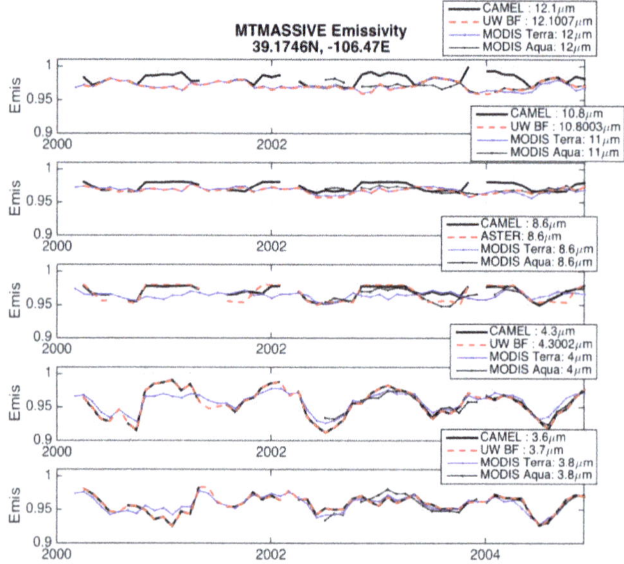

Figure 11. The CAMEL dataset is currently available for the years 2003–2015, and makes use of the Aqua MODIS data as input to the 13 hinge-point product. The dataset is now extended to 2000, and uses the Terra MODIS data for the months of 2000 through December 2002. Time series of the CAMEL, ASTER GEDv4, UW BF (UW BF), and MODIS emissivity are shown, and demonstrate consistency between the Aqua and Terra products over the Rocky Mountain case site.

The CAMEL emissivity products also include the MOD10 snow fraction, the ASTER NDVI, and a quality flag for each pixel. The CAMEL quality flag is determined based on the quality flags of the two input datasets (see Table 3). It ranges from 1 as the best quality to 4 as the least confident quality product. The quality values 2, 3, and 4 indicate when either UW BF, ASTER GEDv4, or both are filled from nearby valid grid cell estimates or from the average value of the neighboring months or yearly data. The quality flag can also be used as a sea/land mask. A zero value means sea or inland water, while a non-zero value is over land.

Table 3. Definition of the CAMEL emissivity quality flag.

Value	Description
0	sea or inland water
1	input UW BF and ASTER GEDv4 data are good quality
2	input UW BF is good quality and ASTER GEDv4 is filled
3	the input UW BF is filled, but ASTER GEDv4 is good quality
4	both the UW BF and ASTER GEDv4 values are filled

The product uncertainty is estimated by a total emissivity uncertainty comprised of three independent components: temporal, spatial, and algorithm variability. Each measure of uncertainty is provided for all 13 hinge points and at every latitude–longitude point. The total uncertainty is calculated from the components as a root square sum. Part 2 of this paper [32] provides detailed information on how each component is determined.

5. Applications

Within the NASA MEaSUREs LST&E project, the surface emissivity product plays a critical role in the estimating of surface skin temperature derived from satellite remote sensing [33]. In particular, the intent of the MEaSUREs LST&E project is to unify the global LST estimates from polar orbiting and geostationary satellites through the use of a common emissivity database. Another important application of the CAMEL emissivity database is for medium range weather forecasting, where variability in land surface emissivity has led to the blacklisting (i.e., neglect) of infrared satellite measurements that are not accurately represented by the forecast model, i.e., due to a lack of knowledge of land surface properties [34]. In numerical weather prediction (NWP), the surface emissivity contributes to both the land surface model and the assimilation of satellite infrared radiance channels.

NWP land surface models attempt to model the diurnal variation of surface air temperature using a broadband radiative balance between incoming short-wave heating and long-wave cooling [35]. Due to a prior lack of spatially and temporally variant global broadband emissivity (BBE) measurements, it has been common practice in land surface models to set BBE as a single constant for all land types. This may lead to systematic biases in the estimated net radiation for any particular location and time. The CAMEL HSR product provides an opportunity to create a monthly mean BBE product [36] by numerical integration over the CAMEL spectrum. Initial investigations show that improved estimates of variations over time and land cover classification are more realistic when using BBE derived from the CAMEL dataset [37].

An accurate emissivity is also required for any application involving calculations of brightness temperatures, such as the assimilation of radiances into global weather (or climate) models. For example, an interface to the HSR emissivity algorithm with the emissivity database was implemented into the European Organization for the Exploitation of Meteorological Satellites (EUMETSAT) Numerical Weather Prediction Satellite Application Facilities (NWP SAF) Radiative Transfer for Television Infrared Observation Satellites (TIROS) Operational Vertical Sounder (RTTOV) model Version 10 (UWIREMIS) and 12 (CAMEL). The implementation, testing, and evaluation of that HSR emissivity module is described in Borbas and Ruston [11]. The RTTOV model [38] is the primary tool used at the Met Office in the United Kingdom (UK) and European Centre for Medium-range Weather Forecasting (ECMWF) for the assimilation of high spectral resolution infrared sounders, including the NASA Aqua AIRS, the EUMETSAT Meteorological Operational Satellite (METOP) Infrared Atmospheric Sounding Interferometer (IASI), and the Suomi National Polar-orbiting Operational Environmental Satellite System (NPOESS) Preparatory Project (S-NPP)/NOAA-20 Cross-track Infrared Sounder (CrIS) sensors. The surface emissivity is used to provide the boundary condition for computing the upwelling radiance from the surface. Spectral errors in surface emissivity can lead to misinterpretation of the IR observations and errors in the derived air temperature or moisture profiles. Table 4 includes a summary of the main differences between the RTTOV Version 10 and Version 12 land surface IR emissivity modules, e.g., the different input datasets and spatial resolution. The CAMEL module kept the original $0.05° \times 0.05°$, while the UWIREMS module was degradated to $0.1° \times 0.1°$ spatial resolution. In addition, the new CAMEL module uses three laboratory datasets, with various numbers of PCs based on surface scene and coverage, as described in Section 3.2.

Table 4. Summary of comparison between the RTTOV10/UWIREMIS and RTTOV12/CAMEL emissivity databases.

	RTOV10/UWIREMIS	RTTOV12/CAMEL
Spatial Resolution:	$0.1° \times 0.1°$	$0.05° \times 0.05°$
Inputs:	MODIS MYD11 (6) MODIS-ASTER Lab	UW BF (10) ATER-GEDv4 (5) MODIS-ASTER Lab
Method:	Baseline-fit conceptual modelPCA regression	Conceptual hinge-points method PCA regression
Laboratory data:	One set of 123 selected MODIS/ASTER	three sets of MODIS/ASTER: 55 general set; 82 general + carbonates; 4 ice/snow
Number of PCs	6	2, 5, 7, or 9 depends on surface types, $ASTER_{8.6}$ emis & NDVI, MOD10 snow fraction
Outputs	Emissivity spectra on 10 hinge points and 417 HSR points, PCA coefficients	Emissivity spectra on 13 hinge points and 417 HSR points, PCA coefficients, uncertainties, NDVI, snow fraction

The geographical locations where significant improvements are expected in RTTOV performance using the CAMEL module in Version 12 versus the UWIREMIS module in Version 10 are over snow/ice surfaces and non-vegetated surfaces including bare soil, sand, and rock (including quartz and carbonates). To ascertain these improvements, IASI observed brightness temperatures were compared to those calculated using the RTTOV UW IR emissivity (UWIREMIS) module based on (1) the UW BF, and (2) the NASA MEaSUREs CAMEL database. The debiased variances over the 3.6–5 μm, 8–9 μm, and 10–13 μm spectral region are calculated and used as the indicator for an improved emissivity estimate. Figures 12 and 13 illustrate an IASI granule at 17:56 UTC on 29 September 2008, where the CAMEL emissivity improves the brightness temperature calculations over the Arabian Peninsula. The full results of the RTTOV simulation study is reported in Part 2 of this paper [32].

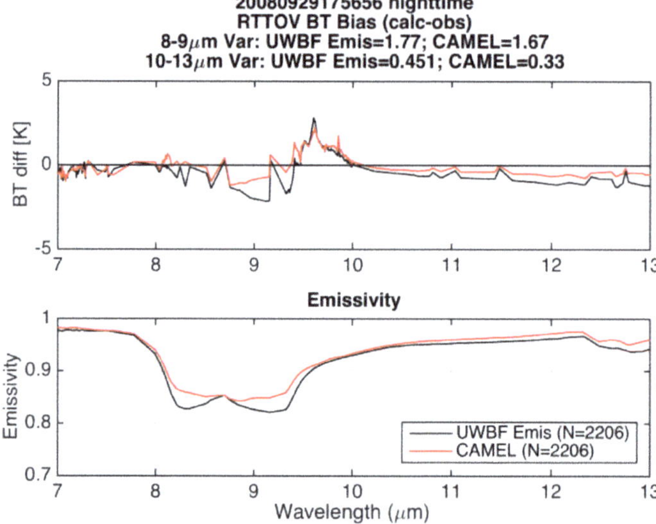

Figure 12. IASI observed brightness temperatures are compared to those calculated using the Radiative Transfer for TOVS (RTTOV) UW IR emissivity module based on the UW BF emissivity database (black) and the CAMEL emissivity database (red) for the granule at 17:56 UTC, on 29 September 2008. The debiased variances are included over the 8–9 μm and 10–13 μm spectral regions.

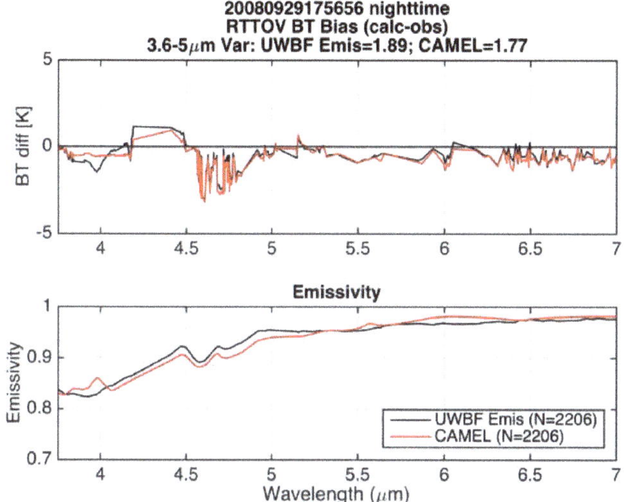

Figure 13. Same as Figure 12 but for the short IR spectral region (between 3.6–7 µm).

A related near-real time application of the CAMEL emissivity is to improve the training sets used to derive the temperature and moisture profiles from the high spectral resolution sounders [39]. The advantage of using the CAMEL database in a retrieval algorithm as a first guess and also in the training phase is demonstrated for the EUMETSAT IASI Algorithm [40]. Similar studies are underway at NOAA for application to the operational CrIS sensor with an emphasis on trace gas retrievals over land [41].

6. Conclusions and Future Plans

The CAMEL database was created by merging the UW MODIS-based emissivity database (UW BF) developed at the University of Wisconsin-Madison, and the ASTER Global Emissivity Dataset Version 4 produced at JPL. The new CAMEL database is integrated to capitalize on the unique strengths of each product's characteristics. The CAMEL ESDR includes monthly global records of emissivity and related uncertainties at 13 hinge points between 3.6 and 14.3 µm, as well as PCA coefficients at 5-km resolution for the years 2000 through 2016. A HSR algorithm has been developed for HSR applications, such as in data assimilation schemes and radiative transfer models that require accurate high spectral resolution emissivity as a first guess for hyperspectral resolution retrieval schemes such as those for AIRS, IASI, and CrIS. This paper describes the 13 hinge-point combination methodology and the high spectral resolution algorithm, and reports on the current status of the dataset. The CAMEL products are evaluated extensively with laboratory measurements over different field sites, with IASI climatological data [42] and using the RTTOV forward model for IASI brightness temperature simulation. Part 2 [32] of this paper provides more information about the results of these validations and evaluations.

The input of the UW BF dataset required for producing CAMEL is the MODIS MOD11C3 monthly mean emissivity products, which includes emissivity at six IR bands. The climate quality of the UW BF dataset is affected by changes over time in the quality of the MOD11 products. In 2017, the processing of the MOD11C3 Col 4.1/5 product was discontinued and replaced by a new Col6 product with some major differences and consequences for CAMEL. The discontinuation of the Col4.1/5 MOD11C3 products and a bug in the Col6 products requires a transition to the new MOD21 emissivity products developed by JPL, which will soon be available. We anticipate that the transition from MOD11 to MOD21 for CAMEL Version 2 will improve the product accuracy and reduce the systematic and time-dependent errors that have been identified in CAMEL Version 1 (V1).

In CAMEL V1, a distinct set of laboratory measurements (including snow and ice spectra) are uniquely assigned to snow-covered areas. The snow emissivity that was derived using the unique snow/ice emissivity spectra is calculated for 5 km × 5 km pixels where the snow fraction (derived from MOD10) is >0.5. This is not optimal for pixels covered partially by snow. In the future, we plan to determine the emissivity as a linear blend of the snow and underlying land emissivity values weighted by the observed snow fraction. This should lead to improvements in high latitude forest regions and in transition zones at mid-latitudes.

The MOD11 emissivity product input to CAMEL uses a day/night algorithm with the assumption that day and night emissivity does not change over a period of a few days. That inherent assumption could introduce an error, such as if, for example, the day/night emissivity changes due to soil moisture, snowmelt, or any rapid changes in surface (e.g., fire). However, this error is very difficult to quantify, and there are no uncertainty estimates of this provided in the original MODC3 products developed by Wan et al. [27]. Masiello et al. [25] used IASI observations, and Li et al. [43] used geostationary satellite data over the Sahara Desert to show that the IR emissivity between 8.7–12 µm has a diurnal variation. The strongest diurnal effect occurs at 8.7 µm, with the highest value at night and the lowest value during the day. The diurnal effect is the weakest at 12 µm, with an opposite feature. The assumption of equal day–night emissivity can result in as much as a 2% emissivity error at 8.7 µm over desert. In future work, we plan to investigate whether there is a diurnal effect in the MODIS product, and how much it contributes to CAMEL uncertainties.

The CAMEL V1 does not yet include an angular dependence to the emissivity. However, laboratory measurement studies [44–46] show an angular variation in thermal IR emissivity in the 8–14 µm spectral band for non-vegetated soils and samples, while homogeneous grass, for example, does not show a strong angular dependence. Garcia-Santos et al. [47] and Ruston et al. [48] found that the emissivity change is small for viewing angles lower than 40°, but at higher viewing angles, the emissivity decreased significantly. Garcia-Santos et al. established a relationship to take into account the viewing zenith angle dependencies for a range of soil types. In contrast to the bare soils, vegetated surfaces can exhibit a "canopy" effect where the emissivity value actually increases along with the viewing angle. Borbas et al. [49], under a EUMETSAT NWP-SAF Associate Scientist Mission, investigated the angular dependence of the IR emissivity using satellite retrieved measurements over the IR spectrum derived from the CrIS as a function of International Geosphere-Biosphere Programme (IGBP) ecosystem types, wavelengths, and seasons. The difference between the mean emissivity at nadir (0°) and the maximum of 60° shows a 6.8% and 8.0% increase at 3.7 µm and 4.3 µm, respectively, during daytime, and 3.2% and 6.4% during nighttime. Over the TIR region, the land surface emissivity does not vary much, so the angular changes are much smaller as well. Decreases of 1.1% and 0.9% in emissivity were observed at 12.1 µm. We are planning to develop a parameterization correction/function for angle dependence, which will be incorporated into the CAMEL algorithm uncertainty based on results from Garcia-Santos et al. [47] and Ruston et al. [48]. Our goal is to provide improved uncertainty estimates for high viewing angles based on these results, which will help promote the use of improved climate quality products for all view angle geometry configurations.

The CAMEL database unifies the infrared emissivity measurements from NASA sensors under the umbrella of a NASA MEASURES LST&E project. The CAMEL dataset is expected to decrease the errors in LST estimation from moderate spatial resolution satellites by providing realistic surface emissivity estimates with global coverage and monthly temporal sampling at 5-km resolution. Applications to NWP short and medium range weather forecasting are in progress, with the potential for climate applications as the emissivity record is extended in time using NOAA satellites.

Author Contributions: E.B., G.H. and R.K. conceived and designed the method; M.F. analyzed the data; S.H. contributed his expertise; E.B. wrote the paper.

Acknowledgments: This research was supported by the NASA grant NNX08AF8A. The MODIS MOD10, MOD11, MYD11, MOD13, ASTER GED Version 4 and CAMEL Version 1 products used in this study are available courtesy of the NASA EOSDIS Land Processes Distributed Active Archive Center (LP DAAC), United States Geological Survey/Earth Resources Observation and Science Center, Sioux Falls, South Dakota, [https://lpdaac.usgs.gov/dataset_discovery/aster].

Conflicts of Interest: The authors declare no conflict of interest.

References

1. Willet, K.; Thorne, P. The ISTI Steering Committee. Progress Report for the International Surface Temperature Initiative. 2011. Available online: http://www.surfacetemperatures.org/progress_reports (accessed on 15 November 2017).
2. Anderson, M.C.; Norman, J.M.; Mecikalski, J.R.; Otkin, J.A.; Kustas, W.P. A climatological study of evapotranspiration and moisture stress across the continental United States based on thermal remote sensing: 2. Surface moisture climatology. *J. Geophys. Res. Atmos.* **2007**, *112*. [CrossRef]
3. Moran, M.S. Thermal infrared measurement as an indicator of plant ecosystem health. In *Thermal Remote Sensing in Land Surface Processes*; Quattrochi, D.A., Luvall, J., Eds.; Taylor and Francis: Abingdon-on-Thames, UK, 2003; pp. 257–282.
4. Zhou, L.; Dickinson, R.E.; Tian, Y.; Jin, M.; Ogawa, K.; Yu, H.; Schmugge, T. A sensitivity study of climate and energy balance simulations with use of satellite-derived emissivity data over Northern Africa and the Arabian Peninsula. *J. Geophys. Res. Atmos.* **2003**, *108*, 4795. [CrossRef]
5. French, A.N.; Schmugge, T.J.; Ritchie, J.C.; Hsu, A.; Jacob, F.; Ogawa, K. Detecting land cover change at the Jornada Experimental Range, New Mexico with ASTER emissivities. *Remote Sens. Environ.* **2008**, *112*, 1730–1748. [CrossRef]
6. Seemann, S.W.; Li, J.; Menzel, W.P.; Gumley, L.E. Operational retrieval of atmospheric temperature, moisture, and ozone from MODIS infrared radiances. *J. Appl. Meteorol.* **2003**, *42*, 1072–1091. [CrossRef]
7. Hulley, G.; Hook, S. *MOD21 MODIS/Terra Land Surface Temperature/3-Band Emissivity 5-Min L2 1 km V006*; NASA EOSDIS Land Processes DAAC: Sioux Falls, SD, USA, 2017. [CrossRef]
8. Hulley, G.; Hook, S. *AG100: ASTER Global Emissivity Dataset 100-meter V003*; NASA EOSDIS Land Processes DAAC; USGS Earth Resources Observation and Science (EROS) Center: Sioux Falls, SD, USA, 2014. [CrossRef]
9. Borbas, E.; Seemann, S.W. *Global Infrared Land Surface Emissivity: UW-Madison Baseline Fit Emissivity Database V2.0*; Space Science and Engineering Center, University of Wisconsin-Madison: Madison, WI, USA, 2007. Available online: http://cimss.ssec.wisc.edu/iremis/ (accessed on 15 November 2017).
10. Hulley, G.; Hook, S. *AG5KMMOH: ASTER Global Emissivity Dataset, Monthly, 0.05 degree, HDF5 V041*; NASA EOSDIS Land Processes DAAC, USGS Earth Resources Observation and Science (EROS) Center: Sioux Falls, SD, USA, 2016. [CrossRef]
11. Borbas, E.E.; Ruston, B.C. *The RTTOV UW IR Land Surface Emissivity Module.* Mission Report NWPSAF-MO-VS-042; EUMETSAT Numerical Weather Prediction; Satellite Applications Facility: Darmstadt, Germany, 2010. Available online: nwpsaf.eu/vs_reports/nwpsaf-mo-vs-042.pdf (accessed on 15 November 2017).
12. Hook, S. *CAM5K30EM: Combined ASTER and MODIS Emissivity for Land (CAMEL) Emissivity Monthly Global 0.05 Deg V001*; NASA EOSDIS Land Processes DAAC, USGS Earth Resources Observation and Science (EROS) Center: Sioux Falls, SD, USA, 2017. [CrossRef]
13. Hulley, G.C.; Hook, S.J. Generating Consistent Land Surface Temperature and Emissivity Products Between ASTER and MODIS Data for Earth Science Research. *IEEE Geosci. Remote Sens. Lett.* **2010**, 1304–1315. [CrossRef]
14. Hulley, G.; Hook, S.J.; Abbott, E.; Malakar, N.; Islam, T.; Abrams, M. The ASTER Global Emissivity Dataset (ASTER GED): Mapping Earth's emissivity at 100 meter spatial resolution. *Geophys. Res. Lett.* **2015**, *42*. [CrossRef]
15. Hulley, G.C.; Hook, S.J. Intercomparison of Versions 4, 4.1 and 5 of the MODIS Land Surface Temperature and Emissivity Products and Validation with Laboratory Measurements of Sand Samples from the Namib Desert, Namibia. *Remote Sens. Environ.* **2009**, *113*, 1313–1318. [CrossRef]

16. Hulley, G.C.; Hook, S.J. The North American ASTER Land Surface Emissivity Dataset (NAALSED) Version 2.0. *Remote Sens. Environ.* **2009**, *113*, 1967–1975. [CrossRef]

17. Vogel, R.; Liu, Q.; Han, Y.; Weng, F. Evaluating a satellite-derived global infrared land surface emissivity data set for use in radiative transfer modeling. *J. Geophys. Res. Atmos.* **2011**, *116*, 11. [CrossRef]

18. Gutman, G.; Ignatov, A. The derivation of the green vegetation fraction from NOAA/AVHRR for use in numerical weather prediction models. *Int. J. Remote Sens.* **1998**, *19*, 1533–1543. [CrossRef]

19. Knuteson, R.O.; Best, F.A.; DeSlover, D.H.; Osborne, B.J.; Revercomb, H.E.; Smith, W.L., Sr. Infrared land surface remote sensing using high spectral resolution aircraft observations. *Adv. Space Res.* **2004**, *33*, 1114–1119. [CrossRef]

20. Abrams, M. (NASA/JPL). Personal Communication, 2018.

21. Wolfe, R.E.; Nishihama, M.; Fleig, A.J.; Kuyper, A.; Roy, P.; Storey, J.C.; Patt, F.S. Achieving sub-pixel geolocation accuracy in support of MODIS land science. *Remote Sens. Environ.* **2002**, *83*, 31–49. [CrossRef]

22. Lin, G.; Wolfe, R.; Kuyper, J.; Yin, Z.; Tan, B. *(Terra, Aqua) MODIS Geolocation Status*; NASA GSFC Code 619; MODIS Science Team Meeting, Calibration Workshop: Greenbelt, MD, USA, 2016. Available online: https://modis.gsfc.nasa.gov/sci_team/meetings/201606/presentations/cal/MODIS_VIIRS_Geolocation_Status_CalWorkshop_Lin_2016.pdf (accessed on 15 February 2018).

23. Moeller, C. (UW-Madison/SSEC). Personal Communication, 2018.

24. Seemann, S.W.; Borbas, E.E.; Knuteson, R.O.; Stephenson, G.R.; Huang, H.L. Development of a global infrared land surface emissivity database for application to clear sky sounding retrievals from multispectral satellite radiance measurements. *J. Appl. Meteorol. Climatol.* **2008**, *47*, 108–123. [CrossRef]

25. Masiello, G.; Serio, C.; Venafra, S.; DeFeis, I.; Borbas, E.E. Diurnal variation in Sahara desert sand emissivity during the dry season from IASI observations. *J. Geophys. Res. Atmos.* **2014**, *119*, 1626–1638. [CrossRef]

26. Baldridge, A.M.; Hook, S.J.; Grove, C.I.; Rivera, G. The ASTER Spectral Library Version 2.0. *Remote Sens. Environ.* **2009**, *114*, 711–715. [CrossRef]

27. Wan, Z.M.; Li, Z.L. A physics-based algorithm for retrieving land-surface emissivity and temperature from EOS/MODIS data. *IEEE Trans. Geosci. Remote Sens.* **1997**, *35*, 980–996.

28. MODIS Land Surface Emissivity/Temperature (MOD11) C4.1 LST Document. Available online: http://landweb.nascom.nasa.gov/QA_WWW/forPage/C41_LST.doc (accessed on 15 February 2018).

29. Wilson, T.; Wu, A.; Geng, X.; Wang, Z.; Xiong, X. Analysis of the electronic crosstalk effect in Terra MODIS long-wave infrared photovoltaic bands using lunar images. *Image Signal Process. Remote Sens.* **2016**, *22*, 100041C.

30. Salisbury, J.W.; D'Aria, D.M. Emissivity of terrestrial materials in the 8-14 μm atmospheric window. *Remote Sens. Environ.* **1992**, *42*, 83–106. [CrossRef]

31. Salisbury, J.W.; D'Aria, D.M. Emissivity of terrestrial materials in the 3-5 μm atmospheric window. *Remote Sens. Environ.* **1994**, *47*, 345–361. [CrossRef]

32. Feltz, M.; Borbas, E.; Knuteson, R.; Hulley, G.; Hook, S.J. The Combined ASTER MODIS Emissivity Over Land (CAMEL) Part 2: Uncertainty and Validation. *Remote Sens.* **2018**, under review.

33. Borbas, E.; Hulley, G.; Knuteson, R.; Feltz, M. MEaSUREs Unified and Coherent Land Surface Temperature and Emissivity (LST&E) Earth System Data Record (ESDR): The Combined ASTER and MODIS Emissivity Database over Land (CAMEL) Users' Guide; 2017. Available online: https://lpdaac.usgs.gov/sites/default/files/public/product_documentation/cam5k30_v1_user_guide_atbd.pdf (accessed on 15 November 2017).

34. Dee, D.P.; Uppala, S.M.; Simmons, A.J.; Berrisford, P.; Poli, P.; Kobayashi, S.; Andrae, U.; Balmaseda, M.A.; Balsamo, G.; Bauer, P.; et al. The ERA-Interim reanalysis: Configuration and performance of the data assimilation system. *Q. J. R. Meteorol. Soc.* **2011**, *137*, 553–597. [CrossRef]

35. Reichle, R.H.; Sujay, V.K.; Sarith, P.P.M.; Randal, D.K.; Liu, Q. Assimilation of satellite-derived skin temperature observations into land surface models. *J. Hydrometeorol.* **2010**, *11*, 1103–1122. [CrossRef]

36. Borbas, E.; Feltz, M.; Knuteson, R. *Broad Band Emissivity Derived from the MEaSUREs CAMEL V001 Database*; UW-Madison, Space Science and Engineering Center: Madison, WI, USA, 2017. Available online: https://doi.org/10.21231/S2PP8H (accessed on 15 November 2017).

37. Feltz, M.; Borbas, E.; Knuteson, R.; Hulley, G.; Hook, S. Global Broadband IR Surface Emissivity Computed from Combined ASTER and MODIS Emissivity over Land (CAMEL). In Proceedings of the American Meteorological Society Joint 21st Satellite Meteorology, Oceanography and Climatology Conference and 20th Conference on Air-Sea Interaction, Madison, WI, USA, 15–19 March 2016. Available online: https://ams.confex.com/ams/21SATMET20ASI/webprogram/Paper296752.html (accessed on 15 February 2018).

38. Saunders, R.; Hocking, J.; Rundle, D.; Rayer, P.; Havemann, S.; Matricardi, M.; Geer, A.; Cristina, L.; Brunel, P.; Vidot, J. RTTOV-12 Science and Validation Report; NWPSAF-MO-TV-41; 2017. Available online: https://www.nwpsaf.eu/site/download/documentation/rtm/docs_rttov12/rttov12_svr.pdf (accessed on 15 February 2018).

39. Weisz, E.; Smith, W.L.; Smith, N. Advances in simultaneous atmospheric profile and cloud parameter regression based retrieval from high-spectral resolution radiance measurements. *J. Geophys. Res. Atmos.* **2013**, *118*, 6433–6443. [CrossRef]

40. August, T. The EUMETSAT operational IASI L2 products and services, from Global to Regional. In Proceedings of the International TOVS Study Conference, Darmstadt, Germany, 29 November–5 December 2017.

41. Gambacorta, A.; Barnet, C.; Wolf, W.; King, T.; Maddy, E.; Strow, L.; Xiong, X.; Nalli, N.; Goldberg, M. An Experiment Using high spectral resolution CrIS measurements for atmospheric trace gases: Carbon monoxide retrieval impact study. *IEEE Geosci. Remote Sens. Lett.* **2014**, *11*, 1639–1643. [CrossRef]

42. Zhou, D.; Larar, M.A.; Liu, X.; Smith, W.; Strow, L.; Yang, P.; Schluessel, P.; Calbet, X. Global Land Surface Emissivity Retrieved From Satellite Ultraspectral IR Measurements. *IEEE Trans. Geosci. Remote Sens.* **2011**, *49*, 1277–1290. [CrossRef]

43. Li, Z.; Li, J.; Li, Y.; Zhang, Y.; Schmit, T.J.; Zhou, L.; Goldberg, M.D.; Menzel, W.P. Determining diurnal variations of land surface emissivity from geostationary satellites. *J. Geophys. Res.* **2012**, *117*, D23302. [CrossRef]

44. Label, J.; Stoll, M.P. Angular variation of land surface spectral emissivity in the thermal infrared: Laboratory investigations on bare soils. *Int. J. Remote Sens.* **1991**, *12*, 2299–2310.

45. Sobrino, J.A.; Cuenca, J. Angular variation of thermal infrared emissivity for some natural surfaces from experimental measurements. *Appl. Opt.* **1999**, *38*, 3931–3936. [CrossRef] [PubMed]

46. Cuenca, J.; Sobrino, A.J. Experimental measurements for studying angular and spectral variation of thermal infrared emissivity. *Appl. Opt.* **2004**, *43*, 4598–4602. [CrossRef] [PubMed]

47. Garcia-Satnos, V.; Valor, E.; Caselles, V.; Burgos, M.A.; Coll, C. On the angular variation of thermal infrared emissivity of inorganic soils. *J. Geophys. Res.* **2012**, *117*, D19116. [CrossRef]

48. Ruston, B.; Weng, F.; Yan, B. Use of a One-Dimensional variation retrieval to Diagnose Estimates of Infrared and Microwave Surface Emissivity Over Land for ATOVS Sounding Instruments. *IEEE Trans. Geosci. Remote Sens.* **2008**, *46*, 393–402. [CrossRef]

49. Borbas, E.; Investigation into the Angular Dependence of IR Surface Emissivity. Mission Report; EUMETSAT NWPSAF-MO-VS-050; 2014. Available online: https://nwpsaf.eu/vs_reports/nwpsaf-mo-vs-050.pdf (accessed on 15 February 2018).

Article

The Combined ASTER MODIS Emissivity over Land (CAMEL) Part 2: Uncertainty and Validation

Michelle Feltz [1,*], Eva Borbas [1], Robert Knuteson [1], Glynn Hulley [2] and Simon Hook [2]

[1] Space Science and Engineering Center, University of Wisconsin–Madison, Madison, WI 53706, USA; evab@ssec.wisc.edu (E.B.); robert.knuteson@ssec.wisc.edu (R.K.)
[2] California Institute of Technology Jet Propulsion Laboratory, Pasadena, CA 91109, USA; Glynn.Hulley@jpl.nasa.gov (G.H.); Simon.J.Hook@jpl.nasa.gov (S.H.)
* Correspondence: michelle.feltz@ssec.wisc.edu

Received: 28 February 2018; Accepted: 16 April 2018; Published: 24 April 2018

Abstract: Under the National Aeronautics and Space Administration's (NASA) Making Earth System Data Records for Use in Research Environments (MEaSUREs) Land Surface Temperature and Emissivity project, a new global land surface emissivity dataset has been produced by the University of Wisconsin–Madison Space Science and Engineering Center and NASA's Jet Propulsion Laboratory (JPL). This new dataset termed the Combined ASTER MODIS Emissivity over Land (CAMEL), is created by the merging of the UW–Madison MODIS baseline-fit emissivity dataset (UWIREMIS) and JPL's ASTER Global Emissivity Dataset v4 (GEDv4). CAMEL consists of a monthly, 0.05° resolution emissivity for 13 hinge points within the 3.6–14.3 μm region and is extended to 417 infrared spectral channels using a principal component regression approach. An uncertainty product is provided for the 13 hinge point emissivities by combining temporal, spatial, and algorithm variability as part of a total uncertainty estimate. Part 1 of this paper series describes the methodology for creating the CAMEL emissivity product and the corresponding high spectral resolution algorithm. This paper, Part 2 of the series, details the methodology of the CAMEL uncertainty calculation and provides an assessment of the CAMEL emissivity product through comparisons with (1) ground site lab measurements; (2) a long-term Infrared Atmospheric Sounding Interferometer (IASI) emissivity dataset derived from 8 years of data; and (3) forward-modeled IASI brightness temperatures using the Radiative Transfer for TOVS (RTTOV) radiative transfer model. Global monthly results are shown for different seasons and International Geosphere-Biosphere Programme land classifications, and case study examples are shown for locations with different land surface types.

Keywords: emissivity; infrared; surface; land; radiation; hyperspectral

---

1. Introduction

Infrared land surface emissivity is an important variable in the estimation of Earth's radiation budget as well as in the retrieval of temperature, water vapor, and other atmospheric constituent profiles from hyperspectral infrared sounders. Under the National Aeronautics and Space Administration's (NASA) Making Earth System Data Records for Use in Research Environments (MEaSUREs) project, a new and improved global land surface emissivity dataset is being made available. This new dataset, termed the Combined ASTER and MODIS Emissivity over Land (CAMEL), combines previously existing satellite emissivity datasets—those from the Moderate Resolution Imaging Spectroradiometer (MODIS) baseline-fit emissivity dataset (BF) developed at the University of Wisconsin–Madison (UW) [1], and the Advanced Spaceborne Thermal Emission and Reflection Radiometer (ASTER) Global Emissivity Dataset v4 (GEDv4) produced at the Jet Propulsion Laboratory (JPL) [2]. CAMEL leverages the additional thermal infrared bands from ASTER which more accurately resolve the thermal infrared region (8–12 μm) and the ability of UW BF to provide information at channels throughout the entire 3.6–12 µm infrared region, thereby combing the strengths of each dataset. The CAMEL methodology and high spectral resolution (HSR) application is described in Part 1 of this paper. An uncertainty product for the CAMEL emissivity dataset is included and is defined on the same monthly, 0.05° resolution. The uncertainty product includes a total uncertainty estimate which is determined by 3 separate components of variability—a spatial, temporal, and algorithm variability component—all of which are included in the uncertainty product file. The CAMEL uncertainty products, which are discussed in greater detail in this paper, are available at the NASA Land Processes Distributed Active Archive Center (LP DAAC) website.

The CAMEL emissivity dataset is being used in the numerical weather prediction community via the European Organization for the Exploitation of Meteorological Satellites (EUMETSAT) Network of Satellite Application Facilities (NWP SAF) Radiative Transfer for TOVS (RTTOV) model Version 12. This model is a fast, radiative transfer model for nadir-viewing passive visible, infrared (IR), and microwave satellites [3]. The CAMEL product is integrated into this model via an IR emissivity module which, like the CAMEL HSR algorithm, creates a high spectral resolution emissivity on 417 infrared spectral channels. A single year of CAMEL data from 2007 is used by the RTTOV Version 12 IR emissivity module. This year is selected for use because it has a lack of observation degradation in terms of calibration from MODIS and is optimal for representing the MODIS record. The use of the CAMEL product within RTTOV's IR emissivity module is an update to the previously used UWIREMIS product, which is available with and leverages the UW-Madison UW BF dataset [4].

This study details the methodology for the CAMEL uncertainty product calculation, and it provides an assessment of the CAMEL emissivity dataset. Comparison of CAMEL with lab data from samples collected at ground sites provides the most important form of validation. However, because this form of validation is only available for case sites and not globally, other forms are used as well. Comparisons with an 8-year average of Infrared Atmospheric Sounding Interferometer (IASI) land surface emissivity derived from measurements are made for specific case sites as well as over the globe [5,6]. In addition, this study makes use of the RTTOV model CAMEL HSR emissivity module to provide an assessment of the CAMEL product. Comparisons of RTTOV simulated IASI hyperspectral IR radiances are made to observations for four focus days. Simulations of IASI radiances are performed using the European Center for Medium Range Weather Forecasting's (ECMWF) forecast model as input into the RTTOV model. Simulations are made using both the CAMEL RTTOV emissivity module as well as a constant 0.98 emissivity, which is a typical simplification in global NWP models. These simulations are then compared to observed radiances, which provide a chance to assess the impact of the difference between the 0.98 constant emissivity and the CAMEL HSR estimate. Statistics are then grouped according to the International Geosphere-Biosphere Programme (IGBP) land classification scheme [7,8] and are used to assess the CAMEL product's accuracy over different surface types.

The paper is structured as follows: Section 2 describes the CAMEL emissivity product inputs. Section 3 outlines the CAMEL uncertainty estimation method in detail and shows results for global maps and lab validation case sites. Section 4 contains the CAMEL product assessments made from comparisons to independent sources and the RTTOV simulated IASI radiances. Section 5 holds the discussion of results and conclusions.

2. Data

2.1. Emissivity

2.1.1. The Combined ASTER MODIS Emissivity Over Land

The CAMEL product is available from 2000 to 2016 at a monthly mean, 0.05° (~5 km) resolution for 13 hinge points within the 3.6–14.3 µm region. It is created using both the ASTER GEDv4 and MODIS UW BF emissivity datasets as input. A standalone CAMEL High Spectral Resolution (HSR) Algorithm included with the dataset is also available, which extends the 13 hinge point CAMEL product to 417 infrared spectral channels using a principal component regression approach. Additional details of the CAMEL dataset are available in Part 1 of this two-part paper series (Borbas et al. 2018). The CAMEL dataset, including the uncertainty product discussed in this paper, is available from the NASA online archive at https://lpdaac.usgs.gov/about/news_archive/release_nasa_measures_camel_5_km_products [9].

2.1.2. Advanced Spaceborne Thermal Emission and Reflection Radiometer Global Emissivity Dataset v4

Housed on NASA's Terra Earth Observing System satellite, ASTER is a cooperative effort between NASA, Japan's Ministry of Economy, Trade and Industry (METI), and Japan Space Systems. The ASTER GED version 4 (v4) product used in this study was produced at JPL and is available at https://lpdaac.usgs.gov/dataset_discovery/aster [2]. The ASTER GEDv4 is available at a monthly, 0.05° resolution over 2000–2016. It is based on the ASTER GEDv3 2000–2008, 100 m resolution climatology which uses emissivity derived using the ASTER Temperature Emissivity Separation algorithm [10]. This climatology is adjusted to vary with the changing land surface (snow cover, vegetation, drought) using a MODIS monthly snow cover (MOD10CM, 0.05° resolution) and monthly NDVI (MOD13C2) product. ASTER GEDv4 has 5 bands located in the thermal infrared region at 8.3, 8.6, 9.1, 10.6, and 11.3 microns. These band placements, three of which are in the Restrahen region, give ASTER the ability to characterize quartz doublet features.

2.1.3. University of Wisconsin Global Infrared Land Surface Emissivity Database

The UW Base-line Fit infrared land surface emissivity dataset is made available by the University of Wisconsin–Madison at 10 wavelengths which were chosen at locations that provide as much characterization of the high resolution infrared spectra as possible. These band placements are at 3.6, 4.3, 5.0, 5.8, 7.6, 8.3, 9.3, 10.8, 12.1, and 14.3 μm. This product is based on the MODIS MOD11C3 emissivity product and is extended to the 10 channels using a conceptual model based on laboratory measurements [1]. Borbas et al. [11] and Borbas and Ruston [4] further extended this emissivity dataset to cover 416 channels from 3.6–14.3 μm using the UW HSR Emissivity Algorithm. This algorithm uses a Principal Component Analyses regression using high spectral resolution laboratory measurements from the ASTER spectral library [12]. Both the UW BF dataset and UW HSR Emissivity Algorithm are available at a monthly, 0.05° resolution from http://cimss.ssec.wisc.edu/iremis/.

2.1.4. Laboratory Validation Data

We chose three different sites to cover a range of land surface types including barren (Namib quartz sand and Yemen carbonate), snow/ice (Greenland), and mixed vegetation (Tucson, Arizona and the Atmospheric Radiation Measurement (ARM) Southern Great Plains (SGP)). Validation data for the Namib site consists of sand samples that were collected over the Namib dunes near Sossussvlei and measured in the lab at JPL using a Nicolet spectrometer [1,2]. The snow and ice validation emissivity was taken from 2006 Greenland ice sheet measurements from University of California, Santa Barbara (UCSB) and from Dozier and Warren [13]. A spline-fit was used to remove high resolution artifacts from the UCSB measurements and data from Dozier and Warren [13] was then used to extend the spectrum to longer wavelengths. Validation data for the ARM SGP site and Tucson, Arizona (AZ) site is based on University of Wisconsin Atmospheric Emitted Radiance Interferometer (AERI) downlooking measurements and coincident airborne data [14]. Measured AERI emissivities of vegetated and non-vegetated surfaces are weighted and combined using a vegetation index that depends on the day of year [15]. The Tucson, Arizona lab data is also a linear combination of AERI measurements. For this case the ASTER Normalized Difference Vegetation Index (NDVI) product is used to define the vegetation fraction used to weight the measurements.

2.1.5. Long Term Infrared Atmospheric Sounding Interferometer Dataset

Monthly, 8-year averages of IASI emissivity derived from measurements are available at 0.25° spatial resolution. For the months of January through May, emissivity is averaged over the years 2007–2014 and for the months of June through December it is averaged over 2008–2015. An inversion scheme which uses both cloudy and cloud-free radiances is used to simultaneously retrieve atmospheric thermodynamic and surface or cloud microphysical parameters [5,6].

2.2. Ancillary Data

2.2.1. European Center for Medium Range Weather Forecasting

European Center for Medium Range Weather Forecasting's (ECMWF) 6 h analyses are used as atmospheric data input to forward model the IASI observed brightness temperature calculations. The ECMWF data has 0.25° (~18 km) horizontal resolution and 91 vertical levels with 0.1 hPa pressure at the model top.

2.2.2. Infrared Atmospheric Sounding Interferometer

IASI Level 1C radiance data is available from EUMETSAT. IASI is a hyperspectral infrared sounder which is flown onboard the Meteorological Operational satellite programme (MetOp) satellites under EUMETSAT and the European Space Agency in the local morning orbit.

2.2.3. Moderate Resolution Imaging Spectroradiometer Land Cover

NASA's Earth Observing System MODIS instrument is used to provide monthly global IGBP land cover classification on a 0.5° × 0.5° grid resolution. The land cover product used is MCD12Q1 and is made available by the Global Land Cover Facility online at http://glcf.umd.edu/data/lc/. More information can be found in [7,8]. There are 17 land cover types as defined by the IGBP classifications.

3. CAMEL Uncertainty Estimates

3.1. Method

Because the inputs to the CAMEL emissivity product do not provide consistent error estimates, it is not possible to use error propagation to estimate the CAMEL uncertainty. When uncertainties are made available for the MODIS emissivity, which is intended for the future MOD21 version, it is planned to include these in combination with the currently available ASTER uncertainties in the CAMEL uncertainty estimates. The current CAMEL uncertainty formula is defined in the following paragraphs and uses coherence analyses to define various components of the total uncertainty. Due to the nature and inherent constraints of the input MODIS and ASTER datasets, CAMEL data is likely to be highly correlated for certain domains, for example where no MODIS data is available in regions of the long polar nighttime.

The CAMEL uncertainties are estimated by combining three separate components—temporal, spatial, and algorithm. Each measure of uncertainty is provided for all 13 hinge points and for every pixel. The total uncertainty is calculated from the components as a root sum square as in Equation (1):

$$\sigma_{total} = \sqrt{\sigma_{spatial}^2 + \sigma_{temporal}^2 + \sigma_{algorithm}^2} \tag{1}$$

The spatial uncertainty component is calculated as the standard deviation of the surrounding 5 × 5 pixel emissivity, which is equivalent to a 0.25° × 0.25° latitude-longitude region. This uncertainty represents the variability of the surrounding landscape and is only provided where the CAMEL emissivity quality flag is non-zero (i.e., sea/ocean is not included). Temporal uncertainty, also defined on a pixel by pixel basis, is defined by the standard deviation of the three bracketing months (e.g., Oct. uncertainty = standard deviation (September, October, November)). Even if emissivity values are not available for all three months as in the case of the starting or ending month of the CAMEL record, an uncertainty is still reported from the closest month.

The algorithm uncertainty is estimated primarily by the differences between the two CAMEL emissivity inputs: the ASTER GEDv4 and UW BF products. Thus, the algorithm uncertainty is a measure of bias in the CAMEL inputs and is not a propagation of the ASTER and UW BF product uncertainties. The intent is to warn the user when differences in the ASTER and UW BF products are comparable to uncertainties due to other sources. To convert the signed difference into a standard deviation, the absolute difference is divided by the square root of 3. This scaling gives the variance associated with the algorithm uncertainty approximately equal weight to the spatial or temporal uncertainties. Table 1 shows the CAMEL, ASTER, and UW BF channel wavelengths, the method for combining the ASTER and UW BF emissivity, and the method for determining the CAMEL algorithm uncertainty. For hinge points 6–9 and 11–13 where ASTER and UW BF report emissivities at nearby frequencies, differences between the ASTER and UW BF emissivity are used to define the algorithm uncertainty as shown in Table 1. Hinge point 10, at 10.8 μm, uses a linear interpolation between the ASTER channels 4 and 5.

Table 1. Bands, inputs, and methodology for generating the CAMEL (Combined ASTER and MODIS Emissivity over Land, combining the Moderate Resolution Imaging Spectroradiometer (MODIS) and the Advanced Spaceborne Thermal Emission and Reflection Radiometer (ASTER)) product and algorithm uncertainty estimates *. Note University of Wisconsin Baseline-Fit (UW BF) is based on the MODIS emissivity product.

CAMEL Channel Number	CAMEL Wavelength [μm]	UWBF	ASTER	CAMEL COMBINING METHOD	ALGORITHM UNCERTAINTY
1	3.6	Y	-	UWBF1	$\mathrm{Abs}(\mathrm{UWBF1} \times [(\mathrm{UWBF6} - \mathrm{ASTER2})/\mathrm{UWBF6}])/\sqrt{3}$
2	4.3	Y	-	UWBF2	$\mathrm{Abs}(\mathrm{UWBF2} \times [(\mathrm{UWBF6} - \mathrm{ASTER2})/\mathrm{UWBF6}])/\sqrt{3}$
3	5.0	Y	-	UWBF3	$\mathrm{Abs}(0.01)/\sqrt{3}$
4	5.8	Y	-	UWBF4	$\mathrm{Abs}(0.01)/\sqrt{3}$
5	7.6	Y	-	UWBF5	0 (Minimal variation)
6	8.3	Y	Y	ASTER 1 + (CAMEL7(UWBF6, ASTER 2) − ASTER 2)	$\mathrm{Abs}(\mathrm{UWBF6} - \mathrm{ASTER1})/\sqrt{3}$
7	8.6	Y	Y	Weighted Mean(UWBF6, ASTER 2)	$\mathrm{Abs}(\mathrm{UWBF6} - \mathrm{ASTER2})/\sqrt{3}$
8	9.1	-	Y	ASTER 3 + (CAMEL7(UWBF6, ASTER 2) − ASTER 2)	$\mathrm{Abs}(\mathrm{UWBF6} - \mathrm{ASTER3})/\sqrt{3}$
9	10.6	-	Y	ASTER 4	$\mathrm{Abs}(\mathrm{UWBF8} - \mathrm{ASTER4})/\sqrt{3}$
10	10.8	-	-	Linear Interpolation(ASTER 4, ASTER 5)	$\mathrm{Abs}(\mathrm{UWBF8} - [(\mathrm{ASTER4} \times 5 + \mathrm{ASTER5} \times 2)/7])/\sqrt{3}$
11	11.3	-	Y	ASTER5	$\mathrm{Abs}(\mathrm{UWBF8} - \mathrm{ASTER5})/\sqrt{3}$
12	12.1	Y	-	UWBF9 but if ASTER5 > UWBF9, UWBF9 + diff(UWBF9, ASTER 5) × w	$\mathrm{Abs}(\mathrm{UWBF9} - \mathrm{ASTER5})/\sqrt{3}$
12 *	12.1	Y	-	if snowfrac > 0.5 UWBF9 + diff(CAMEL10, UWBF8)	$\mathrm{Abs}(\mathrm{UWBF9} - \mathrm{ASTER5})/\sqrt{3}$
13	14.3	Y	-	UWBF10 but if ASTER 5 > UWBF9, UWBF9 + diff(UWBF9, ASTER 5) × w	$\mathrm{Abs}(\mathrm{UWBF10} - \mathrm{ASTER5})/\sqrt{3}$
13 *	14.3	Y	-	if snowfrac > 0.5, CAMEL12	$\mathrm{Abs}(\mathrm{UWBF10} - \mathrm{ASTER5})/\sqrt{3}$

* For more details on the CAMEL Combining Method, see Part 1 of this paper series. In the table w represents a weight factor. CAMEL channels 12 and 13 have separate combing methods based on whether the pixel is defined as snow or non-snow.

For CAMEL hinge points 1–5 which cover 3.6 to 7.6 μm, no ASTER data is available to use in the uncertainty calculation. CAMEL hinge points 1 and 2 use a fractional difference from hinge point 7 due to the correlations between these window regions. Hinge points 3–5 are located in the water vapor absorption region so emissivity does not change much. Thus, hinge points 3 and 4 assume that there is a 1% constant difference. For hinge point 5 at 7.6 μm it is assumed that there is no variation. This is based upon a previous study [1] which found that 7.6 μm was the channel that showed the least amount of variation in a set of 123 laboratory emissivity spectra.

A quality flag is provided for the total uncertainty as defined in Table 2. The quality flag is zero where no CAMEL data is available (over ocean), one where it is determined to be of good quality, and two where it is determined to be unphysical and should not be used. To determine which total uncertainty values are unphysical (i.e., have a 'total_uncertainty_quality_flag' = 2), unphysical uncertainties are first identified in the 3 uncertainty components (i.e., the temporal, spatial, and algorithm uncertainty), though the components' flags are not provided to users due to file size restraints. Spatial and temporal uncertainty values are flagged unphysical if they are located in the 99.9th percentile, while algorithm uncertainty values are determined as unphysical if the ASTER and BF differences prior to having their absolute values taken are in the 0.1st or 99.9th percentiles. Total uncertainty values are then flagged as unphysical if any of the 3 components uncertainties are flagged as unphysical, so that up to 0.4% of the total uncertainty estimates could be marked as unphysical.

In addition to the 13 hinge point CAMEL product, the following analyses (illustrated in Figures 3 and 4) show total uncertainty estimates for CAMEL HSR spectra. While a CAMEL HSR uncertainty is not provided with the official product, it is calculated here to illustrate the sensitivity of the HSR algorithm to the 13 hinge point CAMEL product. To obtain the HSR uncertainty, the 13 hinge point total uncertainty is both added to and subtracted from the monthly emissivity and then input into the CAMEL HSR algorithm. The HSR uncertainty is then estimated as half the difference between the HSR emissivity produced by adding and subtracting the 13-hinge point uncertainty from the monthly emissivity value.

Table 2. Definition of CAMEL emissivity uncertainty quality flag.

Value	Description
0	Ocean or no CAMEL data available
1	Good quality
2	Unphysical uncertainty

3.2. Results

Figure 1 shows a global map of the CAMEL 8.6 μm emissivity for January 2007. Located within the quartz doublet region, this channel's emissivity highlights the location of surface types dominated by quartz, for example the Sahara Desert, Namib Desert, and inland Arabian Peninsula. The associated CAMEL quality flag (qf), defined in Table 3, is also shown in Figure 1 and shows where no CAMEL emissivity values are reported over sea/inland water (qf = 0), where the input UW BF and ASTER data are good quality (qf = 1), the input UW BF is good quality and ASTER is filled (qf = 2), the input UW BF is filled but ASTER is good quality (qf = 3), and both the UW BF and ASTER values are filled (qf = 4). The distinct line over which the quality flag changes at ~55°N latitude is an artifact of the day/night emissivity method of the NASA MOD11C product which is used as input to the UW BF product [16]. A lack of MODIS data is also seen over land in the ~0–20°S tropical belt in regions dominated by persistent cloudiness. The north-to-south stripes over Siberia are caused by the cloudiness over the whole month in the ASTER dataset. For these cases, both datasets use filling techniques from the average of the neighboring months or annual data.

Table 3. Definition of CAMEL quality flag.

Value	Description
0	Ocean or inland water
1	input UW BF and ASTER data are good quality
2	input UW BF is good quality and ASTER is filled
3	the input UW BF is filled but ASTER is good quality
4	both the UW and ASTER values are filled

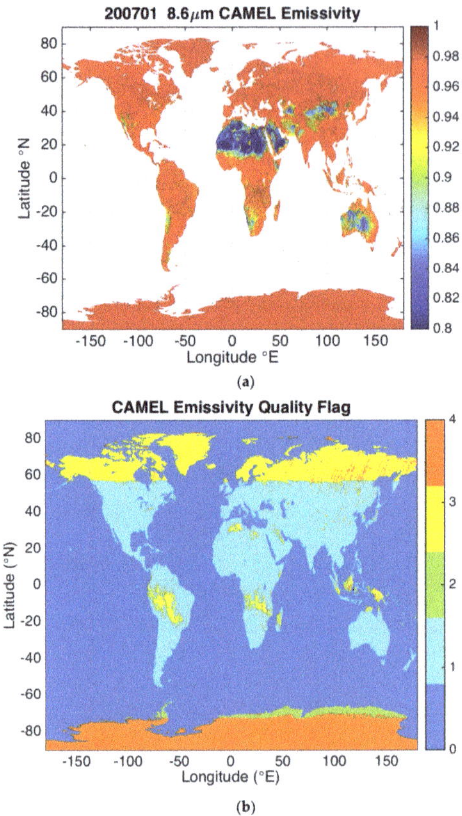

Figure 1. Combined ASTER and MODIS Emissivity over Land (CAMEL) (**a**) 8.6 μm emissivity and; (**b**) quality flag for January 2007. Quality flag is defined as in Table 3. The University of Wisconsin Baseline-Fit (UW BF) has fill values north of ~55°N congruent with the prevailing winter nighttime of January at those latitudes. This is because the MOD11C3 product used as input to UW BF is dependent on a day–night algorithm that requires a day and night observation close in time, and there is not enough daytime data in the Northern Hemisphere winter.

The CAMEL uncertainty product for the 8.6 μm channel during January 2007 is shown in Figure 2. The total uncertainty is shown with its quality flag and three separate uncertainty components. Qualitatively it is apparent how the total uncertainty is a combination of the three components, with the largest contributions coming from the algorithm component. The amount each component contributes to the total value varies with season. For example, the month of July in general has lower algorithm uncertainty and the spatial and temporal uncertainty has increased values north of ~60° latitude which are more readily identified as contributing to the total uncertainty. A line of larger uncertainty values is seen along the ~55°N latitude region in Asia, which is a pattern present in all 3 uncertainty components, and is attributable to the change of CAMEL input data (no UW BF input) as shown in Figure 1's quality flag.

A comparison of the 8.6 μm spatial uncertainty to the monthly mean emissivity in Figure 1 shows areas of spatially uniform emissivity have low spatial uncertainties, for example densely vegetated regions of the Congo Forest, and persistent snow/ice covered regions of Antarctica and areas over Northern Hemisphere winter snow covered areas (northern Arctic Circle). Due to the large difference in the 8.6 μm emissivity between quartz and non-quartz surfaces, even a small mixing of surface types can yield larger spatial uncertainty values. The map of temporal uncertainty reveals that desert regions such as the Sahara, and more permanently snow-covered regions like Antarctica and the Arctic Circle are characterized by low temporal uncertainty. The northeastern United States is marked by a higher, ~0.01 temporal uncertainty likely due to changes in snow cover over the winter months, and areas around the Indus River Valley and northern India are marked by higher uncertainties likely due to changing precipitation and meltwater levels. Such patterns in the uncertainty values reveal physical characteristics and changes of the Earth's surface. The algorithm uncertainty, which is proportional to the UW BF 8.3 µm minus ASTER 8.6 µm emissivity, shows the largest values of over 0.02, mostly over semi-arid and arid regions that exhibit the largest variability in emissivity for this band. This is due to larger differences between retrieved emissivity in this band from the ASTER and MODIS input products. However, because of ASTER's greater number of bands in this region (3), the derived CAMEL emissivities should be of higher accuracy than the original UW BF value, which only has one band in this region. Uncertainty values over northern Russia and Canada show larger values of ~0.01 where snow cover signatures vary between the ASTER and the UW BF products.

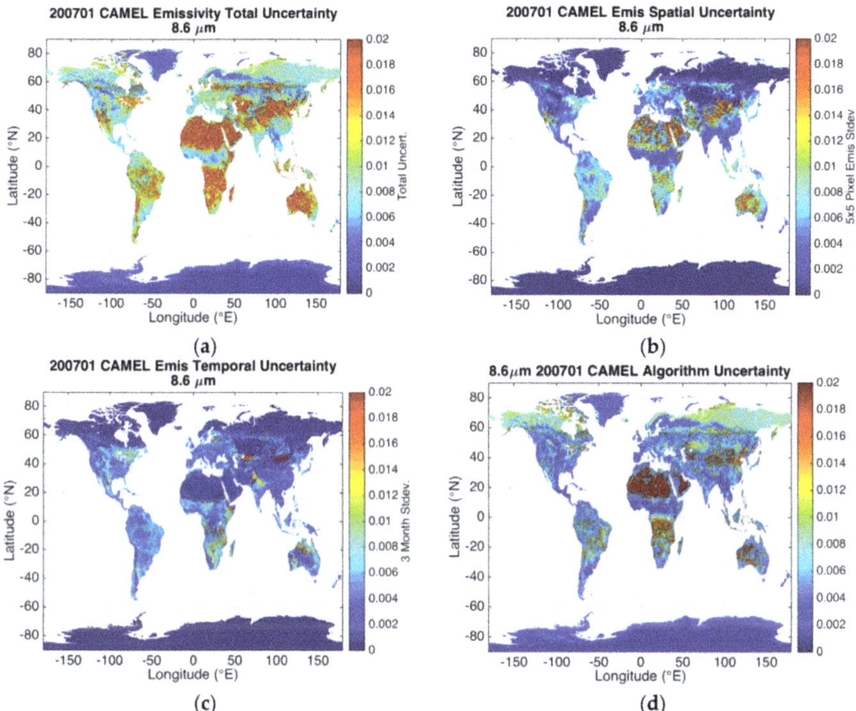

Figure 2. *Cont.*

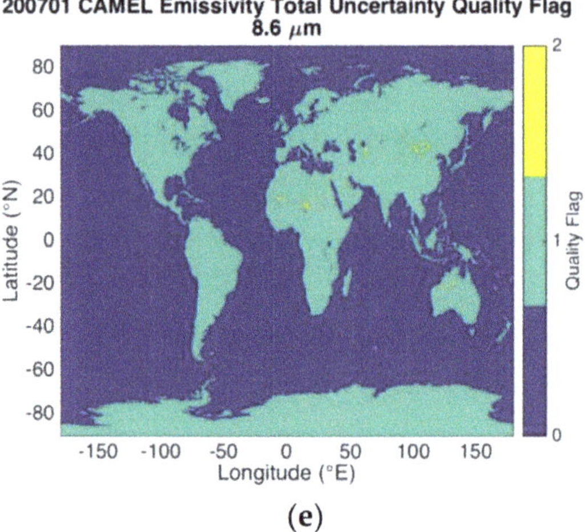

(e)

Figure 2. January 2007 CAMEL 8.6 um emissivity uncertainty (**a**) total estimate; (**b**) spatial component; (**c**) temporal component; (**d**) algorithm component; and (**e**) total uncertainty quality flag which is defined in Table 2.

4. Validation with Laboratory Spectra

4.1. Field Sites

Figure 3 shows the HSR and 13 hinge point CAMEL products overlaid with the UW BF, ASTER, and lab measurement emissivities for four validation sites. The total HSR uncertainty estimate for the CAMEL HSR product is overlaid in the top panel (plotted as red dotted lines), and the CAMEL 13 hinge point total uncertainty with its 3 components are shown separately in the bottom panel.

The first site in Figure 3a shows emissivity spectra comparisons over the Namib desert during January 2007. The quartz doublet feature can clearly be seen at 8.5 and 12.5 μm since the desert sands consists mostly of quartz and hematite. In the top panel, the CAMEL HSR product closely captures the depth of the quartz doublet within the 8–9 μm region. The middle panel shows how the CAMEL emissivity is defined using ASTER bands within this 8–9.3 μm window. The difference between the ASTER and UW BF product around 9 μm is reflected in the CAMEL algorithm uncertainty shown in the bottom panel. With an exception for wavelengths less than ~5 μm where the temporal uncertainty is the largest component, the total uncertainty is dominated by the algorithmic uncertainty.

The second site shown in Figure 3b is for Tucson, Arizona in May 2007. The spectrometer data for this case was obtained from AERI in situ measurements of a characteristic creosote site, where the Creosote bush is the dominant vegetation type [14]. The ASTER NDVI is used as a vegetation fraction to form a weighted emissivity measurement from bare and vegetated spectra. This site shows very close agreement between ASTER, UW BF, and CAMEL products. This close agreement, in addition to the lack of temporal variation of the surface characteristics for wavelengths under 11 μm, produces a total CAMEL uncertainty that is mostly driven by the spatial variability of the region. While the CAMEL HSR product closely resembles the UW BF HSR emissivity across most of the spectrum, the CAMEL HSR product is more accurately able to capture the 8–9.3 μm subtle quartz feature seen in the lab validation data.

The third site in Figure 3c shows results for a site in Greenland, Summit Station, for January 2007. The top panel shows that the CAMEL HSR emissivity more accurately fits the lab data for smooth snow than does the UW BF, particularly for snow features in the 11–14 μm region. Snow spectral features are better captured because the CAMEL HSR for snow is derived from a unique set of principal component (PC) coefficients derived from laboratory spectra of different snow grain sizes, ice, and water, more details of which can be found in Part 1. While the CAMEL HSR uncertainty is around 0.01 for wavelengths over 11 µm, the 13 hinge point CAMEL uncertainty is confined well under 0.01 and is dominated by algorithm variability.

The last case site example in Figure 3d is for Yemen in January 2007. Lab measurements are derived from the ASTER spectral library and illustrate the primary carbonate features that are seen over regions of Yemen and Oman at the southern tip of the Arabian Peninsula. The CAMEL 13 hinge point product is unable to resolve the narrow 6–7 µm main carbonate feature because MODIS does not have any bands covering that region; however, the CAMEL HSR emissivity is able to capture the feature since a unique set of PC coefficients for carbonate minerals are used. The CAMEL is an improvement upon the UW BF HSR in this region since the UW BF uses only one set of PC coefficients for all surface types. Though the lab data does not lie within the bounds of the CAMEL HSR uncertainty within the ~6–7 and 11–11.5 µm regions, it is important to note that the lab spectra are a pure carbonate (dolomite) and CAMEL is representative of a 5 km pixel which likely contains a mixture of material, not purely carbonate. Additionally, the HSR uncertainty does appropriately reflect the increased difference between the CAMEL HSR emissivity and the lab validation data. Like the Namib Desert site, the Yemen desert site also has a higher CAMEL uncertainty in the shortwave region due to an increased temporal variability but otherwise is dominantly influenced by the algorithm uncertainty component.

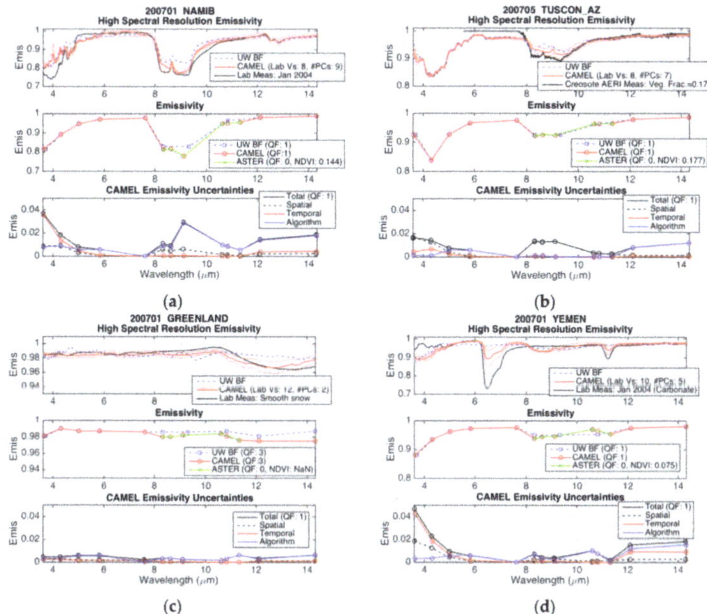

Figure 3. Emissivity at (**a**) January 2007 Namib Desert (−24.25N, 15.25E); (**b**) May 2007 Tucson, AZ (32.01N, −110.77E); (**c**) January 2007 Greenland (72.57N, −38.45E); and (**d**) January 2007 Yemen (19.15N, 55.57E) case sites. HSR emissivity from UW BF (blue dashed), lab measurements (black), and CAMEL (red solid with dotted uncertainty) are shown in top row panels. UW BF (blue squares, dashed), ASTER (green x's, solid), and CAMEL (red circles, solid) emissivity are shown in middle row panels. CAMEL total (black solid), spatial (black dash), temporal (red), and algorithm (blue) uncertainty components are shown in bottom panels. Legend in middle and bottom panel include quality flags (QF) for the different products as well as ASTER normalized difference vegetation index (NDVI). The version number (Vs) of the laboratory dataset and the number of principal components (#PCs) used in the CAMEL HSR Algorithm are included in the legend of the top panels.

Each of the four case sites represents a different surface type and uses a different number of PCs in the CAMEL HSR emissivity construction. The Namib Desert and Tucson, AZ cases represent surfaces characterized by quartz. Within the quartz signature region located at ~8–10.5 μm, CAMEL and the lab measurements agree within 0.05 for the Namib case and within 0.025 for the partially vegetated Tucson, AZ case. At wavelengths greater than 10.5 μm, CAMEL and the lab data agree within 0.01 for both sites, while for the Namib SW region from 3.6–6 μm, CAMEL has departures of ~0.1 magnitude away from the lab measurements. This could be explained by surface roughness and terrain elevation changes present in the satellite observation but not in the lab measurement. The Greenland case site illustrates the CAMEL snow/ice emissivity performance: for wavelengths greater than or less than 8 μm, CAMEL and the lab data agree within 0.005 and 0.01 respectively. The Yemen site characterizes the carbonate surfaces and shows the 6.5 μm spectral feature is included in the CAMEL dataset but the lab data differs by up to 0.15, with the lab measurement lying outside of the CAMEL HSR uncertainty bounds. The reduced spectral contrast in the CAMEL estimate could be explained by surface roughness or sparse vegetation not accounted for in the lab measurement.

4.2. CAMEL Intercomparison with IASI Emissivity

This section shows a comparison of CAMEL to the 8-year averaged IASI HSR emissivity dataset, which is derived from IASI radiance observations and uses a retrieval method based on Principle Component Analysis (PCA) of lab spectra [5]. To facilitate the comparison at the 13 CAMEL hinge points, the CAMEL 13 hinge point product is first resampled to a lower spatial resolution by applying a 5 × 5 pixel running average over the CAMEL grids to match the 0.25° × 0.25° spatial resolution of the IASI dataset. CAMEL 8-year averaged monthly emissivities are then computed corresponding to the years over which the IASI dataset is averaged. For the months of January through May the years 2008–2015 are averaged, and for June through December the years of 2007–2014 are averaged. To compare the HSR CAMEL product to the HSR IASI emissivity, the individual CAMEL HSR spectra are simply computed and averaged over the proper years. Since the CAMEL HSR 8-year monthly averages actually represent a 0.05° × 0.05° resolution, the HSR uncertainty estimate, which includes a measure of the surrounding 0.25° × 0.25° standard deviation, is used to provide a measure of the potential error from the spatial resolution mismatch (i.e., of the CAMEL 0.05° × 0.05° to the IASI 0.25° × 0.25°).

Figure 4 shows the monthly 8-year averages of CAMEL HSR emissivities compared against (i) laboratory validation data (where it is available) and (ii) the IASI HSR dataset for various case sites. The 8-year averaged HSR emissivities are overlaid in the top panels, including individual CAMEL spectra for each of the years averaged, and the bottom panels show the 8-year average differences bounded by the CAMEL uncertainty estimate. At the top of Figure 4, the Namib and Tucson, AZ desert cases show emissivity differences between the CAMEL and lab emissivities of less than 0.05 for wavelengths greater than 5 μm. Within the quartz spectral region, CAMEL agrees more closely with the lab measurements than the IASI emissivity. For the Greenland case site, differences between the lab and CAMEL emissivities are less than 0.01. The IASI emissivity reflects a subtle quartz signature which is absent in CAMEL. This case shows the importance of the selection of laboratory measurements for the PCA method. For snow or ice-covered cases the CAMEL HSR Algorithm includes a separate set of laboratory measurements of snow and ice spectra only. The Yemen case site shows the 8-year average CAMEL emissivity having a ~0.15 maximum difference from the lab measurements around the 6.5 μm carbonate spectral feature. While CAMEL does not capture the full extent of this feature as shown by the lab measurements, it is an improvement upon the UW BF and IASI emissivity, as the feature is present, but lower in magnitude. For the ARM SGP site, the CAMEL and IASI 8-year averages have close resemblance across the spectrum with the exception of the 8–9.5 μm region where IASI resolves the sharper features of the quartz signature. Differences between the CAMEL and lab emissivity are generally less than 0.025 with a small exception in the SW region at 5 μm. The last case site (no lab validation data is available) over Mt. Massive, located in the Colorado Rocky Mountains, is shown for the month of November. The overlaid November CAMEL HSR estimates for the years 2007–2014 located in the top panel reveal the year to year variation which is captured by the monthly CAMEL product. In some years, HSR emissivities reflect snow/ice signatures (version number of the laboratory dataset: Vs = 12 and #PC = 2), while others reflect more quartz-like signatures of bare rocks (Vs = 8 and #PC = 7), reflecting the dynamics of annual snow cover and snow-melt across the region.

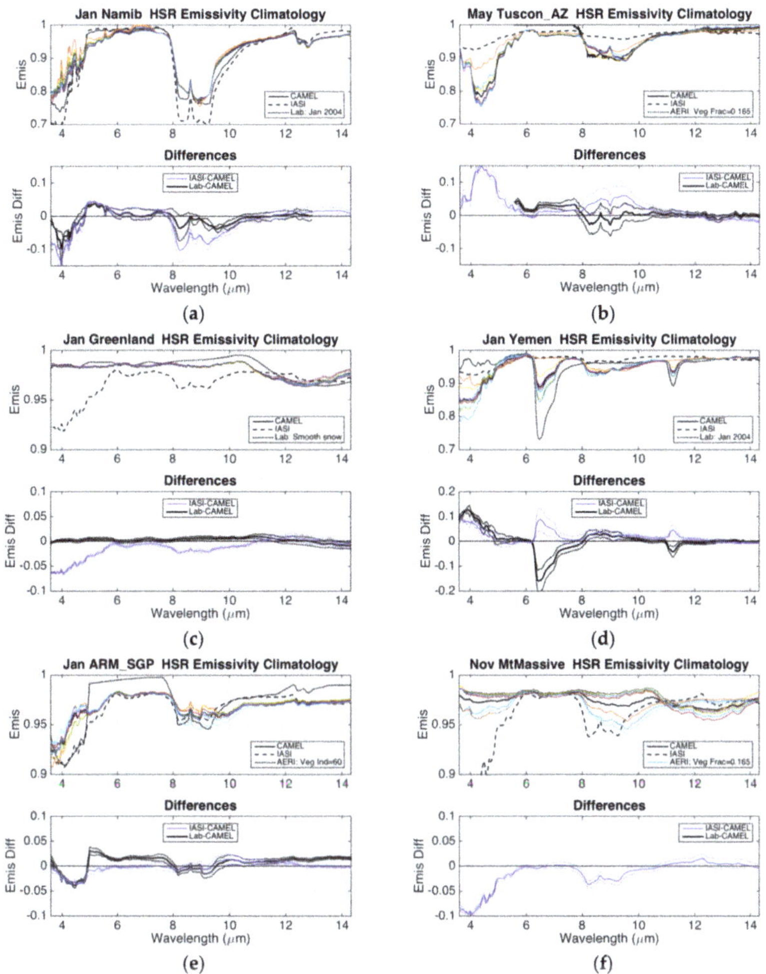

Figure 4. High spectral resolution (HSR) emissivity for (**a**) January Namib (−24.25N, 15.25E); (**b**) May Tucson, AZ (32.01N, −110.77E); (**c**) January Greenland (72.57N, −38.45E); (**d**) January Yemen (19.15N, 55.57E); (**e**) January Atmospheric Radiation Measurement, Southern Great Plains (ARM SGP) (36.60N, −97.48E); and (**f**) Nov Mt. Massive (39.17N, −106.47E) case sites. Top panels show CAMEL 8-year averaged (solid black), lab measurement (dotted black), and Infrared Atmospheric Sounding Interferometer (IASI) 8-year averaged (dashed black) HSR emissivity spectra overlaid on top of the individual CAMEL spectra for the years over which the aforementioned average was computed (colored). Bottom panel shows differences of the IASI minus CAMEL averages (blue solid) and lab minus CAMEL emissivity (black solid) which are both bounded by the CAMEL HSR uncertainty (dotted lines).

Figure 5 shows global maps of the differences between the CAMEL and IASI 8-year averages for the months of January and July at 4 different wavelengths. The global maximum magnitude of differences is larger for the shortwave and 8–9 μm channels (e.g., the 3.6 and 8.6 μm channels), where differences exceed 0.1 in select places (e.g., Sahara Desert, Australia, Tibetan Plateau), typically in regions whose surfaces are barren. Comparison of the January and July results reveals a seasonal dependence in the differences for various regions and channels, seen most notably at high latitudes where snow cover varies with the seasons. In general, a stable emissivity over deserts and polar ice (Greenland and Antarctica) is expected, so the large IASI variations over deserts and polar ice is unphysical and synthetic. The IASI emissivity over permanent snow cover contains quartz signature artifacts as discussed previously. The 10.8 μm channel shows slight changes from January to July and does show select areas where CAMEL emissivity is higher than IASI such as over the Tibetan Plateau, but in general CAMEL emissivities are lower than IASI at 10.8 μm. The 12.1 μm channel shows that for latitudes south of ~40°N where there is no snow cover, that the IASI minus CAMEL 8-year average difference is qualitatively similar between January and July and is largely positive, though it is quite small, being confined to under 0.03 magnitude.

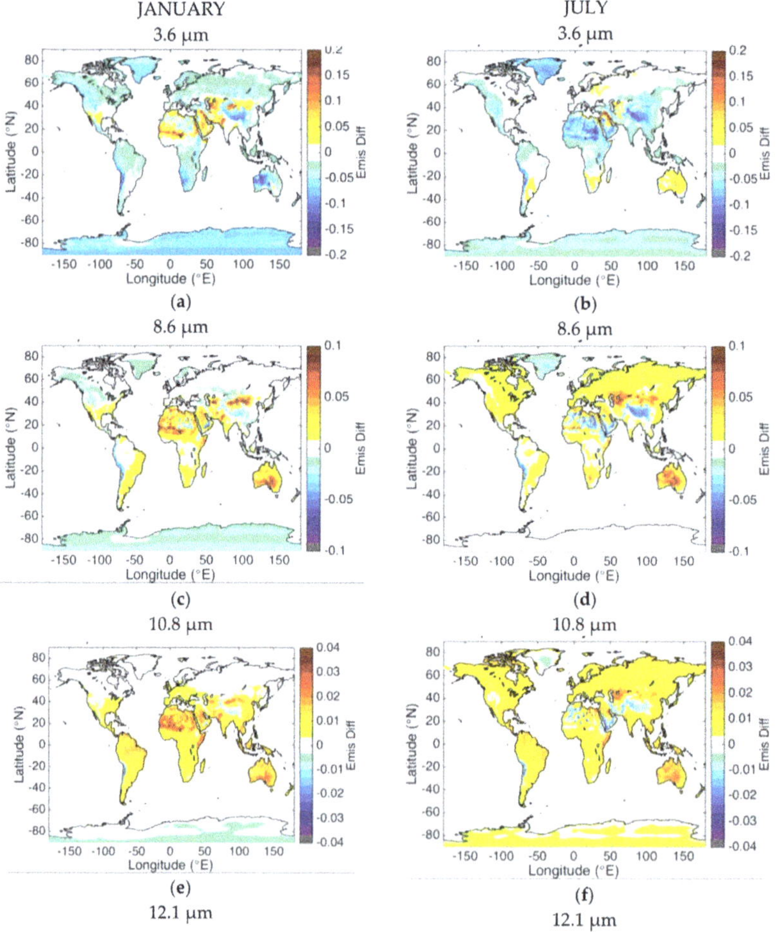

Figure 5. *Cont.*

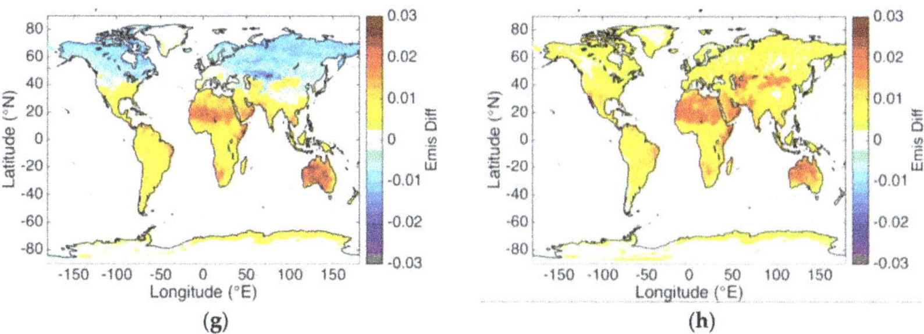

Figure 5. Difference of IASI minus CAMEL 8-year average emissivities at 0.25° × 0.25° spatial resolution for (**a**) January 3.6 μm; (**b**) July 3.6 μm; (**c**) January 8.6 μm; (**d**) July 8.6 μm; (**e**) January 10.8 μm; (**f**) July 10.8 μm; (**g**) January 12.1 μm; (**h**) July 12.1 μm. The years for which the 8-year averages are calculated are 2008–2015 for January and 2007–2014 for July. Colorbar limits change from channel to channel.

Statistics of the 8-year averaged IASI minus CAMEL emissivity over IGBP land classification categories are shown in Figure 6 for the months of January and July. Top panels show the average IASI and CAMEL emissivities and the bottom panel shows the average IASI minus CAMEL difference. Differences are generally less than a magnitude of 0.02 in emissivity and could be due to multiple issues including differences of spatial resolution between the IASI and MODIS/ASTER instruments and differences between the IASI emissivity and CAMEL algorithms. While differences could also be due the assumption of equal emissivity over day and night, further investigation is needed on this topic. Largest differences between the CAMEL and IASI products are seen in the barren/sparse vegetation category, where they exceed 0.03 in magnitude. The IASI product has larger differences between January and July than CAMEL does for the barren/sparse 8–11 μm region. But the CAMEL product has larger differences between January and July in the grassland 8–11 μm region. The CAMEL temporal variations over temperate grasslands and stable emissivity over deserts is as expected due to changes in phenology that affect vegetation cover.

4.3. RTTOV IASI Brightness Temperature Calculations

4.3.1. Methodology

As an additional assessment of the CAMEL HSR product, IASI brightness temperatures (BT) are calculated using the RTTOV model and CAMEL HSR emissivity module for 616 selected channels (National Oceanic and Atmospheric Administration selection) for four focus days representing the four seasons including 15 January, 14 April, 15 July, and 29 September of 2008. Similar calculations are performed for a constant with wavenumber 0.98 emissivity spectrum to assess the impact of the difference between the CAMEL dataset and commonly used default 0.98 emissivity value. Both sets of calculated BTs are then compared to IASI observed radiances for a subset of cases over land (coastlines not included) and for clear sky conditions. The clear sky mask that is used is the MAIA cloud mask [17], which selects scenes with less than 95% cloud cover. Simplifications to the BT calculations include setting the satellite zenith and azimuth and sun zenith and azimuth angles to zero, not applying the snow fraction, and using no trace gas inputs besides water vapor. ECMWF 6 hourly, 0.25° spatial resolution, analyses are used as input into the RTTOV model. The model analyses are bilinearly-interpolated to the IASI observations, specifically the center of the IASI fields of view. To minimize the effect of rapid skin temperature variations on the comparisons, only nighttime cases (solar zenith angles greater than 85°) are used and a two-hour time restriction between IASI observations and ECMWF analyses time steps is applied.

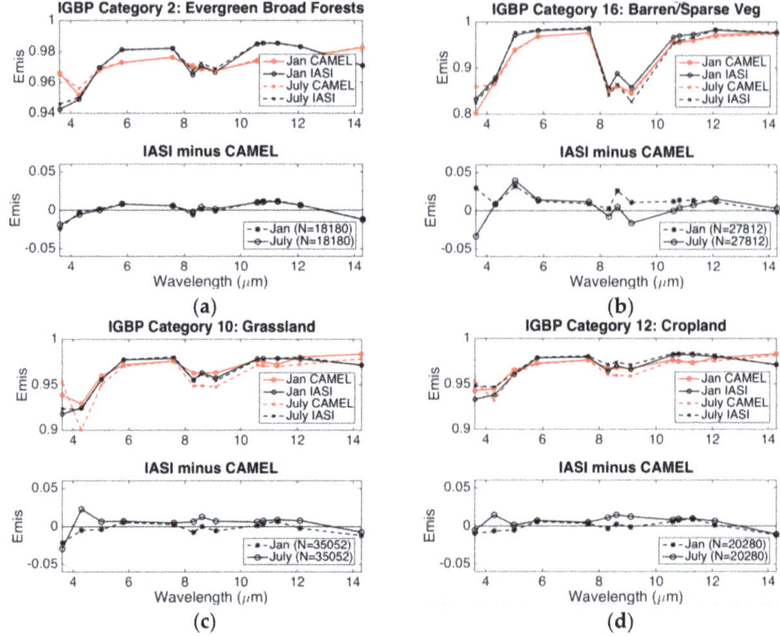

Figure 6. Top panels show IASI (black) and CAMEL (red) 8-year average emissivity averaged over the (**a**) evergreen broadleaf forest; (**b**) barren/sparse vegetation; (**c**) grassland; and (**d**) cropland IGBP land classifications for the months of January (solid line, circle markers) and July (red line, 'X' markers). Bottom panels show the associated averaged IASI minus CAMEL again for January (black dashed 'X' markers) and July (black solid circle markers). Number of samples, N, noted in bottom panel's legend.

4.3.2. Results

Differences between the calculated and observed IASI BTs are computed over the 616 channel set and separated by land classification type using the IGBP categories. The improper skin temperature determination can cause a bias in the BT calculation. This bias between ECMWF and IASI has a magnitude typically around 1–2 K in the Northern Hemisphere and 2–3 K in the Southern Hemisphere. To avoid it the de-biased variances of the calculated minus observed BTs differences for the 3.6–5, 8–9, and 10–13 μm spectral regions are used as indicators for a better emissivity estimate. To compute the de-biased variances for each spectrum, the IASI BT calculation minus observation difference is found and the result is then fitted to the 10.5–13 μm window region by subtracting the calculation minus observation 10.5–13 μm spectral domain average from the calculated spectrum. This de-biasing using the window region is assumed to only account for differences in skin temperature, though it's possible that the de-biasing could account for other errors of lesser magnitudes such as inaccurate surface emissivity characterization and errors induced from the inclusion of cloudy scenes in the analysis. Three separate spectral variances over the 3.6–5, 8–9, and 10–13 μm regions are then computed for the window-fitted BT difference. Using this method, there are 3 separate variances computed for each IASI calculation. The de-biased variance statistics corresponding to the CAMEL and 0.98 emissivity are compared for different time and space domains by taking averages of the variances over the specific regions.

Figure 7 shows mean CAMEL and 0.98 constant emissivity spectra and calculated minus observed IASI BT biases for a nighttime granule over Egypt. The 8.5 μm map of the calculated CAMEL minus observed BTs illustrates the spatial distribution and magnitude of the differences. This example shows the very large improvement of over 5 K when using CAMEL emissivity as opposed to the constant 0.98 emissivity value within the 8–10 µm region Reststrahlen band region. De-biased variance values over the three spectral regions can be found in the upper panel and show lower, improved values for the CAMEL dataset. The CAMEL calculated minus observed BT bias has increased magnitudes in spectral regions where atmospheric gases strongly absorb, in particular the 4.5–5 µm carbon monoxide and carbon dioxide region, the 6–8 µm water vapor region, and 9.6–10 µm ozone absorption region. Figure 8 shows a second nighttime example of statistics from a granule over the coast of Antarctica. Though not many samples are included in this granule, results are representative of snow and ice scenes. The BT bias corresponding to the CAMEL emissivity calculations is greater than the 0.98 emissivity bias from approximately 10–10.8 µm, however, it is less within the ~11–13 µm region where most of the spectral variation in snow/ice emissivity occurs. The magnitude of the BT bias difference is just under 1 K and represents how the CAMEL emissivity changes with respect to the 0.98 constant emissivity in the bottom panel. While the 8–9 µm de-biased variance is greater for CAMEL than the 0.98 constant, the 10–13 µm value is overall decreased.

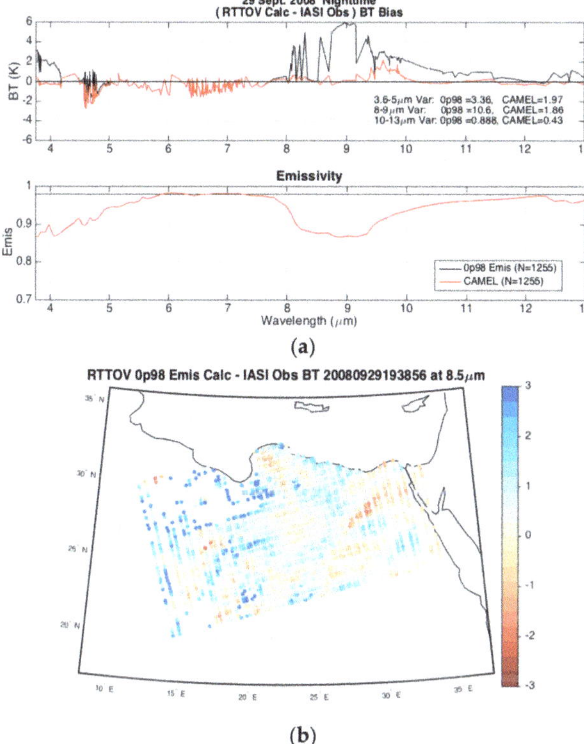

Figure 7. (**a**) Calculated constant 0.98 (black) and CAMEL (red) emissivity brightness temperature (BT) minus observed IASI BT bias (top panel) and mean emissivities (bottom panel) for a 29 September 2008 nighttime granule over Egypt. Number of samples, N, shown in legend. Debiased variances listed for three spectral regions; (**b**) Map of 8.5 µm calculated CAMEL minus observed IASI BTs differences in Kelvin.

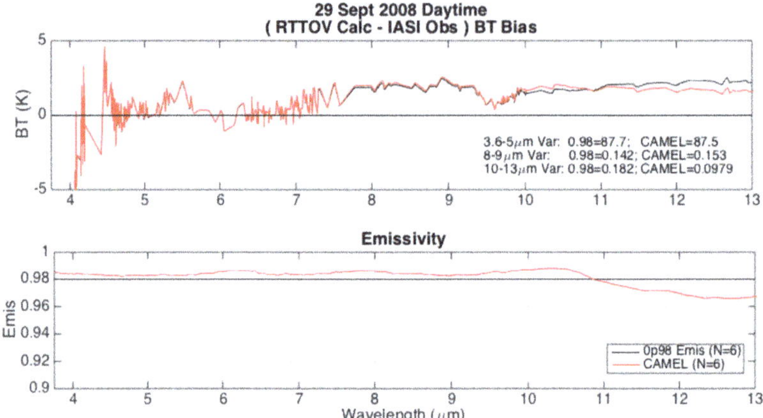

Figure 8. Same as Figure 7a except for daytime granule located over the coast of Antarctica.

Figure 9 shows nighttime statistics of the calculated minus observed IASI BTs for various IGBP land classification categories for the 29 September 2008 case day. While these examples are not comprehensive of all days and seasons, some general characteristics can be seen. For all categories shown, the CAMEL calculated BT is seen to generally be closer to the observed IASI BT than the 0.98 emissivity calculated BT. Land classification categories with emissivity spectra close to 0.98, such as the evergreen broad-leaf forests and croplands show the smallest bias difference between the CAMEL and 0.98 emissivity BT calculations. Largest differences are seen in the barren/desert land category around 8–9.5 µm, where the emissivity spectra have prominent quartz spectral features and reach values less than 0.7, as expected. The CAMEL calculations also show better agreement with the observations than the 0.98 emissivity calculations within the 3.6–4.2 µm region. Improved agreement around these wavelengths is seen for all other land categories shown as well. Comparison of the de-biased variance statistics for the CAMEL and 0.98 emissivity also confirms that for all land categories shown the CAMEL HSR emissivity is a spectral improvement upon the constant 0.98 emissivity within the 3.6–5 and 8–9 µm region. The 10–13 µm region de-biased variance statistics show neutral improvement using the CAMEL emissivity for all categories shown except for the open shrub and barren/desert land category where reductions in variance are significant.

To summarize the global RTTOV calculated minus observed statistics over the 4 case days, the CAMEL and 0.98 emissivity de-biased variances are listed for the 8–9 µm spectral region in Table 4. While variances for the 10–13 and 3.6–4.2 µm regions were calculated in a similar fashion, the results are not listed here and are only discussed qualitatively. The 8–9 µm region was selected to be shown since this spectral region is the region of largest expected improvement in comparison to the previous UWIREMIS. The numbers shown in Table 4 are averages of de-biased variances for the focus days and IGBP land classification categories shown. The corresponding number of samples are illustrated in Figure 9.

Table 4 results show in general that CAMEL is an improvement upon the 0.98 constant emissivity for this spectral region. For all categories with over 2000 samples, the CAMEL de-biased variance is lower than the 0.98 emissivity de-biased variance. The open shrub and barren/desert land categories again show the largest percent decreases in de-biased variance from the 0.98 emissivity to CAMEL. The CAMEL de-biased variance is decreased on average for the four days to ~41% and ~21% of the 0.98 emissivity value respectively for the open shrub and barren/desert land categories.

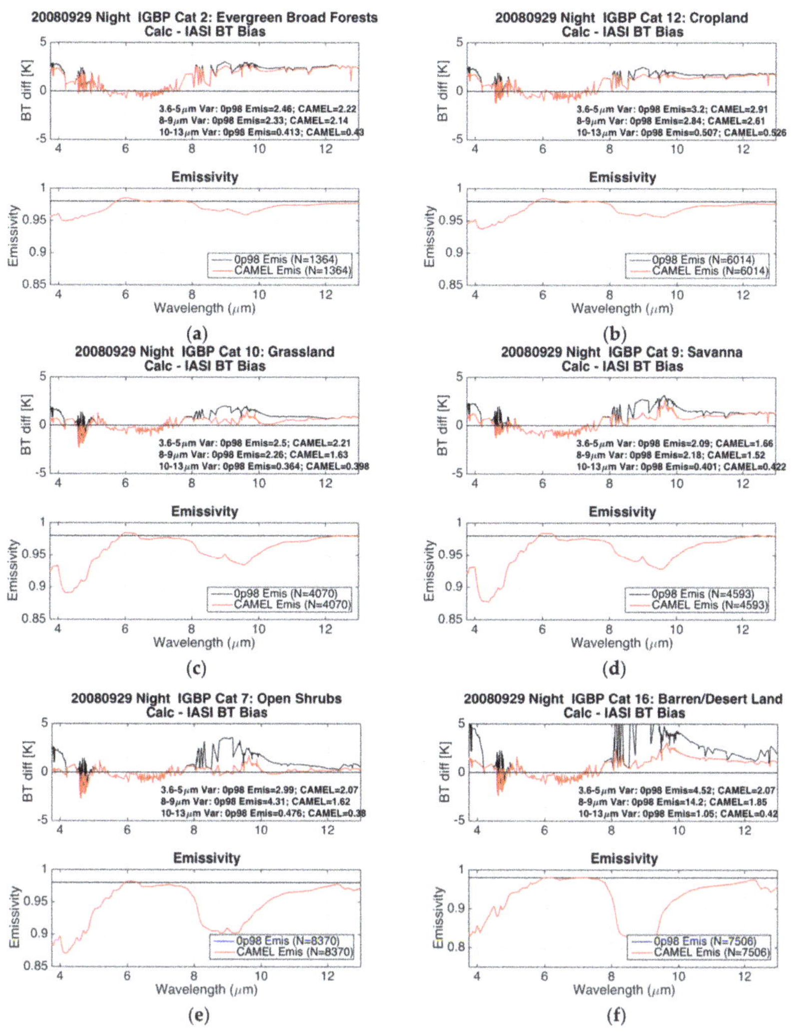

Figure 9. Nighttime RTTOV calculated minus observed IASI BT biases (**top panels**) for a constant 0.98 emissivity (black) and the CAMEL emissivity dataset (red) with associated mean emissivity spectra (**bottom panels**) for 29 September 2008 for the (**a**) evergreen broad-leaf forest; (**b**) cropland; (**c**) grassland; (**d**) savanna; (**e**) open shrub; and (**f**) barren/desert land IGBP categories. De-biased variances (explanation in text) are used as the better emissivity estimate and are shown in figures. The number of samples, N, are listed in figure legends. Note that the bottom panel *y*-axis limits change for the desert/barren land category.

In addition to the 8–9 μm region, the 3.6–5 μm region of CAMEL also better matches the IASI observations than does the 0.98 constant emissivity. De-biased variances for CAMEL are overall lower than those for the 0.98 constant emissivity in the 3.6–5 μm region, with the exceptions being for cases with a low (e.g., less than 100) number of samples. The open shrub and barren/desert categories showed the most CAMEL improvement—the de-biased variances for CAMEL are decreased respectively to be ~79% and ~56% of the 0.98 emissivity de-biased variance.

The last spectral region assessed, the 10–13 μm region, does not show an obvious improvement with the use of the CAMEL. For the average of the four case days, more land classifications show an increase, rather than a decrease, in the CAMEL de-biased variances. Improvement is seen, however, in the barren/desert land and open shrub categories, with the average CAMEL variances being 52% and 80% of the 0.98 variance respectively. The deciduous needle forest and mixed forest categories have the largest average increases of 10–13 CAMEL de-biased variances with respect to the 0.98 values, with 15% being the largest increase.

Table 4. Variance over the 8–9 μm region of de-biased IASI calculated minus observed BTs are averaged by IGBP land classification categories for four focus days and shown below in units of Kelvin for nighttime cases. De-biased variances for BTs calculated using a constant 0.98 emissivity and the CAMEL HSR emissivity module are shown. The corresponding number of samples are shown in Figure 10.

IGBP Category	15 January 2008		14 April 2008		15 July 2008		29 September 2008	
	0.98	CAMEL	0.98	CAMEL	0.98	CAMEL	0.98	CAMEL
1: Evergreen Needle Forests	3.48	3.48	1.35	1.45	3.48	3.38	2.89	2.86
2: Evergreen Broad Forests	5.21	4.95	5.76	5.49	6.19	5.81	2.33	2.14
3: Deciduous Needle Forests	1.98	1.98	5.99	6.02	7.91	7.49	1.31	1.29
4: Deciduous Broad Forests	2.22	2.23	1.92	2.06	3.03	2.84	4.37	4.44
5: Mixed Forests	1.97	1.96	1.24	1.3	5.34	4.96	2.73	2.64
6: Closed Shrubs	2.27	1.98	3.61	3.57	3.62	3.2	2.87	2.41
7: Open Shrubs	4.8	2.57	4.16	1.48	4.83	1.89	4.31	1.62
8: Woody Savanna	3.01	2.88	3.92	3.59	4.67	4.21	3.08	2.67
9: Savanna	3.76	2.93	3.77	3.11	4.68	3.75	2.18	1.52
10: Grassland	2.85	2.34	2.9	2.17	2.56	1.92	2.26	1.63
11: Wetland	1.83	1.84	1.24	1.3	3.63	3.4	2	1.78
12: Cropland	2.71	2.51	2.19	1.81	3.47	3.12	2.84	2.61
13: Urban Area	2.67	2.68	1.77	1.9	4.39	4.22	3.78	3.71
14: Crop Mosaic	3.34	3.15	2.62	2.45	3.9	3.55	3.1	2.75
15: Antarctic/Permanent Snow	1.25	1.3	1.16	1.2	1.46	1.45	1.57	1.72
16: Barren/Desert	6.96	2.28	13.4	2.16	10.3	2.62	14.2	1.85

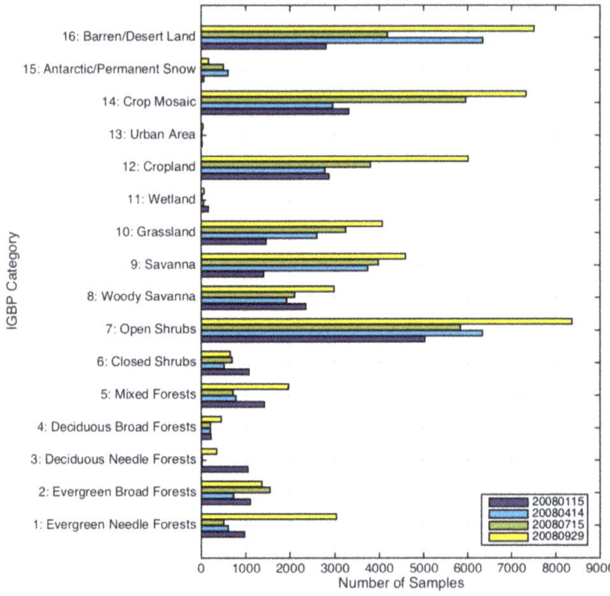

Figure 10. Number of samples for de-biased variances listed in Table 4.

5. Conclusions

The NASA CAMEL product reports monthly global land surface emissivity at 13 hinge points between 3.6–14.3 µm and provides software which uses a PCA approach to compute 417 channel high spectral emissivity spectra from the 13 hinge point product for the same wavelength region. The uncertainty provided with the product is available at the 13 hinge point resolution for every month and grid point and is a combination of 3 separate components of variability—a spatial, temporal, and algorithm component. This work presented the methodology behind the uncertainty estimates included in the product and a comprehensive validation of the CAMEL product which included laboratory measurements, product intercomparisons, and radiative transfer calculations.

Comparisons of CAMEL to validation data, the 8-year averaged IASI dataset, and the previous UW BF emissivity show an improvement of CAMEL in the characterization of snow, quartz, and carbonate spectral features. The monthly and inter-annual variations of the emissivity reveal dynamic land surface changes, for example, snow in the mountains. Comparison of the CAMEL 13 hinge point product to the 8-year averaged IASI dataset at 0. 25° spatial resolution showed maximum differences were located in the shortwave 3.6 µm region, where they reached over 0.15 magnitude in emissivity for some geographic locations, typically dry desert regions. Differences at 8.6 µm were primarily less than 0.1 and at wavelengths of 10.8 µm or larger they were predominantly under 0.02.

Assessment of the CAMEL RTTOV emissivity module was made using IASI brightness temperature observations as a reference. Calculated minus observed BT biases were computed for both the CAMEL emissivity and a constant 0.98 emissivity spectra, a commonly used simplification for a land surface emissivity estimate. The use of a de-biased spectral variance over three spectral regions was used as a measure of emissivity spectral accuracy. The average of the de-biased variances was found over IGBP land classification schemes for four days which represented four seasons. Land classifications which showed improvement with the use of the CAMEL emissivity for all three spectral regions were the open shrub and barren/desert land categories. In general, the 3.6–5 and 8–9 µm region statistics showed the CAMEL to be a better emissivity estimate than the constant 0.98 value, while the 10–13 µm region showed an improvement only for barren or partially barren surfaces. As expected, the percentage decreases in the 3.6–5 and 8–9 µm four day-averaged variances were larger than the percent increases for the 10–13 µm region when comparing CAMEL with respect to the 0.98 emissivity.

This two-part paper has described the methodology, applications, uncertainty, and validation of the first release of the CAMEL dataset available from a NASA MEASUREs project at https://lpdaac. usgs.gov/dataset_discovery/measures/measures_products_table/cam5k30em_v001. Future work will include version updates to improve some known issues, for example, snow fraction in mixed scenes, and the development of a 16-year CAMEL climatology and covariance for use in NWP applications.

Acknowledgments: The ASTER GED version 4 (v4) product used in this study was produced at JPL and is available courtesy of the NASA EOSDIS Land Processes Distributed Active Archive Center (LP DAAC), United States Geological Survey/Earth Resources Observation and Science Center, Sioux Falls, South Dakota, [https://lpdaac.usgs.gov/dataset_discovery/aster]. This work was done under grant number: NNX08AF8A.

Conflicts of Interest: The authors declare no conflict of interest.

References

1. Seemann, S.W.; Borbas, E.E.; Knuteson, R.O.; Stephenson, G.R.; Huang, H.-L. Development of a Global Infrared Land Surface Emissivity Database for Application to Clear Sky Sounding Retrievals from Multispectral Satellite Radiance Measurements. *J. Appl. Meteorol. Climatol.* **2008**, *47*, 108–123. [CrossRef]
2. Hulley, G.C.; Hook, S.J.; Abbott, E.; Malakar, N.; Islam, T.; Abrams, M. The ASTER Global Emissivity Dataset (ASTER GED): Mapping Earth's emissivity at 100 meter spatial scale. *Geophys. Res. Lett.* **2015**, *42*, 7966–7976. [CrossRef]

3. Saunders, R.; Hocking, J.; Rundle, D.; Rayer, P.; Havemann, S.; Matricardi, M.; Geer, A.; Cristina, L.; Brunel, P.; Vidot, J. *RTTOV-12 Science and Validation Report*; EUMETSAT NWP SAF, NWPSAF-MO-TV-41; EUMETSAT: Darmstadt, Germany, 2017.

4. Borbas, E.E.; Ruston, B.C. *The RTTOV UWiremis IR Land Surface Emissivity Module*; EUMETSAT NWP SAF, NWPSAF-MO-VS-042; EUMETSAT: Darmstadt, Germany, 2010.

5. Zhou, D.K.; Larar, A.M.; Liu, X. MetOp-A/IASI observed continental thermal IR emissivity variations. *IEEE J. Sel. Top. Appl. Earth Obs. Remote Sens.* **2013**, *6*, 1156–1162. [CrossRef]

6. Zhou, D.K.; Larar, A.M.; Liu, X.; Smith, W.L.; Strow, L.L.; Yang, P.; Schlussel, P.; Calbet, X. Global land surface emissivity retrieved from satellite ultraspectral IR measurements. *IEEE Trans. Geosci. Remote Sens.* **2011**, *49*, 1277–1290. [CrossRef]

7. Friedl, M.A.; Sulla-Menashe, D.; Tan, B.; Schneider, A.; Ramankutty, N.; Sibley, A.; Huang, X. MODIS Collection 5 global land cover: Algorithm refinements and characterization of new datasets. *Remote Sens. Environ.* **2010**, *114*, 168–182. [CrossRef]

8. Channan, S.; Collins, K.; Emanuel, W.R. *Global Mosaics of the Standard MODIS Land Cover Type Data*; University of Maryland and the Pacific Northwest National Laboratory: College Park, MD, USA, 2014.

9. Hook, S. Combined ASTER and MODIS Emissivity for Land (CAMEL) Uncertainty Monthly Global 0.05Deg V001 [Data set]. *NASA EOSDIS L. Process. DAAC* **2017**. [CrossRef]

10. Gillespie, A.; Rokugawa, S.; Matsunaga, T.; Steven Cothern, J.; Hook, S.; Kahle, A.B. A temperature and emissivity separation algorithm for advanced spaceborne thermal emission and reflection radiometer (ASTER) images. *IEEE Trans. Geosci. Remote Sens.* **1998**, *36*, 1113–1126. [CrossRef]

11. Borbas, E.E.; Knuteson, R.O.; Seemann, S.W.; Weisz, E.; Moy, L.A.; Huang, H.-L. A high spectral resolution global land surface infrared emissivity database. In Proceedings of the Joint 2007 EUMETSAT Meteorological Satellite Conference and the 15th Satellite Meteorology & Oceanography Conference of the American Meteorological Society, Amsterdam, The Netherlands, 24–28 September 2007.

12. Baldridge, A.M.; Hook, S.J.; Grove, C.I.; Rivera, G. The ASTER spectral library version 2.0. *Remote Sens. Environ.* **2009**, *113*, 711–715. [CrossRef]

13. Dozier, J.; Warren, S.G. Effect of viewing angle on the infrared brightness temperature of snow. *Water Resour. Res.* **1982**, *18*, 1424–1434. [CrossRef]

14. Knuteson, R.O.; Best, F.A.; DeSlover, D.H.; Osborne, B.J.; Revercomb, H.E.; Smith, W.L. Infrared land surface remote sensing using high spectral resolution aircraft observations. *Adv. Space Res.* **2004**, *33*, 1114–1119. [CrossRef]

15. Tobin, D.C.; Revercomb, H.E.; Knuteson, R.O.; Lesht, B.M.; Strow, L.L.; Hannon, S.E.; Feltz, W.F.; Moy, L.A.; Fetzer, E.J.; Cress, T.S. Atmospheric Radiation Measurement site atmospheric state best estimates for Atmospheric Infrared Sounder temperature and water vapor retrieval validation. *J. Geophys. Res. D Atmos.* **2006**, *111*, D09S14. [CrossRef]

16. Wan, Z.; Li, Z.-L. A physics-based algorithm for retrieving land-surface emissivity and temperature from eos/modis data. *IEEE Trans. Geosci. Remote Sens.* **1997**, *35*, 980–996. [CrossRef]

17. Lavanant, L. *MAIA AVHRR Cloud Mask and Classification*; Meteo-France: Lannion, France, 2002.

Article

Dynamic Inversion of Global Surface Microwave Emissivity Using a 1DVAR Approach

Sid-Ahmed Boukabara [1,*], Kevin Garrett [1] and Christopher Grassotti [2]

[1] NOAA/NESDIS Center for Satellite Applications and Research, College Park, MD 20740, USA;
    Kevin.Garrett@noaa.gov
[2] University of Maryland Earth Science System Interdisciplinary Center, College Park, MD 20740, USA;
    Chris.Grassotti@noaa.gov
* Correspondence: Sid.Boukabara@noaa.gov; Tel.: +1-301-683-3516

Received: 1 March 2018; Accepted: 23 April 2018; Published: 27 April 2018

Abstract: A variational inversion scheme is used to extract microwave emissivity spectra from brightness temperatures over a multitude of surface types. The scheme is called the Microwave Integrated Retrieval System and has been implemented operationally since 2007 at NOAA. This study focuses on the Advance Microwave Sounding Unit (AMSU)/MHS pair onboard the NOAA-18 platform, but the algorithm is applied routinely to multiple microwave sensors, including the Advanced Technology Microwave Sounder (ATMS) on Suomi-National Polar-orbiting Partnership (SNPP), Special Sensor Microwave Imager/Sounder (SSMI/S) on the Defense Meteorological Satellite Program (DMSP) flight units, as well as to the Global Precipitation Mission (GPM) Microwave Imager (GMI), to name a few. The emissivity spectrum retrieval is entirely based on a physical approach. To optimize the use of information content from the measurements, the emissivity is extracted simultaneously with other parameters impacting the measurements, namely, the vertical profiles of temperature, moisture and cloud, as well as the skin temperature and hydrometeor parameters when rain or ice are present. The final solution is therefore a consistent set of parameters that fit the measured brightness temperatures within the instrument noise level. No ancillary data are needed to perform this dynamic emissivity inversion. By allowing the emissivity to be part of the retrieved state vector, it becomes easy to handle the pixel-to-pixel variation in the emissivity over non-oceanic surfaces. This is particularly important in highly variable surface backgrounds. The retrieved emissivity spectrum by itself is of value (as a wetness index for instance), but it is also post-processed to determine surface geophysical parameters. Among the parameters retrieved from the emissivity using this approach are snow cover, snow water equivalent and effective grain size over snow-covered surfaces, sea-ice concentration and age from ice-covered ocean surfaces and wind speed over ocean surfaces. It could also be used to retrieve soil moisture and vegetation information from land surfaces. Accounting for the surface emissivity in the state vector has the added advantage of allowing an extension of the retrieval of some parameters over non-ocean surfaces. An example shown here relates to extending the total precipitable water over non-ocean surfaces and to a certain extent, the amount of suspended cloud. The study presents the methodology and performance of the emissivity retrieval and highlights a few examples of some of the emissivity-based products.

Keywords: emissivity; variational retrieval; surface parameters

---

1. Introduction

Passive microwave measurements have been used extensively in the past to monitor surface parameters such as the Sea-Ice Concentration (SIC) and ice age over ice-covered surfaces [1,2], the snow depth or Snow-Water Equivalent (SWE) over snow-covered land [3,4], the soil moisture over snow-free land surfaces [5], and the wind speed over the ocean [6], to list just a few examples. These inversions usually rely on brightness temperature measurements with, sometimes ad-hoc, corrections to account for the atmospheric and cloud contamination. The methods generally make use of a limited number of channels, mostly pure window channels, in order to avoid atmospheric impacts as much as possible. However, the measurement, even if properly cleared from the atmospheric effects, is still radiometrically a mixture of the emissivity and the skin temperature. These two parameters are indeed intertwined from the radiometric point of view. Notably though, the signal of the surface properties is most directly contained in the radiometric emissivity. An alternative approach to retrieving surface properties is suggested in this study, in which the surface parameters, such as SIC and SWE, are directly inverted from the emissivity spectrum of the surface in question, not from the brightness temperatures or their combinations. The emphasis is on the use of the term *spectrum* in this case, since the intent is to truly use not only the absolute values of the emissivities, but also the whole spectral shape, in order to retrieve the surface parameters.

Many approaches for deriving surface emissivity rely on the simplified microwave radiative transfer (RT) equation inverted to solve the emissivity term analytically:

$$\varepsilon = \frac{T_b - T_u - T_d \Gamma}{(T_s - T_d)\Gamma},\tag{1}$$

where the emissivity (ε) is a function of the observed brightness temperature at the satellite, T_b, taking into account the skin temperature, T_s, the upwelling and downwelling brightness temperatures (T_u and T_d, respectively), and the atmospheric transmittance, Γ. In this case, as for brightness temperature-based retrieval of geophysical parameters, methodologies must be adopted to remove observations contaminated by clouds and precipitation [7–9]. The derived emissivity depends on radiative transfer calculations to compute the transmittance, and therefore, is sensitive to errors from the input temperature and water vapor profiles as well as the skin temperature and can be large, especially if a forecast is used. Additionally, the accuracy of the emissivity depends on other factors unaccounted for, including the surface roughness.

The usage of a restricted number of channels, the removal of observations affected by clouds and precipitation, and the lack of consideration of errors in a priori information when deriving emissivity and other surface parameters are argued to be limiting factors. In this study, we adopt the use of all channels in order to have a better picture of the spectral shape of the emissivity. One of the key factors for the success of the proposed approach is the ability to distinguish the different signals impacting the brightness temperatures, in order to accurately isolate the emissivity signal from which the surface parameters will be derived. In order to distinguish all of the parameters impacting radiances, one logical approach is to use all channels available that have the potential to inform us about the state of these parameters.

In this study, the main objective is to focus on the retrieved emissivity quality assessment and to highlight some applications derived from it. In Section 2, we give an overview of the Microwave Integrated Retrieval System (MiRS) emissivity retrieval approach (the overall approach is described in [10]), and in Sections 3 and 4, we thoroughly assess the performance of the emissivity retrieval (qualitatively and quantitatively). In Section 5, we highlight a few examples of parameters derived using these emissivities. This constitutes an indirect assessment of the emissivity product itself. For brevity, this study will not present details about the assessment of the emissivity-based products which may be found in other publications ([11] for example, for sea-ice).

2. Overview of the Retrieval Approach

The MiRS-based retrieval of surface parameters consists of two distinct steps. The first step aims to isolate the emissivity signal in the brightness temperatures from the rest of the signals impacting them, such as those from skin temperature or from the atmosphere. This is performed using a one-dimensional, variational-based physical approach (1DVAR) where the simulated radiances and their Jacobians are computed using the Community Radiative Transfer Model (CRTM) [12]. The goal of the iterative 1DVAR is to find a solution that, when fed to a forward model, will produce simulated radiances that closely match the measurements within a level of uncertainty (e.g., radiative transfer error plus instrument noise).

The iterative solution is initiated from a climatological first guess. Figure 1 illustrates this approach where, during this first step the system retrieves the state vector including the emissivity spectrum and the skin temperature (representing the surface), along with the atmospheric parameters represented by temperature, moisture, cloud, rain and ice profiles. A solution is found by minimizing the cost function, and the retrieved state vector when input to the forward operator, fits the radiometric observations (e.g., convergence). Further mathematical basis of the MiRS 1DVAR may be found in [10].

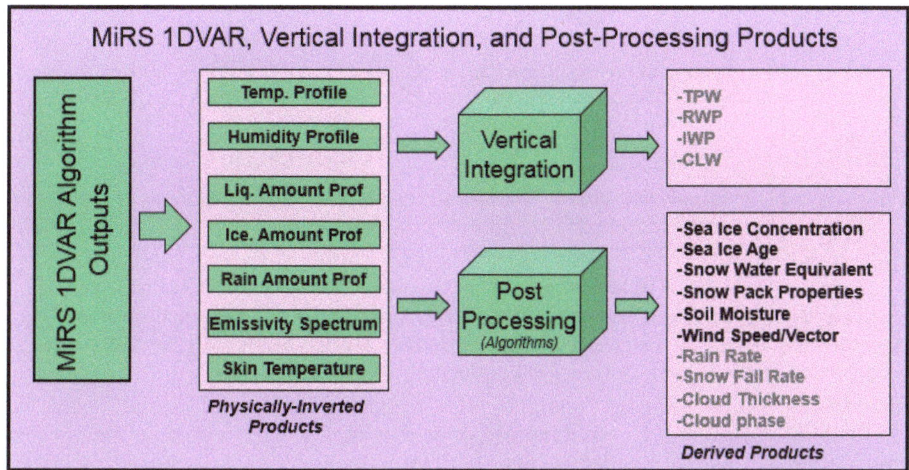

Figure 1. Products from the Microwave Integrated Retrieval System (MiRS) two-steps retrieval process. Products listed in bold under "Derived Products" are generated from the emissivity spectrum. Soil moisture and wind speed/vector are not current operational products. 1DVAR: one-dimensional, variational-based physical approach.

The retrieved state vector is then used in the second step of MiRS, which is called the Vertical Integration and Post-Processing (VIPP) step, also illustrated in Figure 1. Here, some products are simply generated by vertical integration of the atmospheric profiles, such as the moisture profile, to derive the Total Precipitable Water (TPW) or the ice profile to derive the Ice Water Path (IWP). Other parameters are generated by using a more sophisticated mechanism. In the case of SIC and SWE, the post-processing stage utilizes a procedure to search a catalog of pre-computed emissivity spectra to closely match the retrieved one. Along with the search of SIC, the ice age (FYI/MYI for First-Year and Multi-Year Ice, respectively) is also catalogued and searched for, given its impact on the emissivity spectrum. Similarly, along with the search of SWE, the effective grain size is also catalogued and searched for. These two byproducts (sea-ice age and effective grain size) are also operational products from MiRS.

It is worth noting that the emissivity could be used directly as a proxy parameter to determine the degree of wetness over land, for instance. In MiRS, however, as mentioned earlier, the emissivity is further processed to produce more readily usable surface parameters, such as SIC and SWE. Figure 2 details this post-processing of the emissivity. It shows that the emissivity spectra as retrieved in the first step are compared to an offline-computed catalog of emissivity spectra for a range of values of the parameters to be derived.

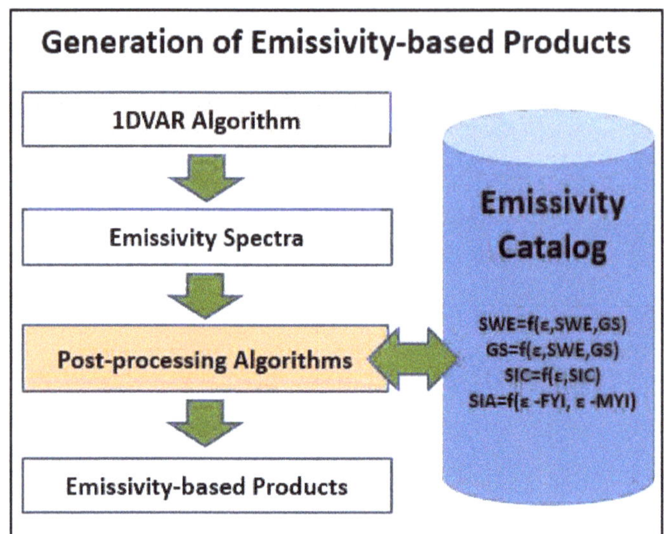

Figure 2. Illustration of MiRS emissivity-based products. The surface characteristics (Sea-Ice Concentration, SIC; Snow-Water Equivalent, SWE) are retrieved along with corresponding secondary parameters (age, grain size respectively), by matching the 1DVAR-retrieved emissivity with pre-computed emissivity spectra stored in a catalog.

The MiRS algorithm is an enterprise algorithm that applies to a multitude of satellites and sensors, provided that the forward operator (CRTM) is applicable to those sensors. Currently, the MiRS algorithm is routinely applied to microwave sensors but efforts are ongoing to validate it for infrared (IR) sensors as well, given that CRTM is applicable to IR sensors. Currently MiRS is applied operationally or routinely to the NOAA and Metop series sounders (AMSU/MHS) as well as the SNPP/ATMS and DMSP SSMI/S series, and research satellites (NASA GPM/GMI, NASA/JAXA AMSR-2, CNES/ISRO Megha-Tropiques SAPHIR). It has recently been extended to process ATMS data from the newly launched NOAA-20 satellite. In this study, we show a variety of emissivity highlights from a number of sensors, but for the performance assessment, we focus on NOAA AMSU/MHS for cross-track type sensors and DMSP SSMI/S for conical type sensors.

3. Emissivity Qualitative Assessment

3.1. Methods

The initial assessment of retrieved surface emissivity was qualitative and focused on assessing the response of the emissivity magnitude to changes in surface type, and the responses of the magnitude and spectral shape to changes in the land surface moisture content. This is an important first assessment since the emissivity-based surface products from MiRS are only as good as the emissivity they are derived from. The main difficulty with the direct validation of emissivity is the lack of a ground-based reference to compare it to. Even if there was a ground-based sensor taking measurements of emissivity, it would have to be made at exactly the same frequency, with similar sensor characteristics (bandwidths for instance) and at the same viewing angle, and it would need to represent the same satellite footprint size and not just a single point. The emissivity variability is simply too high within the footprint size of a polar-based sensor to rely on single point measurements for assessment. This is particularly true over non-ocean surfaces. We are therefore left, for the purpose of assessing our emissivities, with two methods: a qualitative assessment, and a quantitative assessment based on analytical emissivities, which will be explained in the next section.

3.2. Examples

The first qualitative assessment of the dynamic emissivity from MiRS is presented in Figure 3 where satellite data from SNPP/ATMS were used to generate global fields of microwave emissivities. The Continental United States (CONUS) region is highlighted in this figure. The emissivity was found to act as expected—ocean-like emissivity (lower values) over water surfaces and land-like emissivity (higher values) over land surfaces. What is important to note is that these retrievals (with emissivity and other parameters inverted simultaneously) have permitted full convergence or fitting of the measurements, giving us confidence that these parameters belong to the space of possible solutions, including over coastal areas. The system is therefore flexible enough to retrieve intermediate-values (or mixed-signal value) of the emissivity when the brightness temperatures are measured over mixed surface types. This is an important point as we will see later, because this allows the dynamic emissivity retrieval to *absorb* the mixed signal present in the measurement, and it allows the other parameters (like TPW, cloud, rain, etc.) to be retrieved seamlessly over ocean and land interfaces. If there was no emissivity in the retrieval state vector to *absorb* this mixed-type signal, the atmospheric parameters would exhibit discontinuities at these coastal regions. The same applies to other surface type transitions (sea–ice edges, snow and snow-free land, river–land, lake–forest, etc.). We have shown, for example, that rain retrieval over land and over coastal transitions is improved due to this approach. The emissivity, retrieved simultaneously with all other parameters, absorbs the coastal signal and therefore, reduces the number of false alarms in the rain retrievals over land and coastal areas significantly [13]. We argue that over these types of highly variable surfaces, it is critical to allow the emissivity to vary dynamically as part of the retrieval, as we did here, in order to provide seamless atmospheric products. This is especially true because the degree of mixing (of the different surface types) depends, to a significant degree, on the shape and size of the measurement footprint; this is highly uncertain a priori, making attempts to account for these dynamic backgrounds through precomputed atlases sub-optimal at best.

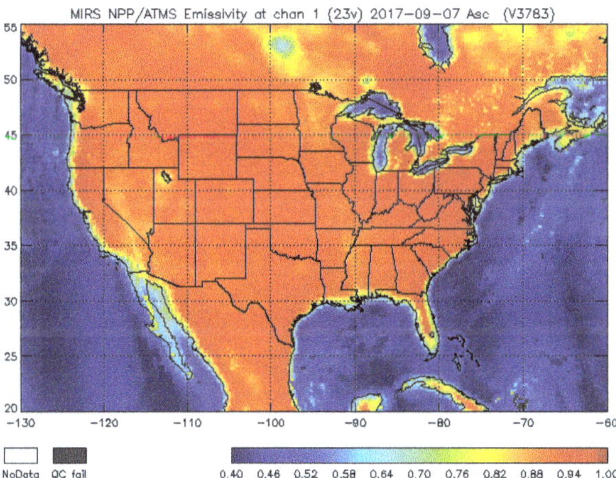

Figure 3. MiRS-based emissivity retrieved using SNPP/ATMS data from 7 September 2017. The emissivity of channel 1 (23 GHz) is shown. The emissivity is behaving as expected, with low values over the ocean, higher values over land, and intermediate values over coastal areas. Over land, lower values are also retrieved over water bodies and rivers. All cases have converged during the retrieval (not shown).

Another qualitative example to emphasize the previous point is given in Figure 4. It represents the emissivity at the 31.4 GHz channel on the Advance Microwave Sensor Unit (AMSU) onboard the NOAA-18 satellite. It shows that the emissivity retrieved from MiRS transitions has low value over oceans and high values over land, as expected, with in-between transitional values on the coast and along water bodies, in this case the Volta Lake in West African Ghana. At the same time, the bottom panel in the same figure shows that convergence (fit to observations) was reached over land, ocean, and mixed-surface areas. The convergence metric shown there is indeed under unity, signaling that the solution found is within noise levels that are consistent with the radiometric measurements.

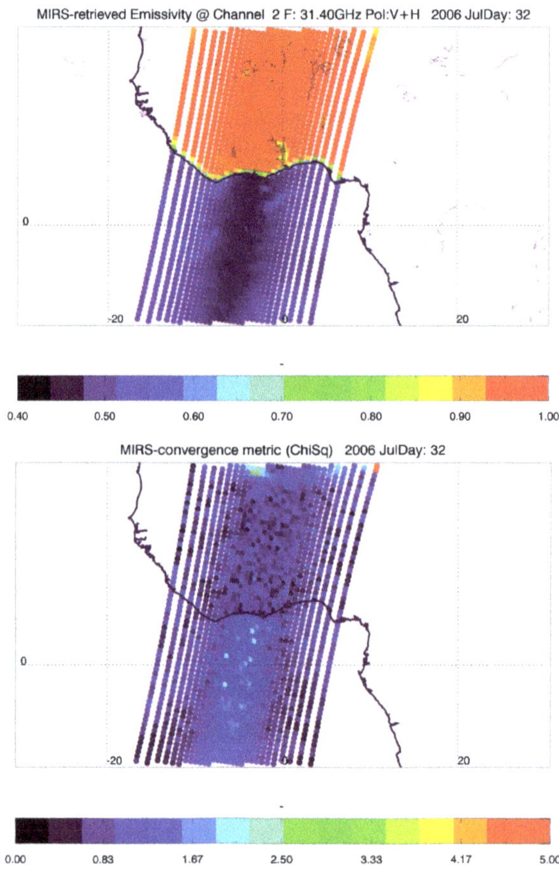

Figure 4. Qualitative assessment of the MiRS-based emissivity retrieval from NOAA-18 Advance Microwave Sensor Unit (AMSU) measurements at the transition of the Sub-Saharan African coast. (The actual emissivity retrieved from MiRS highlighting the low-to-high transition from ocean emissivities to land emissivities, as expected (**top**), all while perfectly converging (**bottom**). Note also the angle dependence of the emissivity over the ocean, as expected, is captured by MiRS emissivities.

A third qualitative assessment of the emissivity performed is illustrated in Figures 5 and 6. It demonstrates the evolution of emissivity (amplitude and spectral shape) over time when a major rain event occurs. One intuitively expects that when such an event occurs over a dry land surface, the rain will wet the surface instantly, altering the emissivity spectrum significantly. After the rain stops and the surface begins to dry out, one would expect that the emissivity would gradually go back to its pre-rain spectral shape. The speed of this cycle depends on many parameters, including the pre-existing soil moisture, the surface type, and atmospheric conditions.

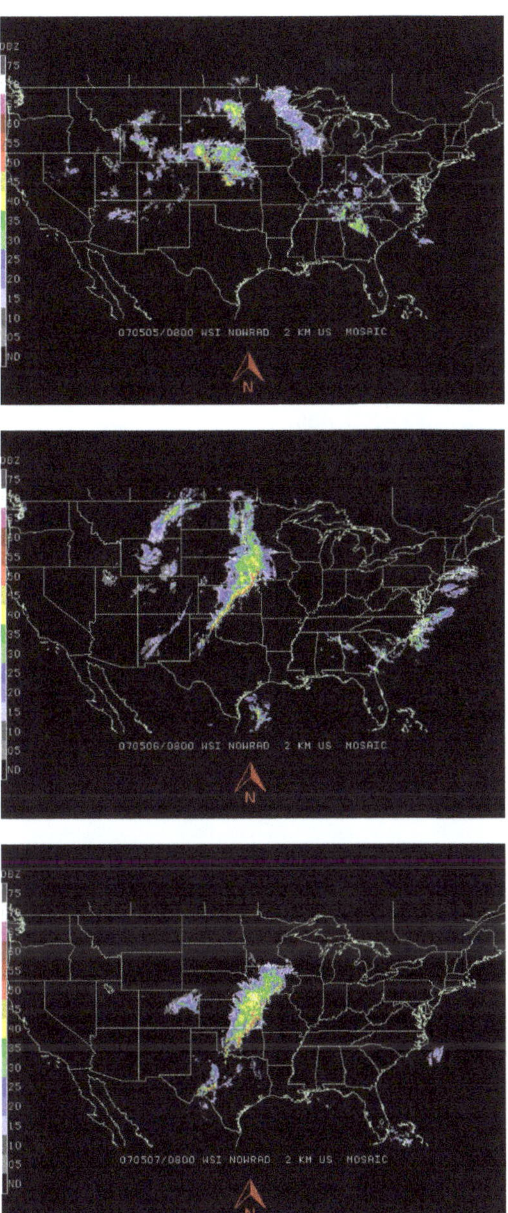

Figure 5. Rain storms in Midwestern US, as captured by NEXRAD radars during the period between 5 May (**top**), 6 May (**middle**) and 7 May (**bottom**), 2007.

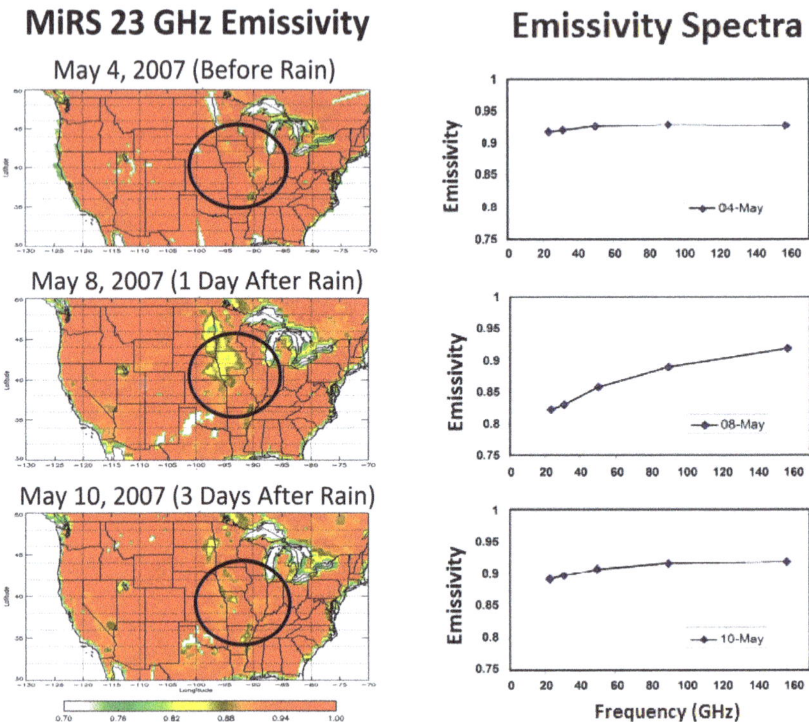

Figure 6. MiRS-based emissivity maps and spectra, before (**top**), shortly after (**middle**) and three days after (**bottom**) the rain events in Midwestern US, shown in previous figure. Circles highlight the areas where the spectra were computed from, corresponding to rain events. The emissivity absolute values and spectra behave as expected, going from their dry land emissivity values and almost flat spectrum before the rain, to much lower values (and water-like spectrum) right after the rain. After a few days of drying, the emissivities go back (although not fully) to their higher values and dry-like flat spectrum.

A significant storm system was recorded due to its wide-spread nature. These storms hit Midwestern US during 5–7 May 2007, as seen from NEXRAD Radar images in Figure 5. The MiRS retrieved emissivity spectra were qualitatively assessed for their behavior before, during, and after these rain storms. This is illustrated in Figure 6. In this figure, both a spatial map and a full spectrum are shown. The emissivity map relates to channel 1 (23.8 GHz) of the NOAA-18 AMSU. It was chosen for its higher sensitivity to the surface. The days of the measurements that were assessed are 4 May 2007 (a day before the rain event hit that area; dry surface); 8 May 2007 (a day after the storm had passed, where the surface was presumably still wet); and 10 May 2007 (three days after the event, where presumably, the surface had dried out). One can see clearly in Figure 6 that both the emissivity values and the spectral shape from the MiRS-retrievals behaved as expected. Indeed, the dry surface emissivity spectrum is almost flat with high values (4 May). This transforms into an almost water-like spectrum on 8 May, when the surface is wet in terms of both the absolute values and the shape of the emissivity spectrum. When the surface returns to its dry state, the emissivity spectrum again resembles the shape and magnitude from before the rain event (10 May).

This physical behavior is quite encouraging, demonstrating that high variability in the emissivity could be well captured, at least qualitatively, in cases of rain events. Note that only convergent points were shown in this case. In these maps, areas of non-convergence can be observed, corresponding to cases where precipitation is occurring far from the area of consideration (circled).

One cannot stress enough the high degree of variability of microwave emissivity over land, not only where rain is involved, but also because of other temporally dynamic phenomena, such as morning dew or snow fall/melt and run off. This is in addition to the temporally-static/semi-static spatial variability of the surface emittance (such as from rocks, mountains, forests, and rivers). For these reasons, it is suggested that the use of atlases, databases or composite averaging of emissivity, should be used with a high degree of caution. Using emissivity from precomputed atlases when performing instantaneous inversions could, in many cases, harm the other parameters. Indeed, this atlas-based emissivity is almost always different from the actual emissivity because of the reasons listed above and because of the dynamic nature of the surface type mixture covered by the sensor footprint. This is true no matter how high of a resolution (spatially or temporally) the atlas is. We argue in this study that emissivity is a highly dynamic variable (and footprint dependent) and should therefore be handled in a dynamic fashion, such as part of the 1DVAR system, especially during and after rainy conditions or other emissivity-altering events, as illustrated above

4. Emissivity Quantitative Assessment

4.1. Methods

Along with the qualitative assessment that we described in the previous section, we performed a more rigorous, quantitative assessment of the MiRS emissivity performances. This assessment was done on a daily basis. Over ocean, the MiRS emissivity is simply compared to fields of emissivities computed using the wind speed from the Numerical Weather Prediction (NWP) analyses, input to an emissivity model (FASTEM) [14]. In this review, however, the quantitative assessment will focus on non-ocean surfaces.

Over non-ocean surfaces (snow, ice and land), the assessment relies on NWP global analyses as input to the CRTM for the computation of a reference emissivity which is computed analytically, as shown by Equation (1). This approach to computing the analytic emissivity was introduced and fully validated by [7,8]. Assuming that NWP analyses can provide accurate temperature and moisture profiles as well as a relatively accurate estimate of skin temperature, the CRTM may be used to compute the total transmittance of the up- and downwelling brightness temperatures. This gives a relatively accurate estimate of the emissivity which can be compared to the MiRS-retrieved emissivity. As mentioned in Section 1, restrictions exist on the observations that can be compared. There are four criteria which need to be satisfied for this analytical emissivity to be valid: (1) $\Gamma \neq 0$ (the equation is not usable for opaque channels); (2) $T_s \neq T_d$ (the solution could be unstable if this condition is not satisfied); (3) the surface is assumed to be specular (for the validity of the simplified RT equation); (4) the atmosphere is assumed to have clear sky and be free of hydrometeors.

Another quantitative assessment methodology relies on inter-comparisons with other independently developed emissivities. This has been the subject of efforts led by others and has been reported in a number of publications [15–17] and therefore, will not be addressed here. Suffice to say that these comparisons highlight that MiRS emissivities are consistent with independently-generated emissivities.

4.2. Examples

An example of the routine monitoring of MiRS retrievals is given in Figure 7, where the emissivity product is shown (the example shown is for emissivity over land). This figure shows two maps corresponding to MiRS retrieval at 50.3 GHz channel (top panel) and NOAA Global Data Assimilation System (GDAS)-based analytic emissivity (bottom). This shows that globally, MiRS-based emissivity retrievals are very similar to analytic emissivity, even if the two methods are different and computed entirely independently. It is worth mentioning that the analytic emissivity computed here obviously cannot be applied in real-time since it depends on NWP analyses. It can only be used as a source of verification and validation.

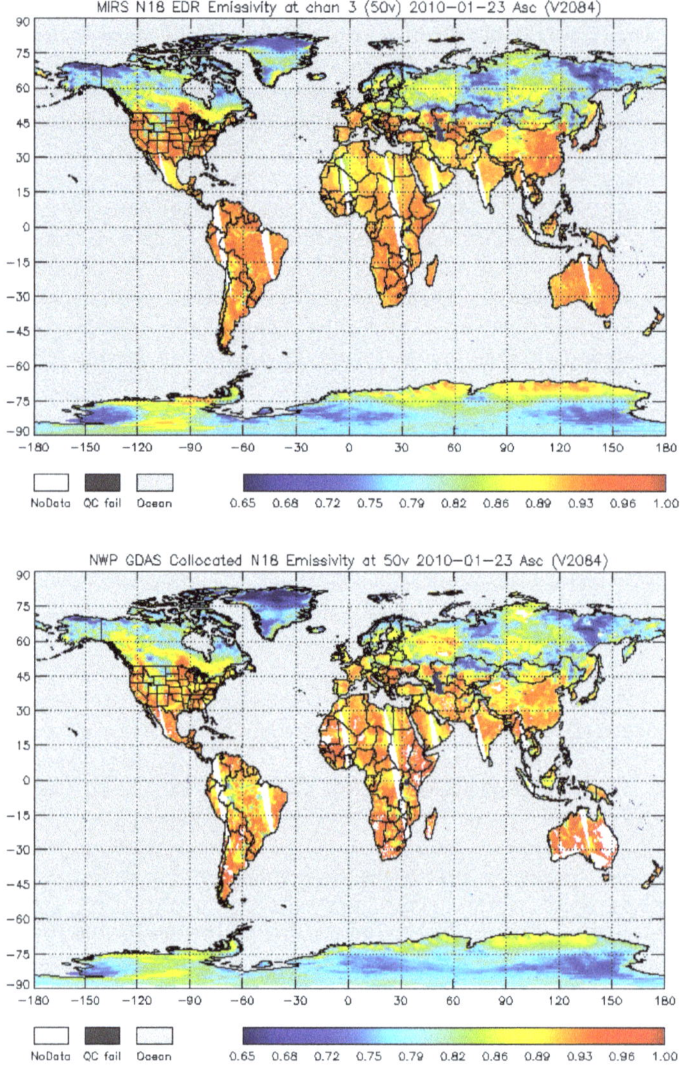

Figure 7. Global fields of non-ocean emissivity for the 50.3 GHz channel, as retrieved by the MiRS algorithm (**top**) and as provided by the analytical emissivity based on the Global Data Assimilation System (GDAS) analysis (**bottom**). The field corresponds to January 23, 2010. See text for more detail and for a discussion on the similarities and differences between the two.

We can see in Figure 7 that the two estimates of emissivities differ in some areas, which could be attributed to (1) the poor accuracy of the skin temperature estimate from GDAS, such as over the high-latitude regions, where snow and ice cover the surface and it is known that skin temperature from NWP is not very accurate, and (2) the non-validity of the assumptions made to compute the analytical emissivity; over rough mountainous regions or over the Amazon forest, the surface is significantly *non*-specular for instance, leading to large uncertainties in the analytic emissivity estimation over those regions.

These differences are further highlighted in Figure 8, which shows the map of the difference between MiRS NOAA-18 retrieved emissivity at 23 GHz and the analytically computed emissivity from GDAS over land surfaces. The scatter plots in same figure provide a snapshot of the emissivity performances using the analytically computed emissivity as a reference, over both land and snow-covered land, separately. They reveal that there is good correlation between the two.

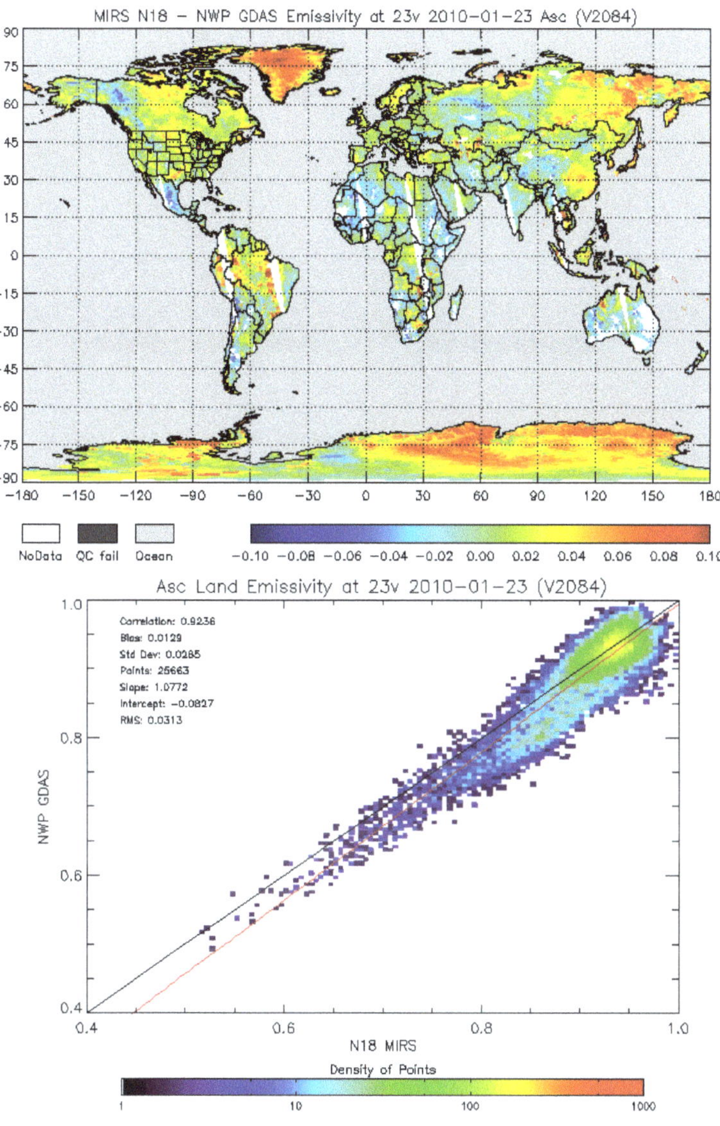

Figure 8. *Cont.*

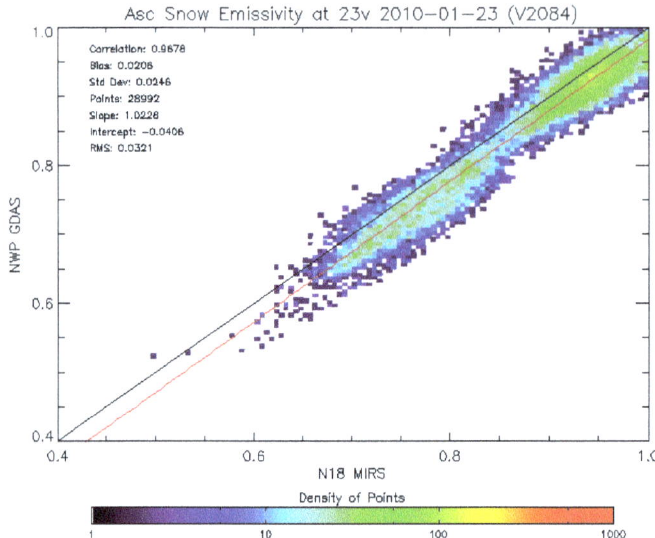

Figure 8. Difference between MiRS-retrieved 23 GHz emissivity over land surfaces from NOAA-18 and analytically computed emissivity from the collocated GDAS analysis on 23 January 2010 (**top**), along with scatter plots and statistics for points over snow-free land (**middle**) and snow-covered land (**bottom**).These results indicate that the MiRS emissivities over land are reasonably accurate with a standard deviation between 2% and 3%, including errors due to collocation, time matching and due to the reference emissivity used.

Tables 1 and 2 compare the MiRS-retrieved emissivities from NOAA-18/AMSU (cross-track scanning geometry) and DMSP-F16 SSMI/S (conical scanning geometry), respectively, using the analytic emissivities as a reference. Recall that emissivities from all channels are retrieved by MiRS, but emissivities for some high-frequency channels are not analytically inverted given the opacity of the atmosphere for those channels. Therefore, only channels in the range from 19 to 50 GHz are compared.

Table 1. Performance of MiRS-based emissivities for three channels from NOAA-18 AMSU, when considering European Centre for Medium-Range Weather Forecasts (ECMWF)-based and GDAS-based analytically computed emissivities as references. Performance is shown in terms of bias (standard deviation) for land, snow and sea-ice surfaces.

Emissivity Reference		Channel 23.8 GHz	Channel 31.4 GHz	Channel 50.3 GHz
Reference: ECMWF-based analytical emissivity	Land	0.015 (0.023)	0.015 (0.023)	0.015 (0.033)
	Snow	0.015 (0.025)	0.003 (0.025)	−0.006 (0.032)
	Sea-Ice	0.0001 (0.026)	−0.010 (0.026)	−0.025 (0.036)
Reference: GDAS-based analytical emissivity	Land	0.013 (0.029)	0.012 (0.026)	0.011 (0.035)
	Snow	0.021 (0.025)	0.008 (0.025)	0.0006 (0.029)
	Sea-Ice	0.0004 (0.026)	−0.012 (0.025)	−0.026 (0.034)

Table 2. Performances of MiRS-based emissivities for three channels (vertical polarization) from DMSP-F16 SSMI/S, when considering ECMWF-based and GDAS-based analytically computed emissivities as references. Performances are shown in terms of bias and (standard deviation) for land, snow and water surfaces.

Emissivity Reference		Channel 19 GHz V	Channel 22 GHz V	Channel 37 GHz V
Reference: ECMWF-based analytical emissivity	Land	0.008 (0.029)	0.009 (0.028)	0.002 (0.027)
	Snow	0.003 (0.039)	−0.002 (0.022)	−0.005 (0.023)
	Sea-Ice	0.001 (0.041)	0.002 (0.019)	−0.001 (0.022)
Reference: GDAS-based analytical emissivity	Land	0.004 (0.026)	0.011 (0.031)	0.002 (0.029)
	Snow	0.002 (0.023)	0.003 (0.025)	−0.002 (0.025)
	Sea-Ice	0.002 (0.017)	0.004 (0.018)	−0.002 (0.018)

The analytical emissivities were generated using both GDAS and the European Centre for Medium-Range Weather Forecasts (ECMWF), with the statistics for biases and standard deviations computed and shown separately for each. For emissivities at AMSU frequencies, the biases and standard deviations had very similar magnitudes compared to both ECMWF and GDAS analytical emissivities, ranging from ~0.015 to ~0.03. This gives us confidence that the statistics are not sensitive to the source of reference used. For emissivity at the SSMI/S frequencies, the same conclusion can be made, with one exception: the 19 GHz channel. At 19 GHz, the standard deviation in emissivity was 0.039 over snow and 0.041 over sea-ice when ECMWF was used as a reference; these values were slightly higher than the values of 0.023 for snow and 0.017 for sea-ice obtained when GDAS was used. This points to some potential discrepancy between ECMWF and GDAS analysis when computing the analytical emissivity over those regions.

An additional important assessment was made to verify the expected angle dependence of the emissivity retrieved by MiRS over surfaces known to have specular features that cause the dependence of emissivity on the viewing angle (from cross-tracking sensors). Figure 9 shows the angle dependence of the emissivity retrieved with MiRS by comparing its scan dependence with those computed from both GDAS and ECMWF analytical emissivities with the AMSU 23.8 GHz channel. The top plot presents the mean emissivities over snow-free land as a function of the scan angle. The MiRS retrieved emissivity appears to compare more favorably to the GDAS emissivity, while the ECMWF-based emissivity demonstrates a negative bias of about 0.01. The bottom figure shows the mean emissivities over snow-covered land as a function of the scan angle. Here, the MiRS-retrieved emissivity contains about a 0.015–0.020 bias relative to both GDAS and ECMWF, which have almost identical means. It is important to note that in MiRS, no bias corrections were applied over land (all non-ocean surfaces) to the brightness temperatures used in this retrieval, which could explain some of the biases found in the emissivity (of either MiRS or analytic emissivities). Angle dependence is also likely a mixture of natural emissivity variation over partially specular surfaces, but also sensor calibration that is not uniform across all angles. This explains, at least partially, the lower emissivities at the scanline edges. The main conclusion that can be drawn from this assessment is that MiRS captures, within reason, the angle dependence, through the analytical emissivity. Keep in mind that these emissivities are not instantaneous—they are averaged over large areas and therefore combine different surface sub-types. For example, over land, forest, desert, bare land, rivers, and urban surface types are included.

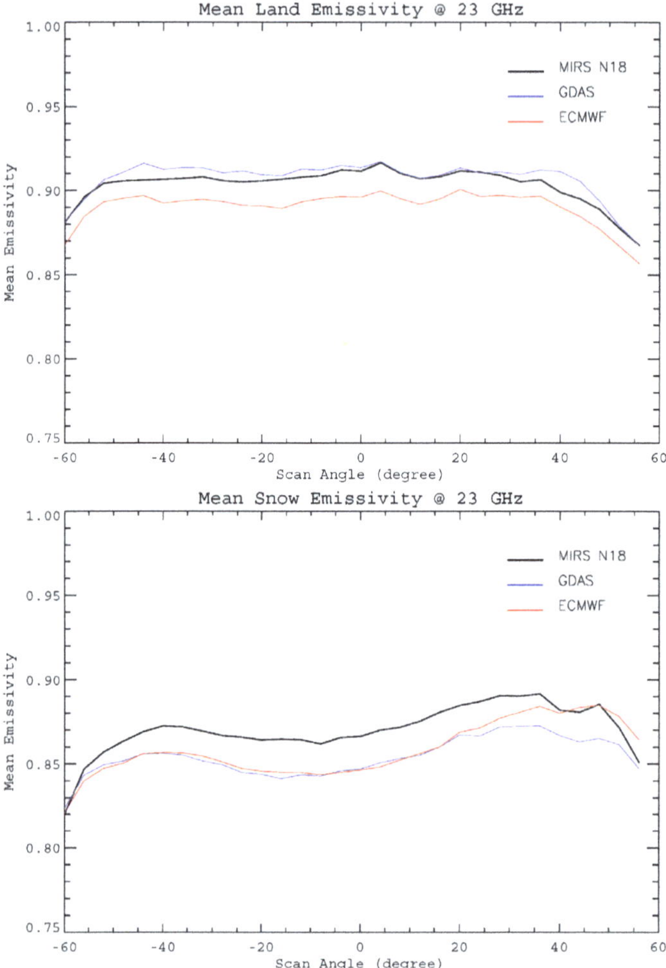

Figure 9. Mean emissivity at 23.8 GHz as a function of scan angle averaged over snow-free land (**top**) and snow-covered land (**bottom**). These correspond to the MiRS NOAA-18 retrieved emissivity, the analytically computed GDAS emissivity, and the analytically computed ECMWF emissivity for 23 January 2010.

5. Examples of Emissivity-Based Products

In this section, we highlight some products that benefit from the emissivity retrieved from MiRS, either directly or indirectly. Some products are indeed generated directly from the emissivity signal (SIC, for instance) based on the post-processing step, as explained above. Other products are generated as part of the 1DVAR system (first step in MiRS), but thanks to the emissivity being part of the same retrieval, these parameters have been expanded to a multitude of other surfaces. Examples of such cases include TPW, which has been extended to land, sea-ice, and snow-covered surfaces. The accuracy of these parameters (those benefiting directly or indirectly from the emissivity dynamic retrieval) presents a tool for the indirect assessment of the emissivity product itself. This is due to direct linkage and simultaneous retrieval; in the context of dynamic simultaneous retrieval such as that employed by MiRS, if an error is found in one parameter, it likely means that a corresponding (or compensating) error exists in the emissivity. In other words, the obtainment of accurate atmospheric parameters retrieved as part of the same retrieval that generates emissivity likely means that the emissivity is also accurate. Similarly, if the SIC generated directly from the emissivity is accurate, it likely suggests that the emissivity is reasonably accurate as well. Note that this section is not aimed at providing a thorough validation of the emissivity-derived parameters, but simply to highlight some of them and point the reader, when appropriate, to publications that contain more thorough validation of specific parameters.

Figure 10 highlights some of the MiRS-based operational cryospheric products, including the SIC and the sea-ice age (FYI or MYI). These are products derived directly from emissivity. These products have been thoroughly validated [11] by intercomparison with established algorithms as well as with ground-truth data from the National Ice Center. They are also routinely generated and monitored for a multitude of sensors (https://www.star.nesdis.noaa.gov/mirs). The fact that SIC, derived from the emissivity, has been found to be valid is an indirect validation of the emissivity itself.

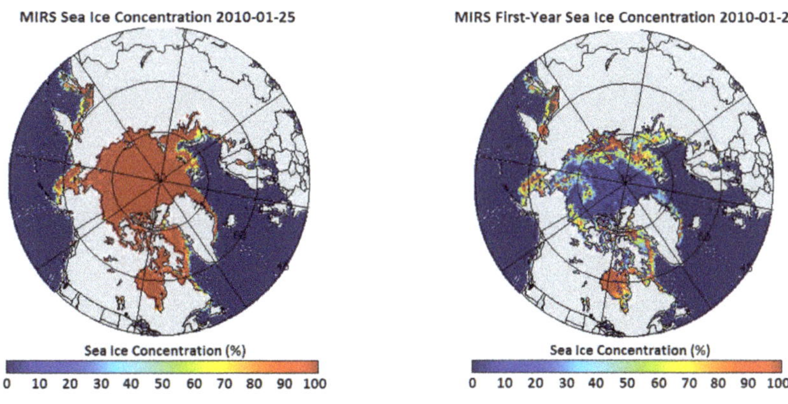

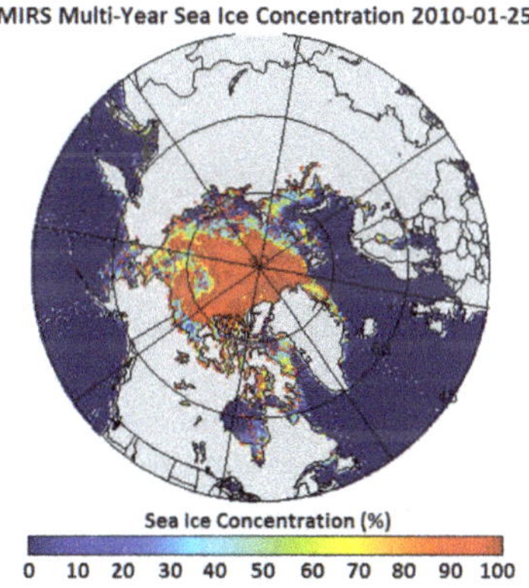

Figure 10. Polar-stereographic image of MiRS NOAA-18 derived Sea-Ice Concentration (**top-left**), First-Year Ice (**top-right**) and Multi-Year Ice (**bottom**) over the Arctic on 25 January 2010, based on MiRS dynamic emissivities. These products were validated extensively in [11].

Another example is shown in Figure 11, which highlights TPW as one of the parameters that benefit indirectly from the simultaneous retrieval of emissivity. Indeed, because the emissivity is dynamically retrieved in the same 1DVAR algorithm as the atmospheric parameters, it is possible to extend the retrieval of TPW over non-ocean surfaces with microwave sensors. The methodology and extensive validation of this process was done in [18]. Figure 11 compares MiRS-based TPW over land to that from the GDAS, and their differences are stratified by zenith angle for snow-free and snow-covered surface types.

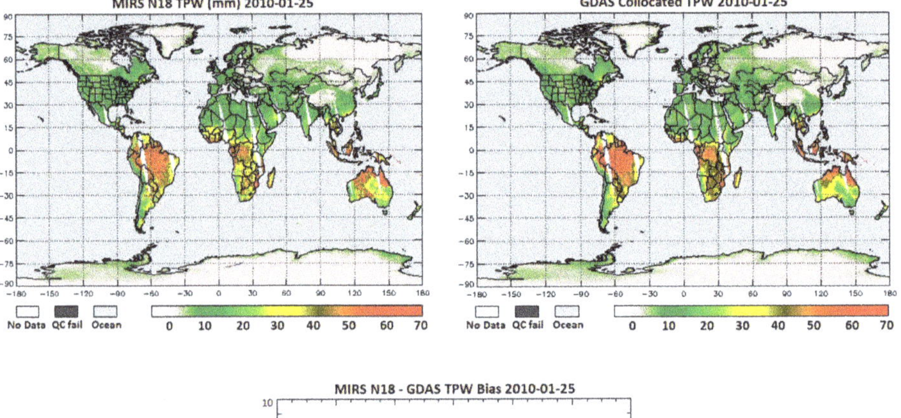

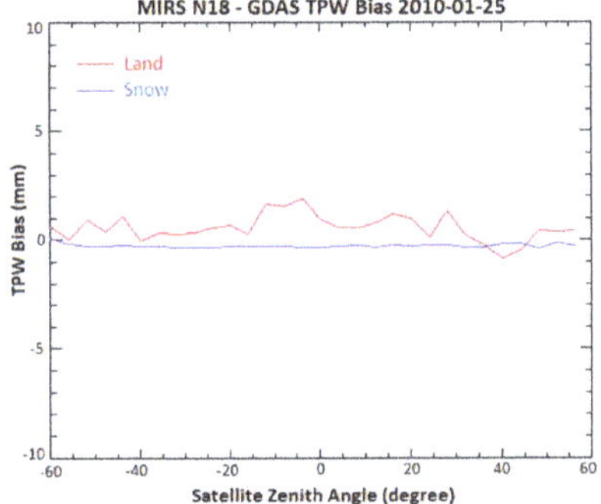

Figure 11. MiRS derived TPW (**top left**) and GDAS TPW (**top right**) for January 23, 2010. The TPW difference as a function of scan angle is shown for land and snow-free land surfaces (**bottom**).

These assessments of MiRS products, for both the cryospheric and total moisture products, benefiting directly or indirectly from the dynamic emissivity, point to the fact that these products have good performances when compared to independent estimates, which indicates that the emissivity that is underlying or allowing their retrieval is also reasonably accurate.

6. Summary and Conclusions

In this study we presented the methodology and validated the performances of a dynamically retrieved emissivity based on a 1DVAR approach called MiRS. We also highlighted the benefit of this dynamic emissivity retrieval for other applications, such as the extension of atmospheric and hydrometeor parameter inversion over difficult (highly variable) backgrounds. This approach can be applied to a multitude of microwave sensors including imagers and sounders, of both cross-track and conical geometries. Previous intercomparison studies have already highlighted that this approach leads to similar results as other independent algorithms, including physical models that simulate emissivities using surface parameters as inputs. The present study attempted to thoroughly validate the performances of MiRS emissivity, using both qualitative and quantitative approaches. The main conclusion from the study is that the emissivity behaves fairly consistently with expectations—spatially, spectrally and temporally—based on current knowledge of emissivity dependencies. Comparisons with analytical emissivities allowed us to quantify the actual uncertainty and precision of retrieved emissivities over a number of surface types and for a number of frequencies. These performances were deemed fairly reasonable (less than 3%). These uncertainties/precisions almost certainly also include errors in the reference data itself, representativeness errors, and collocation errors, so one could argue that the actual emissivity errors (from MiRS and the measurements alone) are more likely within a 1% error margin for the window channels and surface types measured. While the analytical emissivity was found to be a good reference for this assessment, it is worth noting its sensitivity to the accuracy of inputs from numerical models (for skin temperature estimate for instance) and the non-universality of its assumptions (such as the assumed *specularity* of the surface) which limits both its accuracy and scope of validity. We emphasize that this dynamically-inverted emissivity from MiRS is inverted as part of the impact of all parameters on measurements, including the surface temperature, atmospheric, and hydrometeor parameters. These other parameters have also been extensively validated (not shown here and not within the scope of this study), which gives even more credence to the validity of the emissivity measures. If a systematic bias or high random error did exist in the emissivity because of its simultaneous inversion with other parameters, it is likely that it would also cause a systematic or higher random error in one or multiple other parameters. This is because for the inversion to be deemed successful, the inversion is constrained by a strict convergence criterion to fit the observations. No such systematic or high random errors were found when validating these atmospheric products.

Perhaps more importantly than the inversion of the emissivity itself, the inclusion of the emissivity in a dynamic inversion system made it possible to extend the retrieval of atmospheric and hydrometeor parameters from microwave sensors to non-traditional surfaces (such as retrieving TPW over land and ice surfaces). It also allowed us to extend the retrieval to mixed-pixel situations (such as coastlines and other surface type boundaries) in a very seamless fashion. This is an important application for those concerned with capturing the evolution of atmospheric rivers, for instance, which usually bring heavy precipitation to coastal and inland regions.

Another natural extension of this effort is the satellite data assimilation of surface sensitive channels, both in microwave and infrared sensors. Indeed, the same variational dynamic approach could be exploited in variational data assimilation systems by extending the data's assimilation state vector to include the emissivity in the analysis process; or by using the MiRS emissivity as a parameter constraint when assimilating surface sensitive channels. This could potentially lead to an increased assimilation acceptance rate, more accurate lower tropospheric sounding, as well as more cloud- and rain-impacted data assimilation. Regarding this last point, we found that proper rain inversion (or assimilating rain-impacted data) from microwave passive data also accounts for the highly variable emissivity in the same process that extracts the rain information (or atmospheric information in general). The same channels that are sensitive to the rain are also sensitive to the surface, and in these situations (either during or after the rain), the surface emissivity's changes are dramatic due to rain impact and are highly dependent on footprint coverage. Accounting for dynamic emissivity as part of the rain retrieval (or rain-impacted data assimilation) is crucial.

Author Contributions: Sid-Ahmed Boukabara conceived and designed the experiments; Kevin Garrett and Chris Grassotti performed the experiments; Sid-Ahmed Boukabara, Kevin Garrett, and Chris Grassotti analyzed the data and results; Sid-Ahmed Boukabara and Kevin Garrett wrote the paper.

Acknowledgments: The authors gratefully acknowledge many present and past contributors to the MiRS project especially Flavio Iturbide-Sanchez, Cezar Kongoli, Quanhua Liu and Banghua Yan for their particular contribution to the emissivity part, as well as past and current sponsors and in particular the NOAA transition-to-operations PSDI program and the JPSS cal/val program.

Conflicts of Interest: The authors declare no conflict of interest.

References

1.   Comiso, J.; Zwally, H.J. Temperature corrected bootstrap algorithm. *Proc. IEEE IGARS Dig.* **1997**, *3*, 857–861.

2. Svendsen, E.; Mätzler, C.; Grenfell, T.C. A model for retrieving total sea ice concentration from a spaceborne dual-polarized passive microwave instrument operating near 90 GHz. *Int. J. Remote. Sens.* **1987**, *8*, 1479–1487. [CrossRef]

3. Derksen, C. The contribution of AMSR-E 18.7 and 10.7 GHz measurements to improved boreal forest snow water equivalent retrievals. *Remote Sens. Environ.* **2008**, *112*, 2701–2710. [CrossRef]

4. Tedesco, M.; Narvekar, P.S. Assessment of the NASA AMSR-E SWE Product. *IEEE J.-STARS.* **2010**, *3*, 141–159. [CrossRef]

5. Owe, M.; de Jeu, R.; Holmes, T. Multisensor historical climatology of satellite-derived global land surface moisture. *J. Geophys. Res.* **2008**, *113*, F01002. [CrossRef]

6. Grody, N.C. Satellite-based microwave retrievals of temperature and thermal winds: Effects of channel selection and a priori mean on retrieval accuracy. In *Remote Sensing of Atmospheres and Oceans*; Deepak, A., Ed.; Academic Press: Cambridge, CA, USA, 1980; pp. 381–410. ISBN 0-12-208460-8.

7. Prigent, C.; Rossow, W.B.; Matthews, E. Microwave land surface emissivities estimated from SSM/I observations. *J. Geophys. Res.* **1997**, *102*, 21867–21890. [CrossRef]

8. Karbou, F.; Prigent, C.; Eymard, L.; Pardo, J. Microwave land emissivity calculations using AMSU-A and AMSU-B measurements. *IEEE Trans. Geosci. Remote Sens.* **2005**, *43*, 948–959. [CrossRef]

9. Yan, B.; Weng, F.; Meng, H. Retrieval of snow surface emissivity from the advanced microwave sounding unit. *J. Geophys. Res.* **2008**, *113*, D19206. [CrossRef]

10. Boukabara, S.-A.; Garrett, K.; Chen, W.; Grassotti, C.; Kongoli, C.; Chen, R.; Liu, Q.; Yan, B.; Weng, F.; Ferraro, R.; et al. MiRS: An All-Weather 1DVAR Satellite Data Assimilation and Retrieval System. *IEEE Trans. Geosci. Remote Sens.* **2011**, *49*, 1–24. [CrossRef]

11. Kongoli, C.; Boukabara, S.-A.; Yan, B.; Weng, F.; Ferraro, R. New Sea-Ice Concentration Algorithm Based on Microwave Surface Emissivities. *IEEE Trans. Geosci. Remote Sens.* **2011**, *49*, 175–189. [CrossRef]

12. Han, Y.; Van Delst, P.; Liu, Q.; Weng, F.; Yan, B.; Treadon, R.; Derber, J. *JCSDA Community Radiative Transfer Model (CRTM)—Version 1*; NOAA Technical Report NESDIS 122; NOAA: Silver Spring, MD, USA, 2006; 33p.

13. Iturbide-Sanchez, F.; Boukabara, S.-A.; Chen, R.; Garrett, K.; Grassotti, C.; Chen, W.; Weng, F. Assessment of a Variational Inversion System for Rainfall Rate over Land and Water Surfaces. *IEEE Trans. Geosci. Remote Sens.* **2011**, *49*, 1–24. [CrossRef]

14. Liu, Q.; Weng, F.; English, S.J. An Improved Fast Microwave Water Emissivity Model. *IEEE Trans. Geosci. Remote Sens.* **2011**, *49*, 1238–1250. [CrossRef]

15. Ferraro, R.R.; Peters-Lidard, C.D.; Hernandez, C.; Turk, F.J.; Aires, F.; Prigent, C.; Lin, X.; Boukabara, S.-A.; Furuzawa, F.A.; Gopalan, K.; et al. An evaluation of microwave land surface emissivities over the continental United States to benefit the GPM-era Precipitation Algorithms. *IEEE Trans. Geosci. Remote Sens.* **2013**, *51*, 378–398. [CrossRef]

16. Norouzi, H.; Temimi, M.; Prigent, C.; Turk, J.; Khanbilvardi, R.; Tian, Y.; Furuzawa, F.A.; Masunaga, H. Assessment of the consistency among global microwave land surface emissivity products. *Atmos. Meas. Tech.* **2015**, *8*, 1197–1205. [CrossRef]

17. Tian, Y.; Peters-Lidard, C.D.; Harrison, K.W.; Prigent, C.; Norouzi, H.; Aires, F.; Boukabara, S.-A.; Furuzawa, F.A.; Masunaga, H. Quantifying Uncertainties in Land-Surface Microwave Emissivity Retrievals. *IEEE Trans. Geosci. Remote Sens.* **2014**, *52*. [CrossRef]

18. Boukabara, S.-A.; Garrett, K.; Chen, W.; Liu, Q.; Yan, B.; Weng, F. Global Coverage of Total Precipitable Water Using a Microwave Variational Algorithm. *IEEE Trans. Geosci. Remote Sens.* **2010**, *48*, 3608–3621. [CrossRef]

Article

The Combined ASTER and MODIS Emissivity over Land (CAMEL) Global Broadband Infrared Emissivity Product

Michelle Feltz [1],*, Eva Borbas [1], Robert Knuteson [1], Glynn Hulley [2] and Simon Hook [2]

[1] Space Science and Engineering Center, University of Wisconsin—Madison, Madison, WI 53706, USA; eva.borbas@ssec.wisc.edu (E.B.); robert.knuteson@ssec.wisc.edu (R.K.)

[2] Jet Propulsion Laboratory, California Institute of Technology, Pasadena, CA 91109, USA; Glynn.Hulley@jpl.nasa.gov (G.H.); Simon.J.Hook@jpl.nasa.gov (S.H.)

* Correspondence: michelle.feltz@ssec.wisc.edu

Received: 16 May 2018; Accepted: 22 June 2018; Published: 28 June 2018

Abstract: Infrared surface emissivity is needed for the calculation of net longwave radiation, a critical parameter in weather and climate models and Earth's radiation budget. Due to a prior lack of spatially and temporally variant global broadband emissivity (BBE) measurements of the surface, it is common practice in land surface and climate models to set BBE to a single constant over the globe. This can lead to systematic biases in the estimated net and longwave radiation for any particular location and time of year. Under the National Aeronautics and Space Administration's (NASA) Making Earth System Data Records for Use in Research Environments (MEaSUREs) project, a new global, high spectral resolution land surface emissivity dataset has recently been made available at monthly at 0.05 degree resolution since 2000. Called the Combined ASTER MODIS Emissivity over Land (CAMEL), this dataset is created by the merging of the MODIS baseline-fit emissivity database developed at the University of Wisconsin-Madison and the ASTER Global Emissivity Dataset (GED) produced at the Jet Propulsion Laboratory. CAMEL has 13 hinge points between 3.6–14.3 μm which are expanded to cover 417 infrared spectral channels within the same wavelength region using a principal component regression approach. This work presents the method for calculating BBE using the new CAMEL dataset. BBE is computed via numerical integration over the CAMEL High Spectral Resolution product for two different wavelength ranges—3.6–14.3 μm which takes advantage of the full, available CAMEL spectra and 8.0–13.5 μm which has been determined to be an optimal range for computing the most representative all wavelength, longwave net radiation. CAMEL BBE uncertainty estimates are computed, and comparisons are made to BBE computed from lab validation data for selected case sites. Variations of BBE over time and land cover classification schemes are investigated and converted into flux to demonstrate the equivalent error in longwave radiation which would be made by the use of a single, constant BBE value. Misrepresentations in BBE by 0.05 at 310 K corresponds to potential errors in longwave radiation of over 25 W/m^2.

Keywords: broadband emissivity; infrared; surface; land; radiation

---

1. Introduction

Infrared surface emission is needed for the calculation of both the Earth's radiation budget and the Earth's surface radiation budget [1]—two variables which are defined as Essential Climate Variables by the Global Climate Observing System (GCOS) [2]. Land surface emissivity is a measure of how closely Earth's land surface acts as a blackbody in emitting radiation to the atmosphere. Infrared land surface emissivity has long been known to contain spectral variations that vary across the globe [3]. These spectral variations are dependent upon vegetation cover, wetness, snow cover, and mineral composition of the soil including its particle size and compactness [4,5]. Knowledge of the temporal, spatial, and spectral variations in land surface emissivity is important for data assimilation of the hyperspectral infrared sounders [6,7]. Uncertainties in land surface model parameterizations in Numerical Weather Prediction (NWP) models can influence the state parameters in the atmospheric boundary layer and the forecast of severe weather [8].

NWP land surface models, e.g., the Noah Land Surface Model in the Weather Research and Forecasting model (WRF), characterize land surface emissivity in terms of constant broadband values which are multiplied by the Stefan-Boltzman constant and the surface temperature to the fourth power to estimate the surface leaving infrared flux [9]. Due to a previous lack of spatially and temporally varying global broadband emissivity (BBE) measurements, the use of simple parameterizations or a single constant, global BBE value was adopted as common practice [10,11]. Such approximations can lead to systematic biases in the estimated net radiation for any particular location and time of year. A study done by Zhou et al. [11] showed that on average over North Africa a decrease in soil emissivity of 0.1 could increase surface air temperature by ~1 K and decrease upward longwave (LW) radiation by 8.1 Wm^{-2}.

Several efforts in the past have outlined methods to produce optimal BBE datasets using satellite infrared measurements [12–19]. Such methods have typically involved regression of BBE to measured narrowband emissivities and/or albedos from satellite instruments such as the Moderate Resolution Imaging Spectroradiometer (MODIS), Advanced Spaceborne Thermal Emission and Reflection Radiometer (ASTER), or the Advanced Very High Resolution Radiometer (AVHRR). Another method that has been used is the classification-based scheme, which defines BBE for various land surface classifications using laboratory based measurements. An example of this is the parameterizations of BBE that have been developed for input into models such as in Wilber et al. [20], where BBE values are provided for different infrared bands and International Geosphere-Biosphere Programme (IGBP) land cover classifications. Work has also been done to determine an optimal BBE wavelength range for computing the most representative all-wavelength net radiation [18,21]. Cheng et al. [18] found that the all-wavelength surface LW net radiation was best approximated using the wavelength range of 2.5–200 microns. They then established that out of a set of wavelength ranges, the BBE spectral range of 8–13.5 microns was optimal for computing the LW net radiation between 2.5 and 200 microns.

Under the National Aeronautics and Space Administration (NASA) Making Earth System Data Records for Use in Research Environments (MEaSUREs) project, a new global, land surface emissivity termed the Combined ASTER and MODIS Emissivity over Land (CAMEL) has been made available and is a part of the Unified and Coherent Land Surface Temperature and Emissivity Earth System Data Record [22–25]. This new dataset combines previously existing satellite emissivity datasets—those from the MODIS baseline-fit emissivity database (BF) developed at the University of Wisconsin-Madison (UW) and the ASTER Global Emissivity Dataset version 4 (GED v4) produced at the California Institute of Technology Jet Propulsion Laboratory (JPL) [26–29]. CAMEL leverages the ability of ASTER GED v4 to accurately estimate infrared emissivity in the thermal infrared region (8–12 μm) and the ability of UW BF to provide information at select hinge points throughout the entire 3.6–12 μm infrared region, thereby combining the strengths of each dataset. The CAMEL product is available for the years 2000–2016 for monthly mean, 0.05°, or ~5 km, spatial resolution for 13 hinge points within the 3.6–14.3 μm region. An uncertainty product is included which contains estimates of the total uncertainty in addition to the spatial, temporal, and algorithm variability for each of the 13 hinge points on the CAMEL 0.05° grid [22,23]. A high spectral resolution (HSR) emissivity algorithm which extends the 13 hinge point product to 417 infrared spectral channels within the same 3.6–14.3 μm region is also available. This CAMEL HSR algorithm provides an update to the previous, similarly constructed UWIREMIS HSR emissivity product which is produced using the UW BF emissivity dataset [26,30].

This study presents broadband emissivity calculations that make use of the MEaSUREs CAMEL emissivity version V001 and the CAMEL HSR algorithm. The BBE dataset described here provides the advantage that (1) it is consistent with the MEaSUREs HSR emissivity and (2) it does not require regression schemes—BBE can be calculated by simple numerical integration over the CAMEL HSR emissivity product. Monthly, ~5 km resolution BBE is calculated over the globe for the years 2000–2016. BBE is computed for two different wavelength ranges—over the 3.6–14.3 μm region which uses the full available CAMEL wavelength region and the 8.0–13.5 μm region which has been determined to be an optimal range for computing the most representative all wavelength, longwave net radiation [18,21]. The physical reasoning for the choice of the 8.0–13.5 μm is described by Ogawa and Schmugge [21]—they showed that outside the 8.0–13.5 μm region, assuming the surface temperature and near-surface air temperature are close, the difference between the surface temperature and brightness temperature of the downwelling atmospheric radiation is small. Thus, they argue only minor contributions to the net longwave radiation come from outside the 8.0–13.5 μm region [21]. This choice has been supported by other studies as well, e.g., Cheng et al. (2013) [18].

The dependence of BBE on skin temperature is demonstrated for five different case sites for both wavelength ranges. Uncertainty estimates of the CAMEL BBE and comparisons to BBE computed from the UWIREMIS product and lab validation spectra are made to characterize the accuracy of the CAMEL BBE product. Global maps of monthly and climatological results are shown for four months representing the four seasons. Lastly, statistics are calculated for various land surface types using the IGBP land cover categories and, BBE differences are then converted into fluxes to demonstrate the equivalent error in LW radiation which would be made by the use of a single BBE value. Section 2 describes the data sources, Section 3 details the BBE calculation method, Section 4 contains results, and conclusions are made in Section 5.

2. Data

The CAMEL product version V001 is used in this study and is available from 2000 to 2016 at monthly, 0.05° (~5 km) resolution for 13 hinge points within the 3.6–14.3 μm region [22–25]. It was created using the ASTER GED v4 and MODIS UW BF emissivity datasets as input. A standalone CAMEL HSR Algorithm which extends the 13 hinge point CAMEL product to 417 infrared spectral channels using a principal component regression approach is available with the dataset, as well as uncertainty estimates which are reported on the 13 hinge points. Additional details of the CAMEL dataset are available in Borbas et al. and Feltz et al. [22,23]. CAMEL is available from the NASA Land Processes Distributed Active Archive Center (LP DAAC) online at the following site: https://lpdaac.usgs.gov/about/news_archive/release_nasa_measures_camel_5_km_products.

NASA's Earth Observing System MODIS instrument is used to provide monthly, global skin temperature and IGBP land cover classification on a 0.5° × 0.5° and 0.25° × 0.25° grid resolution, respectively. Specifically, the MOD11C3 skin temperature product, which is used in the calculation of BBE, is made available online by the NASA and United States Geological Survey (USGS) LP DAAC at the following site: https://lpdaac.usgs.gov/dataset_discovery/modis [31]. The land cover product MCD12Q1 is made available by the Global Land Cover Facility online at http://glcf.umd.edu/data/lc/ and more information can be found in Friedl et al. and Channan et al. [32,33]. There are 17 land cover types as defined by the IGBP classifications, and these are used to characterize BBE behavior over different surface types.

3. Methods

The CAMEL broadband emissivity is calculated by using the HSR emissivity algorithm which extends the 13 hinge point CAMEL product to 417 contiguous spectral channels between 3.6 and 14.3 μm. The following equation is then used [18]:

$$\varepsilon_{BB} = \frac{\int_{v_1}^{v_2} \varepsilon_v B_v(T_s) dv}{\int_{v_1}^{v_2} B_v(T_s) dv}, \tag{1}$$

where ε_v is the CAMEL monthly, HSR emissivity product and B_v is the Planck function at wavenumber v and temperature T_s, which is the land surface temperature defined by the average of the day and night monthly MOD11C3 MODIS products. If no MODIS temperature is available, a default value of 290 K is used. There is anticipated to be minimal error due to this approximation of skin temperature—this is based upon previous studies which have asserted that BBE is largely insensitive to variations of temperature that are within the typical range experienced on Earth [21], as well as analyses and discussions included below. A quality flag for the CAMEL BBE is defined that describes where the input temperature was taken from and is shown in Table 1. As mentioned, BBE values are calculated for two different wavelength ranges—the 3.6–14.3 µm region which covers the full extent of the CAMEL emissivity spectrum, and the 8.0–13.5 µm region. Lastly, CAMEL BBE uncertainty estimates are computed using a combination of the CAMEL HSR uncertainty product and Equation (1). Specifically, BBE is calculated for the CAMEL HSR product with both the uncertainty added and subtracted from it, and the difference of these is then divided by two to produce the uncertainty estimate.

Table 1. CAMEL broadband emissivity quality flag definition.

Value	Description
0	Good—BBE between 0.8–1.0
1	Good—no MODIS data, 290 K used as default skin temperature
2	BBE outside 0.8–1.0 range
3	BBE calculation failed
4	No BBE calculation—no CAMEL coefficients available
5	No BBE calculation—sea or inland water

4. Results

4.1. CAMEL Broadband Emissivity

Examples of the CAMEL 13 hinge point and HSR products with uncertainty for the month of January 2007 are shown overlaid in Figure 1 for five case study sites that represent different land surface types. The Namib Desert site is dominated by the quartz mineral (with traces of hematite), for which the spectral feature is clearly shown within the 8–9.5 µm region. The Yemen site represents a carbonate surface minerology type, and such areas are found primarily on the southern tip of the Arabian Peninsula. The Atmospheric Radiation Measurement (ARM) Southern Great Plains (SGP) site is located in Lamont, Oklahoma and represents a grass-covered, farmland surface type. The Greenland case study is representative of snow and ice-covered surfaces, and lastly, the Congo Forest site represents a moist, broadleaf forest of the subtropics/tropics. The overlaid emissivities in each panel of Figure 1 show how the CAMEL HSR product, through correlation with laboratory spectra, is able to add physical spectral variations in emissivity not apparent in the 13 hinge point product [22,23].

Figure 2 shows how the January 2007 CAMEL BBE for the five sites varies with temperature for both the 8.0–13.5 and 3.6–14.3 µm ranges. The site with the highest sensitivity to skin temperature is the Namib Desert site, for which the 8.0–13.5 µm BBE decreases by ~0.0014 for every 10 K increase in skin temperature. Overall the 3.6–14.3 µm BBE varies less with skin temperature than does the 8.0–13.5 µm BBE. The small, ~0.002 difference in BBE between the 220 and 320 K skin temperatures seen at the Greenland site implies that the errors produced by the use of a 290 K default skin temperature would lead to negligible errors for snow and ice-covered regions, which are the regions that would be most poorly represented by a 290 K skin temperature.

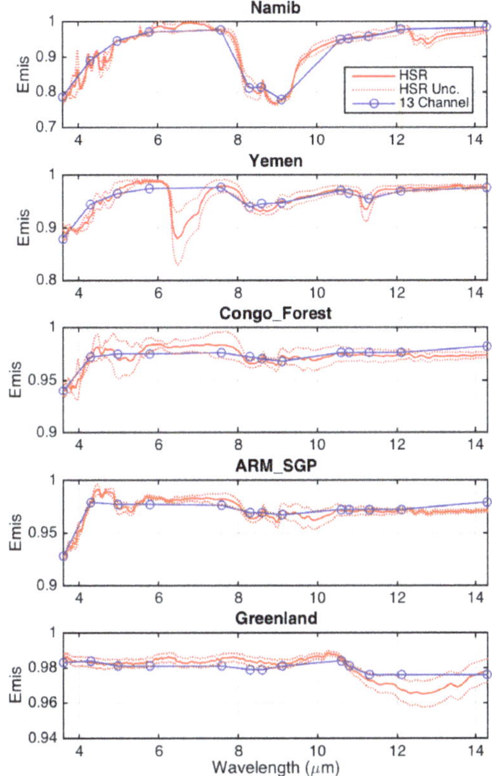

Figure 1. CAMEL emissivity for the 13 hinge points (blue open circles), 417 HSR channels (red line), and HSR uncertainty (red dotted lines) for January 2004 over the (from the top to bottom panel) Namib Desert (quartz), Yemen (carbonates), Congo (forest), Atmospheric Radiation Measurement Southern Great Plains (mixed agriculture), and Greenland (ice/snow) sites. Note changes in y-axis limits between panels.

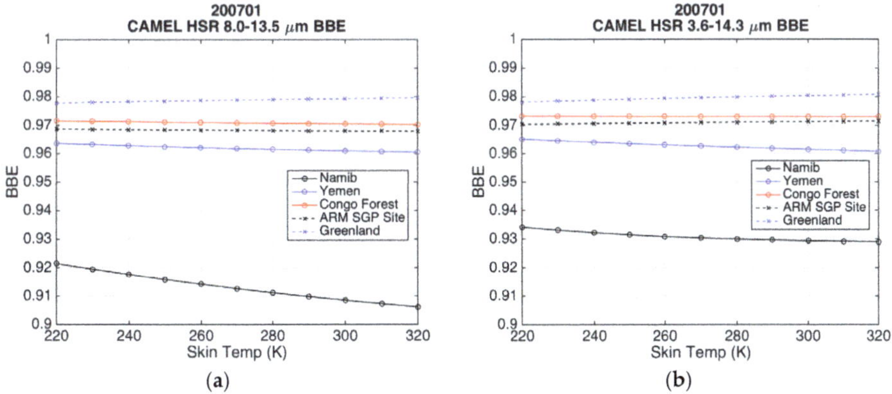

Figure 2. January 2007 CAMEL high spectral resolution (HSR) broadband emissivity (BBE) dependence on skin temperature for 5 different case study sites for (**a**) the 8.0–13.5 μm region and; (**b**) the 3.6–14.3 μm region.

Uncertainty of the CAMEL BBE, in addition to BBE computed from CAMEL, lab validation spectra, and UWIREMIS product are shown for the five case sites in Table 2. These BBE are computed for the 8.0–13.5 μm region with the exception for the Namib site, which is computed over 8.0–13.0 μm, since the lab data for that site does not extend beyond 13 μm. Uncertainty is largest for the Namib site at 0.0112 and smallest for the Greenland site at 0.0038. The CAMEL and lab BBE are in agreement within the CAMEL uncertainty for the Yemen, Congo, and ARM SGP site, and CAMEL BBE is less and more than the lab for the Greenland site and Namib sites respectively. For all sites except for Greenland, CAMEL BBE is in better agreement with the lab validation BBE than the UWIREMIS.

Table 2. CAMEL 8.0–13.5 μm BBE Uncertainty and Validation at Case Sites.

Site	CAMEL	CAMEL Uncertainty	Lab Data	UWIREMIS
Namib **	0.9049	0.0112	0.8934	0.9182
Yemen	0.9609	0.0087	0.9616	0.9580
Congo	0.9716	0.0059	0.9767	0.9673
ARM SGP	0.9684	0.0045	0.9720	0.9667
Greenland	0.9775	0.0038	0.9831	0.9798

** For Namib, lab data is not available for wavelengths larger than 13 μm, so BBE is computed over 8.0–13.0 μm for each the CAMEL, lab validation, and UWIREMIS.

Time series of the CAMEL 8.0–13.5 and 3.6–14.3 μm BBE for the five case sites shown in Figure 1 as well as for a seasonally snow-covered site in the Rocky Mountains (Mt. Massive) are shown in Figure 3. (Note *y*-axis scale change.) Overlaid on the CAMEL BBE time series is BBE computed from the UWIREMIS HSR emissivity product. Generally, the difference between the 8.0–13.5 and 3.6–14.3 μm BBE estimates is small, under 0.01 for both CAMEL and UWIREMIS. An exception exists for the Namib Desert site, where differences are seen to be just over 0.01 in BBE between the wavelength ranges. Differences between the CAMEL and UWIREMIS BBE vary over time with the newer CAMEL dataset showing more long-term stability. The Namib site shows an overall temporally constant difference of ~0.01 between the CAMEL and UWIREMIS BBE; however, the Yemen site shows a large increase in the CAMEL minus UWIREMIS BBE over the 2000–2016 time period. This decrease in the UWIREMIS product is attributed to the MODIS product input to the UWIREMIS HSR algorithm, which is known to degrade over this time period due to intrinsic algorithm issues. The CAMEL BBE for this site is more stable due to the inclusion of ASTER GED v4 emissivity data that is an input for the CAMEL product. This site example, as well as others like the Greenland site where sporadic variations are seen in UWIREMIS BBE between 2003 and 2007, imply that the CAMEL HSR emissivity is an improvement upon the UWIREMIS product. While the Greenland and Namib Desert sites show the CAMEL BBE to be quite constant over time, obvious seasonal dependencies are seen in both the ARM SGP and Rocky Mountain site CAMEL BBE. This is expected, as the Northern Hemisphere mid-latitudes experience large changes in snow and vegetation cover over the year. Changes in BBE over season are seen to be ~0.01 and ~0.02 for the ARM SGP and Rocky Mountain site respectively. Such regions would suffer from the greatest amount of error from the use of a temporally constant BBE value in modeling purposes.

Figure 4 shows the global CAMEL BBE results as maps for the month of January 2007 for both the 8–13.5 and 3.6–14.3 μm regions. The wider 3.6–14.3 μm range produces larger BBE magnitudes but is otherwise qualitatively similar to the 8.0–13.5 μm range. The Saharan Desert, barren regions of Australia, and Tibetan Plateau have markedly lower BBE values, while snow and ice-covered regions have the largest BBEs. The bottom panel shows a cutout of the 8.0–13.5 μm BBE over northeast Africa and the Arabian Peninsula and highlights the detail that the 5 km resolution CAMEL product is able to pick up, e.g., the Nile River and distinct changes in surface mineralogy within the Sahara Desert. For the remainder of the analyses in this paper, the 8–13.5 μm range is used to illustrate the BBE results.

As previously mentioned, broadband emissivity values used in some land surface or radiative models are set to a single, constant value over the globe and over time. To illustrate the magnitude and spatial distribution of error in BBE that is made by such an approximation, Figure 5 shows maps of 8.0–13.5 µm January 2007 CAMEL BBE minus various example constants, ranging from 0.96 to 0.98. Even for the constant 0.97, which appears to globally offer the closest match for the January BBE, differences of over 0.05 are seen in various localized areas as well as over the majority of the Saharan Desert and Arabian Peninsula where differences exceed 0.08.

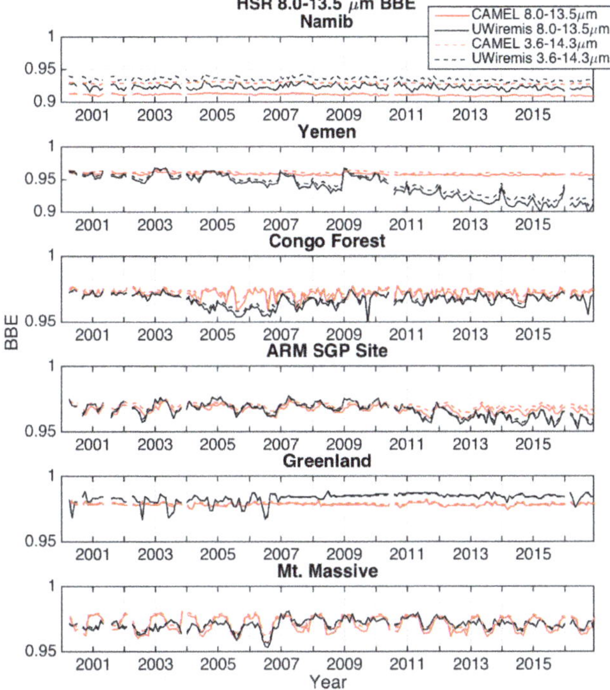

Figure 3. Time series of monthly average 0.5° × 0.5° CAMEL (red) and UWIREMIS (black) 8.0–13.5 µm (solid) and 3.6–14.3 µm (dashed) BBE for five sites (see Figure 1 for single month of corresponding emissivity spectra).

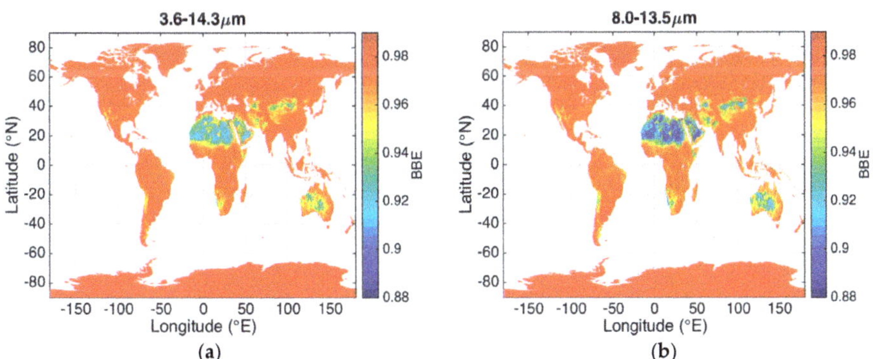

(a)   (b)

Figure 4. *Cont.*

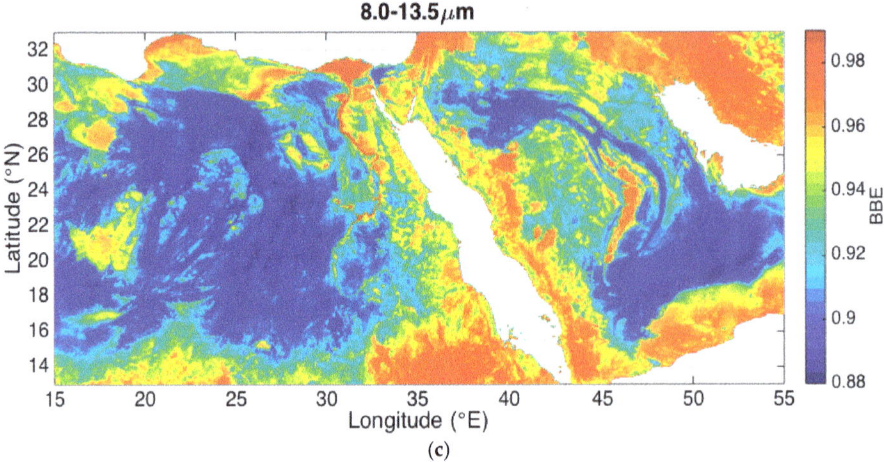

Figure 4. January 2007 CAMEL BBE for (**a**) 3.6–14.3 μm and; (**b**) 8.0–13.5 μm with (**c**) cutout over the eastern Sahara Desert and Arabian Peninsula for the 8.0–13.5 μm BBE.

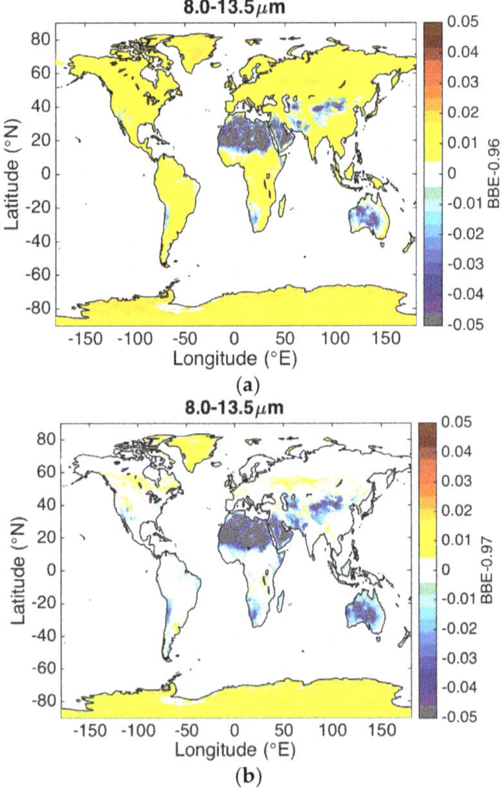

Figure 5. *Cont.*

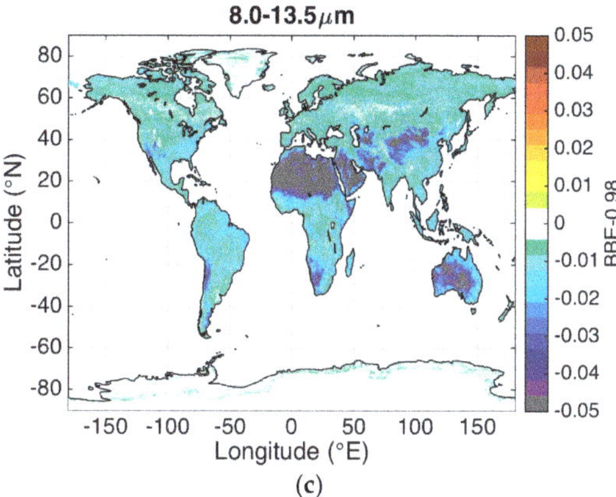

Figure 5. CAMEL 8.0–13.5 μm January 2007 BBE minus (**a**) 0.96; (**b**) 0.97 and; (**c**) 0.98.

Monthly climatologies of the CAMEL 8.0–13.5 μm BBE over the years 2000–2016 are calculated and shown for 4 months that represent different seasons in Figure 6. While BBE does not change dramatically from season to season, there are noticeable changes over the Northern Hemisphere polar and mid-latitudes due mostly to vegetation phenology and snow freeze/melt cycles. Figure 7 shows the standard deviation of the CAMEL BBE over the entire the 2000–2016 record and highlights the regions of highest variability over time. In contrast to the differences between the subpanels in Figure 6 which show the average conditions for single months, Figure 7 shows the variability over all of the individual months of the CAMEL record. Regions of increased standard deviations include the Sahel and southern tip of Africa (likely due to intensive biomass burning over the Sub-Saharan Africa region), Australia, and the Steppes of Central Asia, where standard deviations reach up to 0.025. Mountainous regions including the Andes on the southwestern South American coastline and the Rocky Mountains of the US show larger variabilities as well, reaching over 0.01 in some areas. These areas would be most poorly represented by a single BBE constant over time.

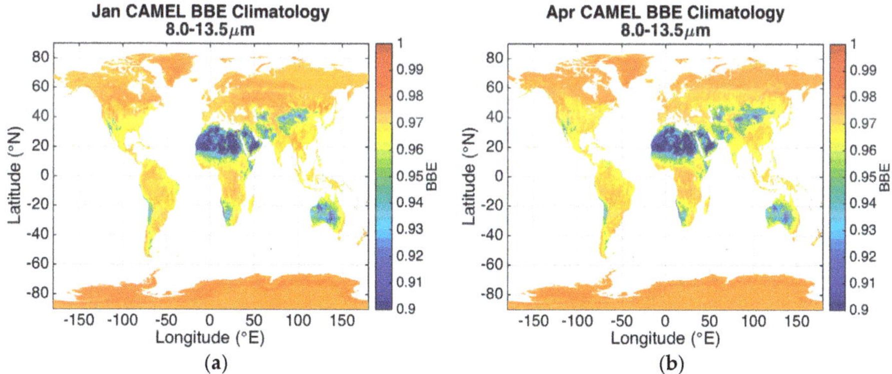

Figure 6. *Cont.*

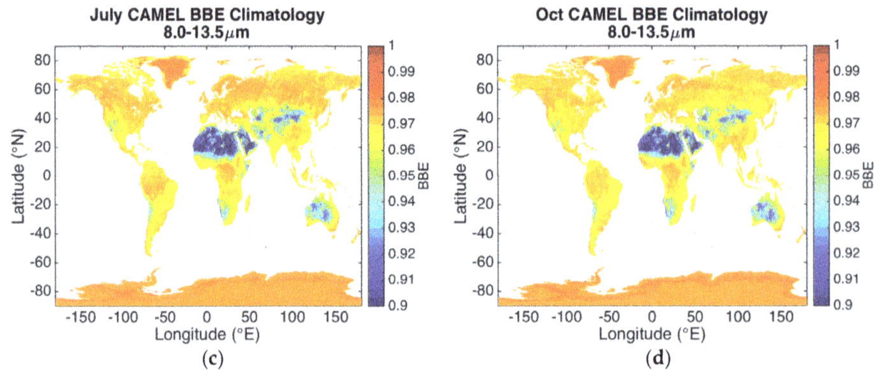

Figure 6. Monthly climatology of CAMEL 8.0–13.5 μm BBE over 2000–2016 for the month of (**a**) January; (**b**) April; (**c**) July and; (**d**) October.

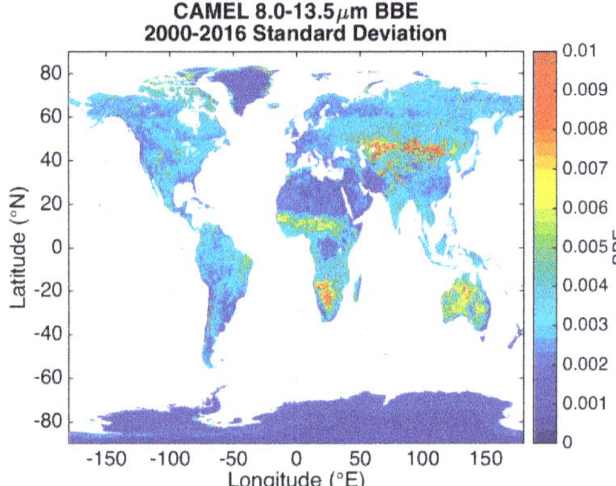

Figure 7. Standard deviation of monthly, 8.0–13.5 μm CAMEL BBE over the years 2000–2016 highlights geographic regions which have larger temporal variations in BBE.

4.2. BBE by Land Type and Climate Regime

To further investigate how BBE changes with surface cover type, statistics are calculated over IGBP land cover classifications listed in Table 3. To calculate the statistics, IGBP categories are mapped to the CAMEL 0.05° resolution latitude and longitude grid. Since the IGBP categories are defined on a 0.25° grid, they are interpolated to the higher CAMEL resolution. Yearly updates of IGBP land cover are available for 2002–2012, so the categories for the year 2012 were used for IGBP statistics over the years 2012–2016.

For each month and land cover category, the mean, μ, standard deviation, σ, and uncertainty of the mean, μ_{unc}, is calculated. Uncertainty of the mean is calculated as follows:

$$\mu_{\text{unc}} = \sigma / \sqrt{(N)}, \tag{2}$$

where N is the number of samples; however, the uncertainty of the mean value is found to be negligible so is not shown in figures.

Figure 8 shows time series of the monthly mean 8.0–13.5 μm BBE for several IGBP categories. Overlaid on the time series are monthly means bounded by monthly standard deviations. Categories with higher changes in mean BBE throughout the year include grassland, cropland, open shrubs, wetland,

and needle forests. This makes physical sense as the grassland, cropland, wetland, and needle forests land cover categories are either dominantly located in the Northern Hemisphere or have significant portions of their coverage in the Northern Hemisphere where land cover changes significantly with the seasons and associated snow coverage. Larger monthly standard deviations, which indicate the land cover types that are not as well represented by a single, constant BBE value, are seen for the barren/sparse vegetation, grassland, and open shrub categories. Closer inspection of various time series, particularly for the wetland, evergreen needle forest and woody savanna, reveals an apparent change in the magnitude of the seasonal oscillation of the time series around the year 2007. This change in the time series behavior is likely attributable to the 2007 version change in the MODIS product which is used as input into the CAMEL emissivity. The update in the MODIS product from Collection 4.0 to 4.1 was made for a change in the way clouds are dealt with in the processing.

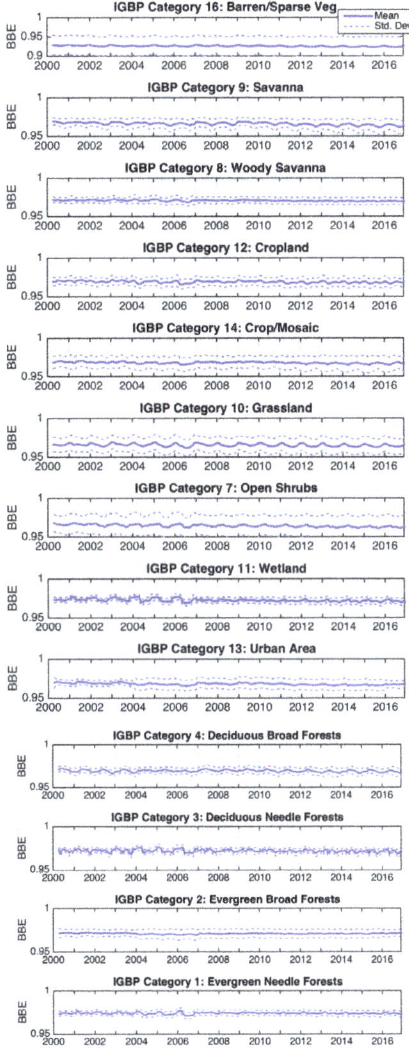

Figure 8. Time series of CAMEL 8.0–13.5 μm BBE statistics by IGBP land cover classifications. Note the *y*-axis limit is different for the top panel.

A summary of the CAMEL 8–13.5 μm BBE over IGBP categories is given in Table 3, which contains the average BBE over IGBP categories for each monthly climatology of the CAMEL product. The associated number of samples are listed in Table 4. For this analysis, the IGBP categories for the year 2009 were chosen for computing statistics for the 2000–2016 CAMEL climatology. While the climatological averaging over multiple months of the CAMEL record mixes BBE values from different land cover categories, as land cover and land use change from year to year, the IGBP statistics for the CAMEL climatology give an approximate snapshot of how the 8.0–13.5 μm BBE changes over IGBP category and by season. Variation in BBE values over the land cover categories is larger than the month to month variation for a specific category by an order of magnitude. The grassland category experiences the largest change in BBE over seasons, but the difference is only 0.005 in magnitude. This is roughly equivalent to the CAMEL BBE uncertainty of the forested Congo and grassland/mixed agriculture of the ARM SGP sites (0.0059 and 0.0045 respectively), whereas the Yemen and Namib sites have larger uncertainties at 0.0087 and 0.0112 respectively and Greenland as a smaller uncertainty at 0.0038. Differences over land categories are seen to be as large as 0.05, larger than any of the site uncertainties. These differences in emissivity are shown as differences in LW radiation for three different skin temperatures in Table 5. The values in Table 5 are computed using the following equation:

$$F = \varepsilon_{BB}\sigma T^4, \tag{3}$$

where σ is the Stefan-Boltzmann constant and is equal to 5.67×10^{-8} W/m^{2}K^4, T is the skin temperature, ε_{BB} is the broadband emissivity (or BBE perturbation), and F is the LW radiation values reported in Table 5. Misrepresentations in BBE by 0.05 correspond to misrepresentations of LW radiation by ~8 to 37 W/m^2 for skin temperatures of 230 to 340 K. If accounting for differences of these magnitudes is proved to be important for various surface modeling endeavors, then Table 3 for the 8–13.5 μm band could be used as a simple parameterization of infrared surface BBE values.

Table 3. Average CAMEL 2000–2016 monthly climatology BBE for the 8–13.5 μm region for IGBP land cover categories.

	Jan	Feb	Mar	Apr	May	June	July	Aug	Sept	Oct	Nov	Dec
1: Evergreen Needle	0.974	0.975	0.975	0.974	0.973	0.973	0.974	0.974	0.973	0.973	0.973	0.974
2: Evergreen Broad	0.97	0.97	0.97	0.97	0.97	0.971	0.971	0.971	0.971	0.971	0.971	0.971
3: Deciduous Needle	0.972	0.972	0.974	0.973	0.97	0.972	0.973	0.972	0.971	0.969	0.972	0.972
4: Deciduous Broad	0.968	0.969	0.969	0.97	0.971	0.971	0.97	0.969	0.969	0.969	0.968	0.968
5: Mixed Forest	0.971	0.972	0.972	0.971	0.971	0.972	0.973	0.972	0.972	0.971	0.97	0.97
6: Closed Shrubs	0.964	0.964	0.964	0.966	0.966	0.964	0.963	0.963	0.963	0.964	0.965	0.964
7: Open Shrubs	0.965	0.965	0.965	0.966	0.964	0.963	0.963	0.963	0.962	0.963	0.964	0.964
8: Woody Savanna	0.971	0.971	0.971	0.971	0.971	0.971	0.971	0.97	0.97	0.97	0.971	0.971
9: Savanna	0.967	0.967	0.967	0.967	0.967	0.967	0.966	0.966	0.965	0.964	0.965	0.966
10: Grassland	0.968	0.968	0.966	0.965	0.965	0.964	0.963	0.963	0.964	0.964	0.965	0.967
11: Wetland	0.973	0.974	0.975	0.974	0.971	0.971	0.972	0.972	0.971	0.971	0.972	0.973
12: Cropland	0.97	0.97	0.969	0.968	0.968	0.968	0.969	0.969	0.969	0.968	0.968	0.969
13: Urban Area	0.968	0.968	0.967	0.967	0.968	0.967	0.967	0.967	0.967	0.967	0.967	0.968
14: Crop/Mosaic	0.968	0.968	0.967	0.966	0.967	0.968	0.968	0.968	0.969	0.968	0.968	0.968
15: Snow & Ice	0.975	0.975	0.976	0.976	0.976	0.975	0.974	0.974	0.975	0.976	0.975	0.975
16: Barren/Sparse Veg	0.927	0.927	0.925	0.925	0.925	0.924	0.924	0.924	0.924	0.925	0.925	0.926

Table 4. Number of samples for the average BBE climatology statistics over IGBP categories shown in Table 2.

IGBP Category	Number of Samples
1: Evergreen Needle Forests	179,000
2: Evergreen Broad Forests	454,500
3: Deciduous Needle Forests	94,500
4: Deciduous Broad Forests	43,400
5: Mixed Forest	492,200

Table 4. *Cont.*

IGBP Category	Number of Samples
6: Closed Shrubs	1400
7: Open Shrubs	1,184,900
8: Woody Savanna	418,000
9: Savanna	361,900
10: Grassland	876,300
11: Wetland	46,000
12: Cropland	507,000
13: Urban Area	3300
14: Crop/Mosaic	277,300
15: Snow & Ice	367,700
16: Barren/Sparse Vegetation	695,300

Table 5. Changes in LW radiation shown for various skin temperatures and changes in BBE.

Representation	Emissivity (-)	Skin Temperature (K)	LW Radiation (W/m^2)
Change between months	0.005	230	0.79
Change between months	0.005	310	2.62
Change between months	0.005	340	3.79
Change in land cover	0.05	230	7.9
Change in land cover	0.05	310	26.1
Change in land cover	0.05	340	37.89

5. Conclusions

This study presented and applied a method for calculating infrared surface broadband emissivity estimates from the recently released MEaSUREs CAMEL version V001 emissivity dataset (https://ezid.lib.purdue.edu/id/doi:10.21231/S2PP8H). In comparison to previous methods that are based upon regression techniques using satellite narrow band channels, this method takes advantage of the CAMEL high spectral resolution emissivity product across the full 3.6–14.3 micron infrared spectrum and directly integrates Equation (1) to obtain BBE estimates. BBE calculations were performed for two wavelength ranges: (1) the full CAMEL 3.6–14.3 μm range and (2) the 8.0–13.5 μm range. The first range was used to take advantage of the full, available high spectral resolution CAMEL spectrum. The second range was selected based on work of previous studies [18,21]. Both Ogawa and Schmugge [21] and Cheng et al. [18] found that from various wavelength ranges studied, BBE over the 8.0–13.5 μm range was optimal for computing all wavelength LW net radiation, which is a key parameter in radiation models and energy budgets.

For the case sites shown, BBE was found to be generally insensitive to skin temperature, with the exception of the Namib Desert site, for which a ~0.0014 decrease in BBE was seen for every 10 K increase in skin temperature. BBE from CAMEL were compared to those from lab validation spectra and were in agreement with them within the CAMEL BBE uncertainty for 3 of the sites—the Yemen (carbonate), Congo (forested), and ARM SGP (mixed agriculture). The Namib (quartz) lab data had a smaller BBE (by 0.0115) and the Greenland (snow/ice) lab data had a slightly larger BBE (by 0.0056) than CAMEL. Comparisons of CAMEL and UWIREMIS 8.0–13.5 μm BBE time series at selected sites showed the CAMEL HSR product to be an improvement upon the UWIREMIS HSR and that changes in BBE of up to ~0.02 can be experienced for a single location over the course of the year.

Differences of the monthly CAMEL BBE product from a single, global value illustrated the error from using single constants over time and space to represent BBE in radiation models. Differences from 0.98, a constant which is known to be used in various models, revealed magnitudes greater than 0.05, which is larger than the CAMEL BBE site uncertainties. Results showed more appropriate constant values could be chosen for specific applications using the global 5 km resolution CAMEL BBE dataset available from the authors upon request [34].

Statistics of CAMEL 8–13.5 μm BBE over IGBP land cover categories were shown as times series and revealed that the land cover categories changed by only ~0.01 in BBE over an annual cycle. Changes in the BBE time series behavior were experienced prior to the year 2007 for some IGBP categories and suggest the influence of a MODIS product version change. IGBP land cover statistics were also calculated for the entire CAMEL records monthly climatologies, and are summarized in Table 3, an example parameterization of BBE which could be used in land surface models. Simple calculations of LW radiation corresponding to changes in monthly, climatological BBE over months (0.005 change in BBE) resulted in changes in LW radiation of 0.79–3.79 W/m^2 for skin temperatures of 230–340 K, and changes over land cover categories (0.05 change in BBE) of 7.9–37.89 W/m^2 for similar skin temperatures. If a single BBE constant were used in land surface modeling, then systematic errors in the LW radiation of these magnitudes would occur.

Author Contributions: E.B., R.K., and M.F. conceived and designed the method, M.F. analyzed the data and wrote the paper, E.B. developed the software and data files, and G.H. and S.H. contributed their expertise.

Funding: This research was funded by NASA grant number NNX08AF8A.

Acknowledgments: Thanks are given to the reviewers of this work for their time and helpful comments.

Conflicts of Interest: The authors declare no conflict of interest.

References

1. Ramanathan, V. The Role of Earth Radiation Budget Studies in Climate and General Circulation Research. *J. Geophys. Res.* **1987**, *92*, 4075–4095. [CrossRef]
2. WMO (World Meteorological Organization). *The Global Observing System for Climate: Implementation Needs GCOS-200*; WMO: Geneva, Switzerland, 2016.
3. Prabhakara, C.; Dalu, G. Remote sensing of the surface emissivity at 9 micron over the globe. *J. Geophys. Res.* **1976**, *81*, 3719–3724. [CrossRef]
4. Salisbury, J.W.; D'Aria, D.M. Emissivity of terrestrial materials in the 3–5 μm atmospheric window. *Remote Sens. Environ.* **1992**, *42*, 83–106. [CrossRef]
5. Norman, J.M.; Divakarla, M.; Goel, N.S. Algorithms for extracting information from remote thermal-IR observations of the earth's surface. *Remote Sens. Environ.* **1995**, *51*, 157–168. [CrossRef]
6. Garand, L.; Buehner, M.; Wagneur, N. Background Error Correlation between Surface Skin and Air Temperatures: Estimation and Impact on the Assimilation of Infrared Window Radiances. *J. Appl. Meteorol. Climatol.* **2004**, *43*, 1853–1863. [CrossRef]
7. Dutta, S.K.; Garand, L.; Heilliette, S. Assimilation of Infrared Radiance Observations with Sensitivity to Land Surfaces in the Canadian Ensemble–Variational System. *J. Appl. Meteorol. Climatol.* **2016**, *55*, 561–578. [CrossRef]
8. Duda, J.D.; Wang, X.; Xue, M. Sensitivity of Convection-Allowing Forecasts to Land-Surface Model Perturbations and Implications for Ensemble Design. *Mon. Weather Rev.* **2017**, *145*, 2001–2025. [CrossRef]
9. Niu, G.Y.; Yang, Z.L.; Mitchell, K.E.; Chen, F.; Ek, M.B.; Barlage, M.; Kumar, A.; Manning, K.; Niyogi, D.; Rosero, E.; et al. The community Noah land surface model with multiparameterization options (Noah-MP): 1. Model description and evaluation with local-scale measurements. *J. Geophys. Res. Atmos.* **2011**, *116*. [CrossRef]
10. Jin, M.; Liang, S. An Improved Land Surface Emissivity Parameter for Land Surface Models Using. *Am. Meteorol. Soc.* **2006**, *19*, 2867–2881.
11. Zhou, L.; Dickinson, R.E.; Tian, Y.; Jin, M.; Ogawa, K.; Yu, H.; Schmugge, T. A sensitivity study of climate and energy balance simulations with use of satellite-derived emissivity data over Northern Africa and the Arabian Peninsula. *J. Geophys. Res.* **2003**, *108*. [CrossRef]
12. Wang, K.; Wan, Z.; Wang, P.; Sparrow, M.; Liu, J.; Zhou, X.; Haginoya, S. Estimation of surface long wave radiation and broadband emissivity using moderate resolution imaging spectroradiometer (MODIS) land surface temperature/emissivity products. *J. Geophys. Res. Atmos.* **2005**, *110*, D11109. [CrossRef]
13. Tang, B.-H.; Wu, H.; Li, C.; Li, Z.-L. Estimation of broadband surface emissivity from narrowband emissivities. *Opt. Express* **2010**, *19*, 185–192. [CrossRef] [PubMed]

14. Ogawa, K.; Schmugge, T.; Jacob, F.; French, A. Estimation of broadband land surface emissivity from multi-spectral thermal infrared remote sensing. *Agron. EDP Sci.* **2002**, *22*, 695–696. [CrossRef]

15. Ogawa, K.; Schmugge, T.; Rokugawa, S. Estimating broadband emissivity of arid regions and its seasonal variations using thermal infrared remote sensing. *IEEE Trans. Geosci. Remote Sens.* **2008**, *46*, 334–343. [CrossRef]

16. Liang, S.; Zhao, X.; Liu, S.; Yuan, W.; Cheng, X.; Xiao, Z.; Zhang, X.; Liu, Q.; Cheng, J.; Tang, H.; et al. A long-term Global LAnd Surface Satellite (GLASS) data-set for environmental studies. *Int. J. Digit. Earth* **2013**, *6*, 5–33. [CrossRef]

17. Ren, H.; Liang, S.; Yan, G.; Cheng, J. Empirical algorithms to map global broadband emissivities over vegetated surfaces. *IEEE Trans. Geosci. Remote Sens.* **2013**, *51*, 2619–2631. [CrossRef]

18. Cheng, J.; Liang, S.; Yao, Y.; Zhang, X. Estimating the optimal broadband emissivity spectral range for calculating surface longwave net radiation. *IEEE Geosci. Remote Sens. Lett.* **2013**, *10*, 401–405. [CrossRef]

19. Cheng, J.; Liang, S.; Yao, Y.; Ren, B.; Shi, L.; Liu, H. A comparative study of three land surface broadband emissivity datasets from satellite data. *Remote Sens.* **2014**, *6*, 111–134. [CrossRef]

20. Wilber, A.C.; Kratz, D.P.; Gupta, S.K. *Surface Emissivity Maps for Use in Retrievals of Longwave Radiation Satellite*; NASA Langley Research Center: Hampton, VA, USA, 1999.

21. Ogawa, K.; Schmugge, T. Mapping Surface Broadband Emissivity of the Sahara Desert Using ASTER and MODIS Data. *Earth Interact.* **2004**, *8*, 1–14. [CrossRef]

22. Borbas, E.; Hulley, G.; Feltz, M.; Knuteson, R.; Hook, S. The Combined ASTER MODIS Emissivity over Land (CAMEL) Part 1: Methodology and High Spectral Resolution Application. *Remote Sens.* **2018**, *10*, 643. [CrossRef]

23. Feltz, M.; Borbas, E.; Knuteson, R.; Hulley, G.; Hook, S. The Combined ASTER MODIS Emissivity over Land (CAMEL) Part 2: Uncertainty and Validation. *Remote Sens.* **2018**, *10*, 664. [CrossRef]

24. Hook, S. *Combined ASTER and MODIS Emissivity for Land (CAMEL) Uncertainty Monthly Global 0.05Deg. V001*; NASA EOSDIS Land Processes DAAC, USGS Earth Resources Observation and Science (EROS) Center: Souix Falls, SD, USA, 2017. Available online: https://lpdaac.usgs.gov/dataset_discovery/measures/measures_products_table/cam5k30uc_v001 (accessed on 15 June 2018). [CrossRef]

25. Borbas, E.E.; Hulley, G.C.; Knuteson, R.O.; Feltz, M.L. MEaSUREs Unified and Coherent Land Surface Temperature and Emissivity (LST&E) Earth System Data Record (ESDR): The Combined ASTER and MODIS Emissivity Database over Land (CAMEL) Users' Guide. 2017. Available online: https://lpdaac.usgs.gov/sites/default/files/public/product_documentation/cam5k30_v1_user_guide_atbd.pdf (accessed on 7 May 2018).

26. Seemann, S.W.; Borbas, E.E.; Knuteson, R.O.; Stephenson, G.R.; Huang, H.L. Development of a global infrared land surface emissivity database for application to clear sky sounding retrievals from multispectral satellite radiance measurements. *J. Appl. Meteorol. Climatol.* **2007**, *47*, 108–123. [CrossRef]

27. Hulley, G.C.; Hook, S.J.; Abbott, E.; Malakar, N.; Islam, T.; Abrams, M. The ASTER Global Emissivity Dataset (ASTER GED): Mapping Earth's emissivity at 100 meter spatial scale. *Geophys. Res. Lett.* **2015**, *42*, 7966–7976. [CrossRef]

28. Borbas, E.E.; Seemann, S.W. Global Infrared Land Surface Emissivity: UW-Madison Baseline Fit Emissivity Database V2.0. 2007. Available online: http://cimss.ssec.wisc.edu/iremis/ (accessed on 15 June 2018).

29. Hulley, G.; Hook, S. *AG5KMMOH: ASTER Global Emissivity Dataset, Monthly, 0.05 Degree, HDF5 V041*; NASA EOSDIS Land Processes DAAC, USGS Earth Resources Observation and Science (EROS) Center: Souix Falls, SD, USA, 2016. Available online: https://lpdaac.usgs.gov/node/1123 (accessed on 15 June 2018). [CrossRef]

30. Borbas, E.E.; Ruston, B.C. *The RTTOV UWiremis IR Land Surface Emissivity Module*; EUMETAT NWP SAF, NWPSAF-MO-VS-042; EUMETSAT: Darmstadt, Germany, 2010.

31. Wan, Z. *MOD11C3 MODIS/Terra Land Surface Temperature/Emissivity Monthly L3 Global 0.05Deg CMG. V006*; NASA EOSDIS Land Processes DAAC, USGS Earth Resources Observation and Science (EROS) Center: Souix Falls, SD, USA, 2015. Available online: https://lpdaac.usgs.gov/dataset_discovery/modis/modis_products_table/mod11c3_v006 (accessed on 15 June 2018). [CrossRef]

32. Friedl, M.A.; Sulla-Menashe, D.; Tan, B.; Schneider, A.; Ramankutty, N.; Sibley, A.; Huang, X. MODIS Collection 5 global land cover: Algorithm refinements and characterization of new datasets. *Remote Sens. Environ.* **2010**, *114*, 168–182. [CrossRef]

33. Channan, S.; Collins, K.; Emanuel, W.R. *Global Mosaics of the Standard MODIS Land Cover Type Data*; University of Maryland and the Pacific Northwest National Laboratory: College Park, MD, USA, 2014.
34. Borbas, E.E.; Feltz, M.L.; Knuteson, R.O. Broad Band Emissivity derived from the MEaSUREs CAMEL V001 Database. UW-Madison Space Science and Engineering Center. Dataset. 2017. Available online: https://ezid.lib.purdue.edu/id/doi:10.21231/S2PP8H (accessed on 15 June 2018). [CrossRef]

Article

Bayesian Bias Correction of Satellite Rainfall Estimates for Climate Studies

Margaret Wambui Kimani *, Joost C. B. Hoedjes and Zhongbo Su

Faculty of Geo-Information Science and Earth Observation, University of Twente,
217 7500 AE Enschede, The Netherlands; j.c.b.hoedjes@utwente.nl (J.C.B.H.); z.su@utwente.nl (Z.S.)
* Correspondence: m.w.kimani@utwente.nl

Received: 16 April 2018; Accepted: 2 July 2018; Published: 5 July 2018

Abstract: Advances in remote sensing have led to the use of satellite-derived rainfall products to complement the sparse rain gauge data. Although these products are globally and some regional bias corrected, they often show substantial differences relative to ground measurements attributed to local and external factors that require systematic consideration. A decreasing rain gauge network inhibits the continuous validation of these products. Our proposal to deal with this problem was to use a Bayesian approach to merge the existing historical rain gauge information to create consistent satellite rainfall data for long-term applications. Monthly bias correction was applied to Climate Hazards Group Infrared Precipitation with Stations (CHIRPS v2) using a corresponding gridded (0.05°) rain gauge data over East Africa for 33 years (1981–2013). The first 22 years were utilized to derive error fields which were then applied to independent CHIRPS data for 11 years for validation. Assessments of the approach's influence on the rainfall estimates spatially and temporally were explored. Results showed a significant spatial reduction of the underestimation and overestimation of systematic errors at both monthly and yearly scales. The reduced errors increased with increased rainfall amounts, hence was less so in the relatively drier months. The overall monthly reduction of Root Mean Square Difference (RMSD) was between 4% and 60%, and the Mean Absolute Error (MAE) was between 1% and 63%, while the correlations improved by up to 21%. Yearly, the RMSD was reduced between 17% and 49%, and the MAE between 13% and 48%, while the increase in correlations was between 9% and 17%. Decreased yearly bias correction corresponded with years of high rainfall associated with El Niño. Results for the assessments of the effectiveness of the Bayesian approach showed that it was more effective in reducing systematic errors related to rainfall magnitudes, but its performance decreased in areas of sparse rain gauge network that insufficiently represented rainfall variabilities. This affected areas of deep convection, leading to minimal overestimation reductions associated with the cirrus effect. Conversely, significant corrections were during years of low rainfall from shallow convections. The approach is suitable for long-term applications where consistencies of mean errors can be assumed.

Keywords: Bayesian bias correction; satellite rainfall; rain gauge; East Africa

1. Introduction

Rainfall data is vital for many applications such as climate studies, water resource management, and agriculture. As its accurate spatial and temporal representations can improve socio-economic planning. Rain gauges provide the most direct representations of rainfall, but their distribution over land are sparse, especially in mountainous areas [1], and being point observations, they lack spatial representativeness [2]. However, they offer useful information in modelling regarding local rainfall processes that are not accurately parameterized by the Global Circulation Models (GCM) [3]. Alternative uses of satellite rainfall products are increasing because of their high spatiotemporal coverage. However, these products often exhibit large discrepancies with ground measurements [4,5], and the errors need to be reduced to make the products more representative of the local rainfall variability. Although some of these products are globally validated [6–8] and some at regional scales [9,10], relatively few efforts have been made to reduce the often-large errors that occur at local scales. Studies [11–13] have found that satellite rainfall products have systematic errors that cause overestimations/underestimations, especially in high elevated areas [13]. Although rain gauge data have low spatial distributions, their direct way of measuring rainfall are still vital as a reference to the local rainfall variability. For better representations of local rainfall processes, the inclusion of all available quality controlled rain gauge data merged with satellite products can enhance the products' future applications. Different methods have been proposed to reduce satellite rainfall estimates errors. A study by [12] applied bias correction using empirical cumulative distribution (CDF) maps on a seasonal basis for hydrological applications in the upper Blue Nile in Ethiopia. To reduce temporal rainfall variability, a seasonal timescale was utilized. However, in high elevated areas, areas near inland water bodies, and those with maritime influences, high rainfall variabilities are experienced. As such, the choice of temporal scale may differ from place to place. It is worth noting that the effectiveness of bias correction on rainfall products may also differ from location to location and consideration of spatial scale is of great importance. Quantile mapping approach was applied by [14] to bias correct rainfall products and they observed that the approach improved estimates in some locations, while it degraded in others. However, it is crucial that the applied method does not necessarily change the product's original rainfall estimates. Therefore, the consistencies of the systematic errors corrected are worthy of consideration. Mateus et al. [15] assessed the performance of two bias correction methods—successive correction method (SCM) and optimal interpolation and qualitative analysis—and visual inspections showed better results by SCM. However, the study noted the limitation of this approach in defining the optimal weight of the error distributions. Elsewhere, ref. [16] evaluated satellite rainfall estimates combined with high-resolution rain gauge data using different bias correction methods based on an additive, multiplicative, and merged scheme approach. The evaluation was carried out on a monthly basis in different rainfall seasons and with different rain gauge networks. The results revealed that the choice of both the temporal and spatial scale of the rain gauge data was vital for adequate bias correction. In their study, the merged scheme showed the best results. Nevertheless, this approach is more suitable for real-time applications and in many areas of the world, the degradation of the rain gauge network is a common problem due to the lack of maintenance. Furthermore, for climate studies and other long-term applications, real-time data is not applicable.

A probabilistic Bayesian approach was applied by [17] on high temporal resolution rain gauge data. Historical rain gauge and satellite data were used to create a satellite estimates–rain gauge data relationship, which is applicable in the absence of real-time rain gauge data. The approach worked well even in areas of low rain gauge distributions, but over corrections were observed in some areas. However, it is understandable that rainfall variability differs from place to place and the impact of rain gauge distribution needs to be determined. It is a fact that over the world, rain gauge distributions are decreasing [18], especially in African countries due to their cost of maintenance. Their availabilities to validate the increasing satellite rainfall products may be affected by inconsistencies caused by the low network. Despite this, they offer useful information on the local rainfall variabilities. Over equatorial East Africa, few studies like [19] have used high-resolution ground data to bias-correct satellite rainfall estimates for hydrological applications. It is crucial to have long-term bias correction because of the accumulation of errors in time and for externally induced errors, particularly in areas that experience high rainfall variability [13].

We proposed a Bayesian approach that could be used with the existing historical rain gauge information to create consistent satellite rainfall data for long-term applications. This approach assumes consistencies in time for the average errors in both datasets. As such, the error weight derived from their climatology is considered to be representative of a given area. In our approach, we converted the probability into independent variables to apply a linear relationship using the least square techniques [20]. This approach is superior to other methods in that it does not always modify the estimates during corrections, but considers the mean error consistencies of the input data in time. Therefore, the corrected satellite estimates approach the uncorrected state in areas of poor rainfall representations arising from sparse rain gauge distribution. This way, the satellite rainfall estimates remain close to the original state in areas of inconsistent rain gauge data. A long-term (1981–2013) temporal scale bias correction was applied to the Climate Hazards Group

Infrared Precipitation with Stations (CHIRPS v2). The product was chosen based on its high spatial resolution and lengthy climatology suitable for climate studies to help end users in planning [21]. Furthermore, a recent study by [13] over East Africa showed a close correspondence of CHIRPS v2, Tropical Rainfall Measuring Mission (TRMM)3B43, and the Climate Prediction Center (CPC) morphing technique (CMORPH) with ground observations. However, in that study, all the satellite products assessed exhibited large biases in high elevated areas. Although CHIRPS is globally bias corrected using some of the rain gauge data used in this study, the data mainly come from the Global Telecommunication System (GTS). The GTS stations are sparse and may therefore not accurately represent the rainfall variability over the region.

This study assessed the performance of the Bayesian approach in reducing systematic errors on CHIRPS v2 rainfall estimates relative to regional gridded rain gauge data. The assessments on the effectiveness of the method were on a monthly and yearly basis. This paper has six sections. Section 1 presents the introduction; Section 2 gives a brief description of the study area and data used, Section 3 describes the Bayesian approach and methods of evaluation; Section 4 presents the results, discussions of the results are in Section 5 while our conclusions are in Section 6.

2. Study Region and Data

The study area in East Africa (Figure 1) extended between 29°E and 42°E, and 12°S and 5°N and covered five countries: Kenya, Uganda, Tanzania, Burundi, and Rwanda. The region shows diverse topography delineated by the embedded elevation map. Two main rainy seasons occur during March, April, and May (MAM) and October, November, and December (OND). These rainy seasons coincide with the overlying of a low-pressure belt of the Inter-Tropical Convergence Zone (ITCZ). The ITCZ migrates from 15°S to 15°N between January and July with characteristics of convective activities that lead to increased precipitation. A third rainfall season occurs in the months of June through to August (JJA) and affects a small part of Western Kenya and Uganda. This season significantly affects water resources within the region and areas around Lake Victoria.

Two monthly rainfall datasets were used in this study and included CHIRPS v2 rainfall estimates and rain gauge data. CHIRPS is a quasi-global dataset developed by the United States Geological Survey (USGS) Earth Resources Observations and Science Centre and the University of California Santa Barbara Climate Hazards Group. It has a spatial resolution of 0.05°, and a daily/pentad/monthly temporal resolution. It uses TRMM multi-satellite precipitation analysis version 7 to calibrate the Cold Cloud Duration (CCD) rainfall estimates. The product covers the area between 50°N and 50°S, and data are available from January 1981 to the near present. Further details of CHIRPS v2 used in this study can be found in [22], and an evaluation of its performance relative to other products in [10].

The gridded (0.05°) rain gauge was from the Intergovernmental Authority on Development (IGAD) Climate Prediction and Application Centre (ICPAC [23]. Although the data includes global telecommunication stations, ICPAC includes data from other stations sourced from the five countries (Kenya, Uganda, Tanzania, Burundi, and Rwanda) (Figure 1). This move was prompted by the decreasing rain gauge distributions, especially in developing countries, partially due to the cost and lack of skilled personnel. In East Africa, the decreasing trend is worrying, and the only solution is to grid the available rain gauge data [24] to preserve their information. It is in this context that the member states of East African countries brought together their available data from all the operational stations of the National Meteorological and Hydrological Services (NMHSs). They interpolated and quality controlled the rain gauge measurements from 284 rainfall stations. They used the GeoCLIM [25] tool with inverse distance weighting (IDW) [26]. The Tamuka Magadzire of the United States Geological Survey (USGS) Famine Early Warning Systems Network (FEWSNET) developed GeoCLIM for rainfall, temperature, and evapotranspiration analysis. The gridded data have been used regionally for hazard and regional rainfall predictions, and recently for the evaluation of satellite rainfall data [13].

Elevation data was downloaded from the Shuttle Radar Topography Mission (SRTM) 90 m Digital Elevation Model (DEM) [27]. The 5° spatial resolution tiles were then mosaicked over East Africa as shown in Figure 1 by using the Geographical Information System (GIS) functionality.

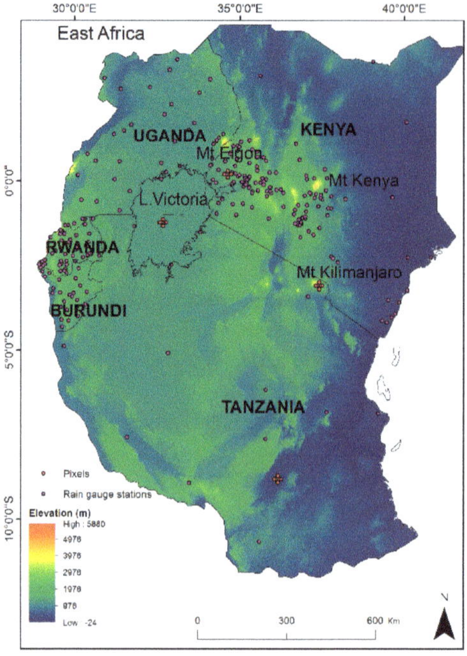

Figure 1. Map of East Africa with the Shuttle Radar Topography Mission (SRTM) 90 m digital elevation model. Rain gauge station distributions used for gridding are highlighted in black. In red are the selected pixels for assessments.

3. Methodology

We first describe the Bayesian method and then explain the training and testing procedures.

3.1. Bayesian Method

A Bayesian method is a probabilistic approach that merges data from different sources [28] to obtain the optimal representative values from the input datasets. It is based on spatial transformation and uses the variances of the input datasets. In this study, it was used to adjust monthly CHIRPS satellite rainfall estimates using the gridded rain gauge data for 33 years (1981–2013) in two steps. First, training data from 22 years (1981–2002) were used to derive bias fields for the multi-annual monthly averages of each month. The monthly averaged bias fields were then used to correct independent satellite rainfall estimates during an 11-year (2003–2013) validation period. The hypothesis of the approach was the temporal consistency of average errors. It was carried out at a $0.05° \times 0.05°$ spatial scale for both datasets, but for compatibility, the CHIRPS data were resampled using the nearest neighbor interpolation [29] to match the georeference of the rain gauge data. The resampling approach is robust in reprocessing algorithms according to this study and has been applied successfully in other areas [30].

3.1.1. Training Period

The Bayes theorem [28] aims to obtain the maximum likelihood of $P(s\,|\,g)$, which is the conditional probability of the satellite estimates (s) given the gridded rain gauge data (g).

$$P(s|g) = \frac{P(s)P(g|s)}{P(g)} \tag{1}$$

where $P(s)$, $P(g|s)$ denotes the probability of satellite data and likelihood function of rain gauge data given by the satellite estimates, respectively. Since the gridded rainfall data distribution is known, $P(g) = 1$, then Equation (1) reduces to Equation (2):

$$P(s|g) = P(g|s)P(s) \tag{2}$$

Following [31], the least squares estimation can be used to simplify the data assimilation problems to linear relationships and Equation (2) is changed from the probabilistic form into independent variables.

Assuming the monthly averaged errors (ε) of the satellite rainfall estimates and the gridded rain gauge data to be unbiased and consistent in time, E is the expected value as in Equation (3).

$$E(\varepsilon_g) = E(\varepsilon_s) = 0 \tag{3}$$

The variances (σ^2) of each dataset can be related to the errors (ε), assuming the errors are uncorrelated (Equations (4) and (5)). σ is the standard deviation described in Section 3.2.

$$E(\varepsilon_g^2) = \sigma_g^2 \tag{4}$$

$$E(\varepsilon_s^2) = \sigma_s^2 \tag{5}$$

Bias-corrected satellite estimates are linearly combined with the gridded rainfall data and the uncorrected satellite rainfall estimates (Equation (6)). The weighing factors, α_g and α_s, are dependent on the respective variances. The higher the variance value, the lower the corresponding weighting factor. Implying that in areas where variance of rain gauge data is high, the bias correction is minimal.

$$\overline{s_c} = \alpha_g \overline{g} + \alpha_s \overline{s} \tag{6}$$

With the overbars denoting the averaged values for each month in the 22 years training dataset, Equation (6) assumes the bias-corrected satellite estimates (in this case CHIRPS) denoted as 's' with a subscript 'c' to be unbiased as their errors are consistent during the training period. The sum of the CHIRPS estimates' weighing factor, α_s, and the gridded rain gauge weighting factor, α_g, equals one (Equation (7)).

$$\alpha_g + \alpha_s = 1 \tag{7}$$

S_c will be the best estimate of g if the weighing factors α_g and α_s are chosen to minimize the mean squared error of the corrected satellite estimates S_c following Equations (8)–(11).

$$\sigma_c^2 = \overline{(s_c - g)} = \overline{\left[\alpha_g(s_c - g) + (1 - \alpha_g)(s_c - g)\right]^2} \tag{8}$$

$$\frac{\partial \sigma_c^2}{\partial \alpha_g} \rightarrow 0 \tag{9}$$

$$\sigma_c^2 = (\overline{s_c} - \overline{g})^2 \tag{10}$$

$$\sigma_c^2 = \alpha_g^2 \sigma_g^2 + (1 - \alpha_g)^2 \sigma_s^2 \tag{11}$$

$$\alpha_g = \frac{\sigma_s^2}{\sigma_g^2 + \sigma_s^2} \tag{12}$$

This leads to

$$\alpha_s = \frac{\sigma_g^2}{\sigma_g^2 + \sigma_s^2} \tag{13}$$

Equations (12) and (13) imply that the weights of the satellite estimates and the corresponding rain gauge data are related to the inverse of their variances. The weighting factors correct the average satellite estimates for each month during the 22 years training period using the linear relationship shown in Equation (6) which can be rewritten as Equation (14)

$$\overline{s_c} = \overline{s} + \frac{\sigma_s^2}{\sigma_s^2 + \sigma_g^2}(\overline{g} - \overline{s}) = \overline{s} + \alpha_g(\overline{g} - \overline{s}) \tag{14}$$

Equation (14) implies that when the variance of the reference (rain gauge) data is very large, that is, $\sigma_g \gg \sigma_s$, then $\sigma_{g \to 0}$ and s_c approaches s, and that when $\sigma_g \ll \sigma_s$ $\sigma_{g \to 1}$ and s_c approaches g.

3.1.2. Testing Period

In this section, the Bayesian approach is described using the error fields derived during training with the monthly data. The CHIRPS and bias-corrected CHIRPS estimates comparison with corresponding rain gauge data was conducted on a monthly and yearly basis. The bias fields were derived from the satellite estimates for each month using Equation (15). The subscript 'i' stands for the time step.

$$Bias = \frac{1}{n} \sum_i^N (s_{ci} - g_i) \tag{15}$$

The bias is then subtracted from the satellite data of each corresponding month (subscript 'i') using Equation (16).

$$s_{ci} = s_i - bias \tag{16}$$

3.2. Evaluation of Bias Corrected CHIRPS Rainfall

Validation of bias correction of CHIRPS rainfall estimates was carried out for 11 years (2003–2013) between the raw and bias-corrected CHIRPS (bc) relating to the gridded rain gauge data on a monthly and yearly basis. Continuous statistics of the correlation coefficient (*cc*), RMSD, standard deviations (σ) (Equations (17)–(19)), MAE (Equation (20)), and mean bias (Equation (21)) were used to quantify their relationships. For visualization, Taylor diagrams [32] and spatial maps were utilized.

$$cc = \frac{\frac{1}{N} \sum_{i=1}^{N}(s_i - \overline{s})(g_i - \overline{g})}{\sigma_s \sigma_g} \tag{17}$$

$$RMSD = \sqrt{\frac{1}{N} \sum_{i=1}^{N}(s_i - g_i)^2} \tag{18}$$

$$\sigma_{g(s)} = \sqrt{\frac{1}{N} \sum_{i=1}^{N}\left(g(s) - \overline{g(s)}\right)^2} \tag{19}$$

where the overbar stands for the respective mean satellite estimates (*s*), the gridded rain gauge datasets (*g*), *g(s)* is either the gridded rain gauge or satellite dataset, and *N* is the number of samples considered.

$$MAE = \frac{1}{n} \sum_{i=1}^{N} |s_i - g_i| \tag{20}$$

$$Mean\ bias = \frac{1}{N} \sum_{i=1}^{N}(s_i - g_i) \tag{21}$$

3.3. Assessments of Bayesian Approach Performance

East Africa has complex terrain, comprising of lakes, mountains, and lowlands. As a result, the high elevated areas and Lake Victoria influence local rainfall variability. As such, further assessments were carried out to determine the performance of the Bayesian approach in eradicating CHIRPS biases on an annual basis. Pixels within the areas of lowest Bayesian performance were assessed using CHIRPS and bias-corrected CHIRPS rainfall estimates relative to rain gauge data. These included areas of large uncorrected overestimations over Mt. Elgon, Southern Tanzania, Lake Victoria, and Mt. Kilimanjaro (Figure 1). Scatter plots and MAE were used to quantify these relationships.

4. Results

4.1. Assessments of Bias-Corrected CHIRPS Estimates

4.1.1. Monthly Assessments

This section describes the monthly assessments of bias correction of CHIRPS rainfall estimates using rain gauge data over East Africa during a validation period of 2003 to 2013. Figure 2 presents the Taylor diagrams displaying the error metrics before (CHIRPS) and after bias corrections (abbreviated with a 'bc') during the wet months of March to May, and October to December. These results showed that the Bayesian approach significantly improved the accuracy of the CHIRPS estimates as indicated by the reduced RMSD and increased correlations for all months. The biases showed seasonality and occurred more during OND when compared to the MAM months. During OND, the orographic processes were more dominant and more challenging for the infrared-based satellite rainfall products [13]. The bias increased with an increase in rainfall magnitudes [11] and were largest in April and November. These are the peak rainfall months of the MAM and OND seasons, respectively. These observations of bias dependence on rainfall amounts concurred with [10] over other parts of Africa.

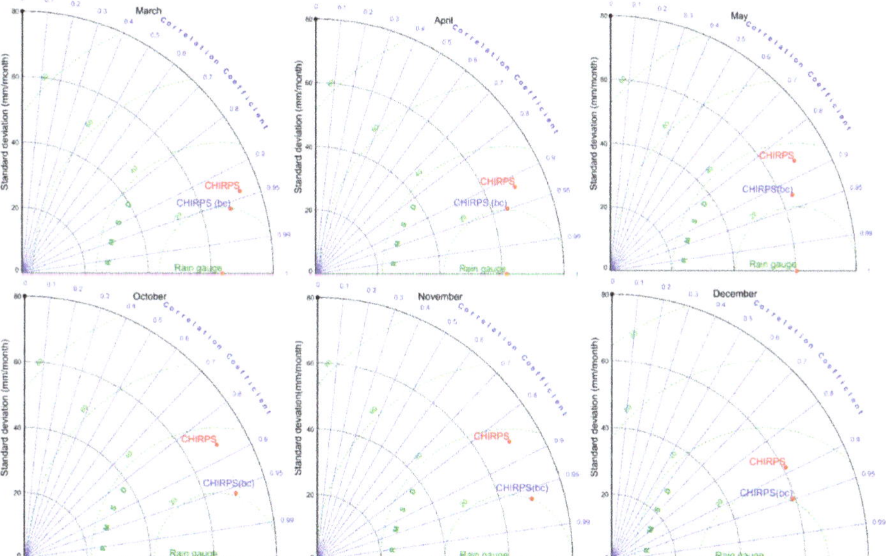

Figure 2. Monthly Taylor diagrams displaying the statistical comparison between the Climate Hazards Group Infrared Precipitation with Stations (CHIRPS) (red) and bias-corrected CHIRPS (bc) (blue) estimates with corresponding rain gauge data (green) as the reference. Shown are the wet months of the rainfall seasons (March–May and October–December) over a period of 11 years (2003–2013). The azimuthal angle represents the correlation coefficient; radial distance is the standard deviation (mm/month), and green contours represent RMSD (mm/month).

Figure 3 presents the results of the relatively dry months of January, February, and June through September and showed a significant reduction of biases. It was evident that there were increased error magnitudes than during the wet months, mainly from June–September. These errors are attributable to orographic processes during the northeast (January through February) and southeast (June through September) monsoon period. The high grounds inhabit moisture influx inland, thereby limiting the rainfall occurrence to the high ground areas. In their study [33,34], associated Turkana Jets that run parallel to the highlands cause rainfall variabilities during the south-east monsoon season.

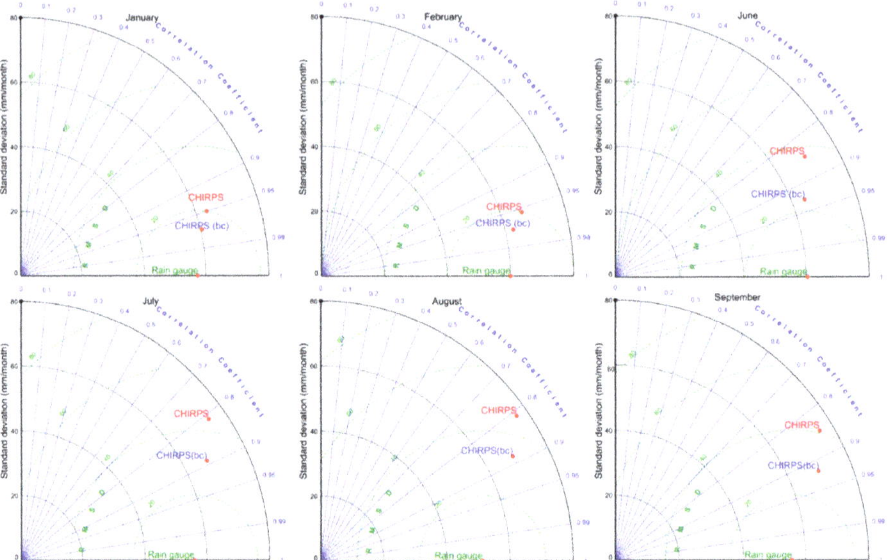

Figure 3. Same as for Figure 2, but for the relatively drier months of January, February, June–August and September.

Figure 4 shows the mean bias (2003–2013) derived from CHIRPS and the bias-corrected CHIRPS (bc) relative to rain gauge data for April and May, August, and November. The three wet seasons are representative of March–May, June–August, and October–December, respectively. The spatial patterns display areas of underestimations and overestimations and indicate the areas of improvement after Bayesian applications. It is evident that increased bias followed areas of the highest rainfall amounts. Consequently, the largest underestimations (negative bias) were observed in highly elevated regions of Mt. Kenya in November, around Mt. Elgon in August, and over the coastal areas bordering the Indian Ocean in May. This confirmed that the CHIRPS monthly estimates underestimated high rainfall amounts [29]. These findings were also in line with [11], where observed systematic errors increased with increased rainfall amounts. Evidently, in April, which is the peak rainfall month of MAM, the mean biases were well distributed and attributable to mixed rainfall regime characteristics. Additionally, overestimations (positive bias) were evident in areas around Lake Victoria, eastern parts of Kenya, and Southern Tanzania in April and May. The overestimations arose from cirrus effects common to infrared based products using a cold cloud temperature threshold as they consider the cold cirrus clouds that occur in deep convections as precipitating [11]. It was evident that the approach adequately reduced overestimations except over Southern Tanzania, which showed sparse rain gauge stations (Figure 1) and may therefore not have well-represented rainfall variability. Similarly, a study by [17] used the probability distribution to adjust satellite rainfall estimates and associated overcorrections for the misrepresentation of rainfall variability in sparse rain gauge areas.

It is recommended that this approach is sensitive to the inconsistencies of the reference data and does not necessarily modify the satellite rainfall estimates.

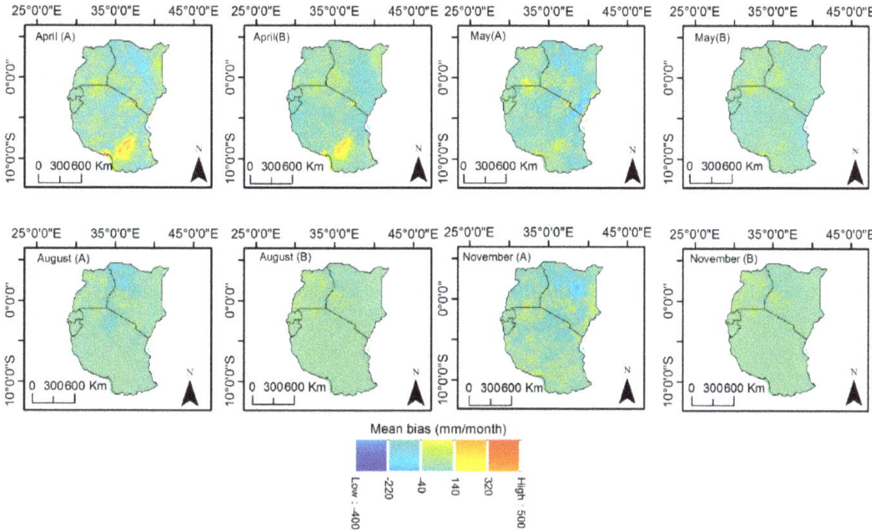

Figure 4. Monthly mean bias derived from averages (2003–2013) of rain gauge data and satellite rainfall estimates of CHIRPS (**A**) and bias corrections (CHIRPS (bc)) (**B**).

Table 1 shows a summary of the monthly error statistics for all months before and after bias corrections and the change of errors in percentages. It was evident the bias-corrected CHIRPS (bc) estimates showed reductions in RMSD and MAE errors and an increase in correlations. One notable observation was the low percentage change in correlations during the relatively dry months. Further observations showed that these months corresponded with the southeast and northeast monsoon months of June through September and December–February, respectively. Therefore, it was evident that the systematic errors locally induced by topographic effects were also externally influenced by the largescale circulations. The low linearity arose from abrupt rainfall variability, hence increasing the average error inconsistencies in time assumed in the Bayesian approach. The overall monthly reduction of RMSD was between 4% and 60%, and the MAE was between 1% and 63%, while the correlations increased up to 21%.

Table 1. Statistics for the monthly spatial evaluation.

Months	CC	CC (bc)	Change (%)	RMSD	RMSD (bc)	Change (%)	MAE	MAE (bc)	Change (%)	Rainfall (mm/month)
Jan	0.95	0.98	2	23.1	13	-45	17.6	9	-49	88.0
Feb	0.96	0.97	2	23.1	15	-37	14.9	10	-31	70.3
Mar	0.87	0.90	3	40.1	31	-22	24.7	23	-9	105.5
Apr	0.47	0.61	14	64.3	48	-25	43.1	28	-35	155.1
May	0.77	0.94	17	49.5	25	-50	36.2	20	-45	103.5
Jun	0.87	0.97	10	22.8	10	-58	15.8	6	-61	38.4
Jul	0.89	0.98	9	20.1	8	-60	12.0	5	-63	31.9
Aug	0.89	0.98	9	24.1	11	-56	14.2	6	-57	38.8
Sep	0.92	0.97	6	19.3	13	-34	13.3	8	-43	39.4
Oct	0.95	0.95	0	21.1	19	-11	16.9	15	-11	87.1
Nov	0.72	0.94	21	41.4	27	-35	29.3	23	-22	129.5
Dec	0.92	0.92	0	27.2	26	-4	20.7	20	-1	127.0

4.1.2. Yearly Evaluations

This section describes the annual spatial evaluation of the CHIRPS and bias-corrected CHIRPS (bc) rainfall estimates relating to the rain gauge data during the validation years (2003–2013). Figures 5 and 6 show the Taylor diagrams displaying the error metrics of the RMSD and correlations and their respective standard deviations. It was evident that in all of the years, the CHIRPS rainfall estimates were adjusted towards the rain gauge data as indicated by the reduced RMSD and increased correlation coefficients. Large differences occurred in the years associated with El Niño (2003 and 2006) and the least during the relatively drier year (2005). Therefore, it was clear that systematic errors accumulated with increased rainfall, especially during anomalous years and it is worth noting that the seasonality over East Africa corresponded with the observed bias. This shows the importance of a long-term correction period that is inclusive of the known external variabilities. A notable observation of increased bias was in the year following El Niño, for example, the years 2004, 2007, and 2010. El Niño occurred towards the end of the year in October through December and increased rainfall in the months following may have arisen from recycled water.

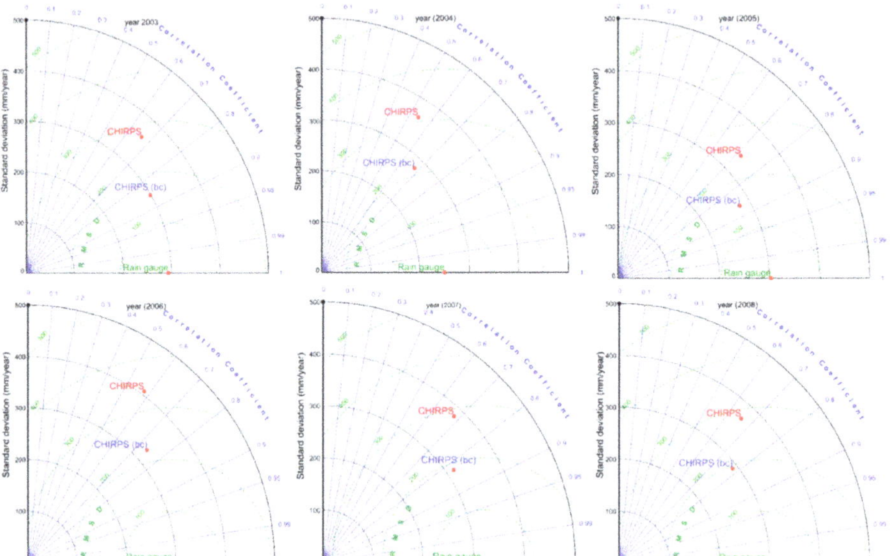

Figure 5. Taylor diagrams displaying the statistical comparison between the CHIRPS (red) and bias-corrected (blue) CHIRPS estimates with corresponding rain gauge data (green) as the reference for years 2003–2008. The azimuthal angle represents the correlation coefficient; the radial distance represents the standard deviation (mm/year), and the green contours represent RMSD (mm/year).

Figure 7 shows the spatial mean bias distribution during the anomalous wet years of 2003 and 2006, the relatively dry (2005), and standard (2008) years. Similar to the monthly analysis, underestimations were evident before bias correction. The Bayesian approach substantially reduced the biases over the high elevated areas of Mt. Kenya in 2006, which was also an El Niño year. Similarly, overestimations were evident around Lake Victoria, Southern Tanzania, near Mt. Elgon, and North-eastern Kenya. Large uncorrected overestimations were observed, particularly in the southern parts of Tanzania and around Mt. Kilimanjaro, which are areas of poor rain gauge network (Figure 1).

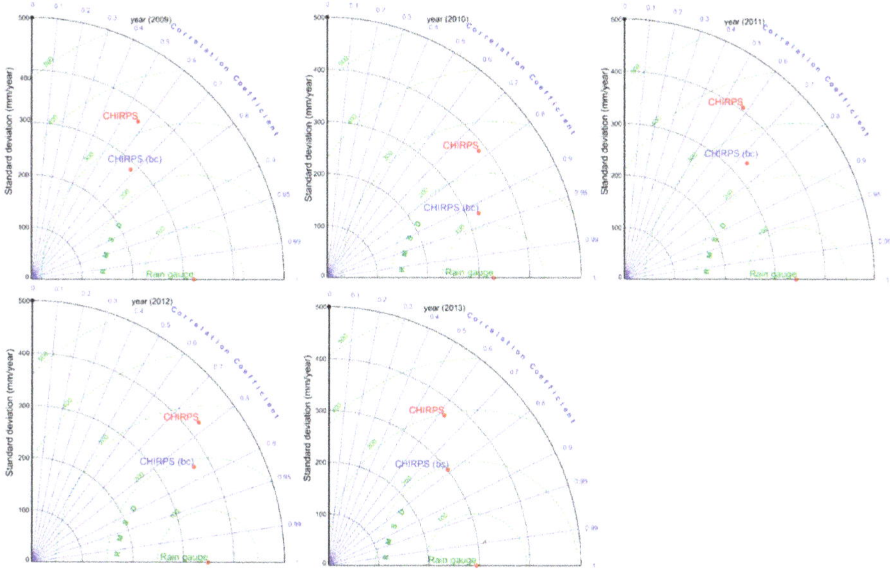

Figure 6. As per Figure 5, but for the years 2009–2013.

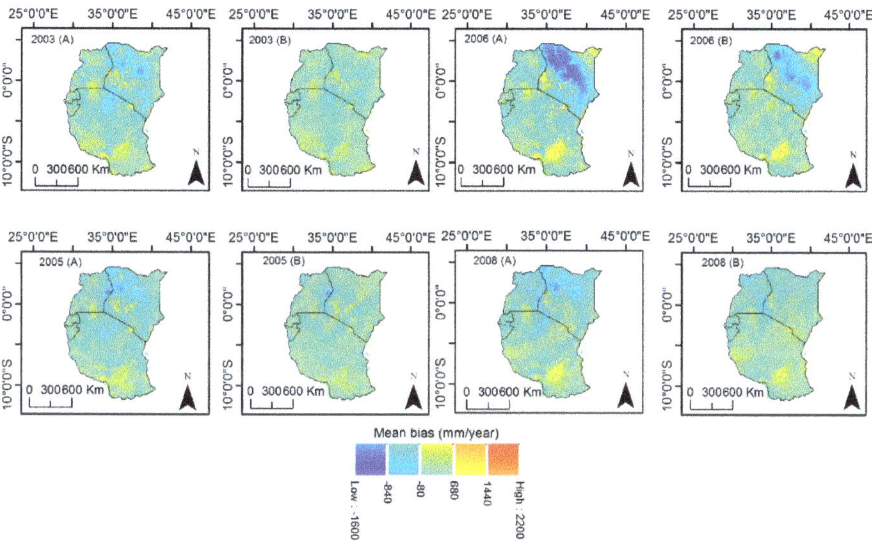

Figure 7. Yearly spatial bias patterns derived from the averages of the anomalous wet years (2003 and 2006), the dry (2005), and standard (2008) of the CHIRPS and bias-corrected CHIRPS (bc) abbreviated as (**A**,**B**), respectively, relative to the rain gauge data.

Table 2 shows a summary of the yearly error statistics before and after bias corrections, and the change of errors in percentages. There was a remarkable decrease in the RMSD and MAE as well as an increase in correlations after bias corrections. Similar to the monthly analysis, the errors corrected showed a dependence on rainfall magnitude, but also on rainfall regime. As such, the years 2003, and 2009, which were also El Niño years, did not correspond to the highest mean rainfall as El Niño occurred during the short rainfall period of OND. However, the years after (2004 and 2010), had higher rainfall that spread to the MAM rainfall season. Similarly, in 2007 and 2010, the years following El Niño, there was increased rainfall and a correspondingly high percentage of errors corrected. It is worth noting that during the relatively dry year (2005), the percentage of RMSD and MAE corrected was high (46%). This can be attributable to the fact that CHIRPS estimates are designed for drought monitoring, hence there are more consistent mean errors during dry years. In [11], it was observed that the systematic errors were seasonal, hence the importance of understanding the rainfall distribution during the year in a given area.

Table 2. Statistics for the yearly spatial evaluation.

Years	CC	CC (bc)	Change (%)	RMSD	RMSD (bc)	Change (%)	MAE	MAE (bc)	Change (%)	Rainfall (mm/year)
2003	0.71	0.88	17	264	145	−45	197	107	−46	847
2004	0.54	0.68	14	277	183	−34	206	131	−36	962
2005	0.72	0.86	15	241	138	−43	178	96	−46	772
2006	0.58	0.74	16	449	324	−28	312	225	−28	1234
2007	0.68	0.83	15	276	155	−44	199	107	−46	951
2008	0.66	0.78	12	270	168	−38	201	123	−39	917
2009	0.57	0.68	11	256	178	−30	195	131	−33	887
2010	0.77	0.92	15	261	134	−49	187	97	−48	943
2011	0.58	0.74	16	348	290	−17	258	224	−13	911
2012	0.78	0.87	9	258	205	−21	191	155	−19	866
2013	0.62	0.79	17	316	264	−17	239	204	−15	847

4.2. Analysis of Bayesian Performance

In addition, pixels within the areas of lowest Bayesian performance were assessed using the yearly CHIRPS and bias-corrected (bc) rainfall estimates relative to rain gauge data. Nine pixels from Mt. Elgon, Southern Tanzania, Lake Victoria, and Mt. Kilimanjaro were considered (Figure 1). Figure 8 shows the scatter plots of the uncorrected CHIRPS rainfall estimates and MAE change in percentage after correction. A linear relationship was evident, indicating that as the CHIRPS rainfall estimates increased, the effectiveness of the approach reduced. Further results also showed that for the 11 years of analysis, overestimations were dominant except over Mt. Elgon, which showed isolated incidences of underestimation. CHIRPS overestimates in areas of high rainfall, and this is likely to be attributable to cirrus effects in those areas of deep convection [13].

Table 3 shows a summary of the yearly statistics for the 11-year validation period on Bayesian performance analysis over Mt. Elgon, Southern Tanzania, Lake Victoria, and Mt. Kilimanjaro. These were the areas of largest uncorrected overestimations. The results showed that over the four regions, the lowest MAE in the uncorrected CHIRPS corresponded to the highest percentage of error corrected. This suggests that the retrieval capability of CHIRPS determines the magnitude of errors corrected due to the average error consistencies. Consequently, the MAE errors corresponded to rainfall magnitude, suggesting overestimations increase in years of high rainfall. As such, the relatively dry year of 2005 had substantial MAE errors corrected. Furthermore, observations showed that although each region received different amounts of rainfall in different years, the percentage change in the MAE showed a similar pattern. An example was in the year 2009 over the Mt. Elgon and Lake Victoria areas, where the lowest MAE in CHIRPS rainfall were 49.4 mm/year and 273.5 mm/year, respectively. Consequently, the corrected MAE were the highest (38% and 44%). Similarly, over Southern Tanzania, in 2007, the lowest MAE in CHIRPS was 175.6 mm/year, and corresponded to the highest percentage (50%) of corrected errors. In Mt. Kilimanjaro in 2005, the lowest MAE error in CHIRPS was 1140.4 mm/year and also coincided with the highest percentage error corrected of 23%. These errors of overestimations were attributable to the cold cirrus effect, which infrared based algorithms including CHIRPS consider as precipitating. The Bayesian approach showed little skill in eradicating these errors, partly due to the low misrepresentation of rainfall variability in areas of poor rain gauge distribution in these areas. As such, the rain gauge data were inconsistent in time.

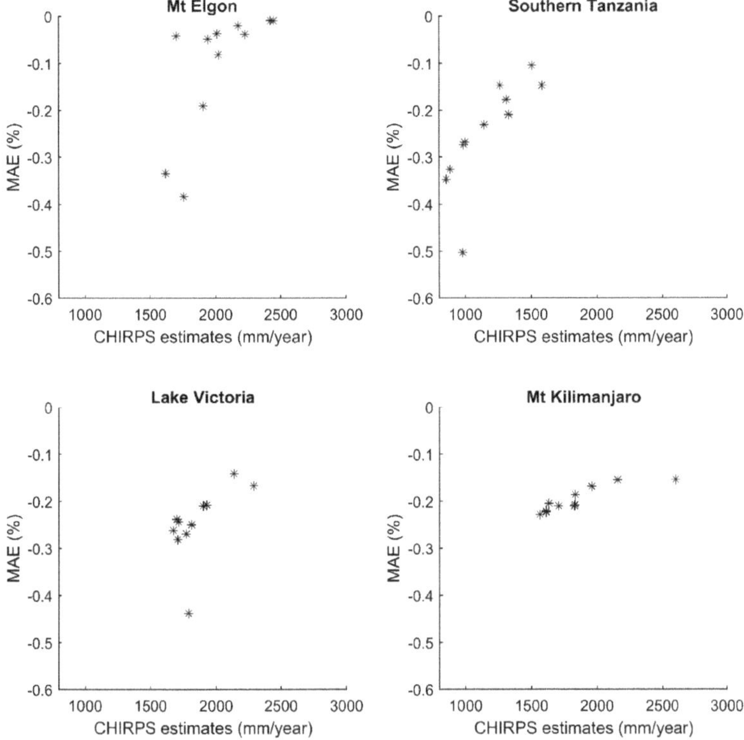

Figure 8. Yearly scatter plots for of pixels in areas of lowest bias correction.

Table 3. Statistics for the yearly Bayesian performance analysis.

	Mt Elgon			Southern Tanzania		
Years	MAE	MAE (bc)	Change (%)	MAE	MAE (bc)	Change (%)
2003	105.2	85.2	−19	270.9	182.6	−33
2004	167.3	160.3	−4	423.0	334.7	−21
2005	56.2	37.4	−33	254.3	166.0	−35
2006	771.7	764.8	−1	601.4	513.0	−15
2007	143.9	137.0	−5	175.6	87.3	−50
2008	113.1	103.9	−8	498.3	410.0	−18
2009	49.4	30.5	−38	383.2	294.8	−3
2010	182.9	176.0	−4	322.6	234.3	−27
2011	347.9	341.0	−2	844.4	756.1	−10
2012	728.5	721.6	−1	329.6	241.3	−27
2013	188.8	181.9	−4	602.8	514.5	−15
	Lake Victoria			Mt Kilimanjaro		
	MAE	MAE (bc)	Change (%)	MAE	MAE (bc)	Change (%)
2003	504.6	384.8	−24	1167.7	907.7	−22
2004	492.6	372.8	−24	1254.7	994.7	−21
2005	459.2	339.4	−26	1140.4	880.5	−23
2006	720.4	600.5	−17	1692.8	1432.8	−15
2007	577.6	457.8	−21	1238.9	978.9	−21
2008	446.3	326.5	−27	1685.8	1425.9	−15
2009	273.5	153.7	−44	1548.5	1288.5	−17
2010	481.2	361.4	−25	1239.0	979.1	−21
2011	851.9	732.0	−14	1399.5	1139.5	−19
2012	572.6	452.8	−21	1272.8	1012.8	−20
2013	427.1	307.3	−28	1178.0	918.0	−22

5. Discussion

This study assessed the Bayesian bias correction on CHIRPS v2 using the regionally gridded rain gauge data provided by the ICPAC over East Africa. Although CHIRPS includes some of the rain gauge data used in this study, they were a small percentage derived mainly from the GTS. The ICPAC gridded dataset incorporates all of the available quality controlled meteorological and hydrological rain gauge data from East African member states. Furthermore, a recent study by [13] on the commonly used satellite products including CHIRPS over the region, showed large discrepancies with ground data, especially in high elevated areas. This study aimed to reduce these errors mainly caused by local effects by using Bayesian methods. The correction was on a monthly scale, which is relevant for long-term applications of 33 years (1981–2013). The usage of the approach assumes the mean errors of the input data are consistent in time. The results from the monthly analysis showed that the approach significantly reduced the systematic errors in all months from January to December with varying magnitudes. The corrected biases were mainly of underestimations affecting the high elevated areas attributable to orographic processes. Overestimations affected the areas around Lake Victoria and low-lying areas to the eastern parts of East Africa and Southern Tanzania. This is attributable to the cirrus effects caused by the assignment by infrared sensors in satellite products of all cold clouds as precipitating. It is suggested the cirrus effect affects areas of poor rain gauge networks areas as they misrepresent rainfall variability. As such, the rain gauge data were inconsistent with time, which is an assumption in this approach, hence the low effectiveness of bias correction over those areas. This observation concurred with that in [35] where at monthly time scales, CHIRPS overestimated low rainfall amounts and underestimated high rainfall amounts. However, the underestimations were associated with local processes, while overestimations were more likely to arise from cirrus effects occurring in deep convection [36]. Consequently, cirrus effects are observed mainly in months of increased rainfall in April and May. The study further revealed that during the southeast and northeast monsoon months of June to September and December to February, low linearity improvements were observed as indicated by corrections change. The inhibition of rainfall further inland from the low-level diffluence of the monsoon winds leads to the dominance of orographic processes [13]. Consequently, the rainfall variability in Eastern Kenya and Southern Tanzania coincided with Turkana low-level jets [33,37], which link the local rainfall variabilities to external influences. These abrupt changes lead to increased rainfall variabilities, hence the reduced linearity assumed in the Bayesian approach.

The yearly analysis showed an increase of the corrected biases attributable to the reduced rainfall variability. Similar to monthly assessments, the underestimations in corrected data were more dominant over the high elevated areas. Similarly, overestimations were observed in areas around Lake Victoria, Mt. Elgon, Eastern Kenya, and Southern Tanzania. These are areas of poor rain gauge distribution, and the misrepresentation of rainfall variabilities led to a reduction in the minimal errors. Consequently, the CHIRPS rainfall estimates approached its uncorrected state. One advantage of the method the incorporation of the error variance in the correction, and the correction only applies when the reference dataset is reliable, and errors are systematic, not random. Although 2006, which was also an El Niño, received the highest rainfall, the percentage of errors corrected was not equivalently high. This suggests that the errors are not only dependent on rainfall magnitude, but on distribution in time. El Niño years occur during the October–December rainfall months, which also have fewer rain days when compared to the March–May rainfall months [38]. Consequently, the years following the El Niño years were observed to have high rainfall and a large percentage of bias were also corrected. This high rainfall after El Niño is suggested to arise during the MAM rainfall months. However, it is worth noting that East Africa has high rainfall variability and the yearly analysis presented here is for the whole region. This observation is essential as the systematic errors occur more with increased rainfall amounts, and also depend on the distribution of rainfall in time and space.

Further annual assessments of the performance of Bayesian approach were conducted on selected pixels centered in areas of large CHIRPS overestimations. Nine pixels from each region over Mt. Elgon, Southern Tanzania, Lake Victoria, and Mt. Kilimanjaro were selected. MAE in percentage change of

CHIRPS and bias-corrected CHIRPS relative to rain gauge data were used to quantify the relationships. Results showed that CHIRPS overestimated in all years (2003–2013) over those areas except for a few incidences of underestimation over Mt. Elgon. Furthermore, the MAE percentage change showed a linear relationship with rainfall magnitudes. An increase in rainfall magnitude corresponded with a decrease in the percentage of MAE change. This showed that there were uncorrected errors during high rainfall periods. It is understandable that cirrus coexists in deep convective systems [36] that are also related to high rainfall amounts. Overestimations in the cirrus effect have also been observed in infrared based products in Eastern Africa [37]. The years 2005, 2007, and 2009 showed the lowest annual rainfall in these regions, and consequently, the highest MAE percentage change. The designation of CHIRPS is for dry conditions; hence the mean errors are consistent in time, hence an increased Bayesian performance. Similar findings in a recent study [38] showed that CHIRPS had low rainfall detection during the wet season. The Bayesian approach showed low skill in reducing errors attributed to cirrus effects in areas of poor rain gauge network due to the misrepresentation of rainfall variability.

6. Conclusions

Rain gauge distributions are on the decline, and satellite rainfall estimates are increasingly used to complement the sparse network. However, rain gauges continue to be used as reference data to validate the incoming satellite products as they offer the most direct information on local rainfall variabilities. This is understandable in developing countries where other ground observations like weather radar are too expensive. However, considering the decreasing trend in rain gauge density, other measures like gridding the available gauge data are preferred. As such, validation exercises are enhanced, which may otherwise become increasingly compromised in the future. This impact may be much more severe for long-term data applications. This study assessed the applicability of the Bayesian approach with an existing historical rain gauge dataset to create consistent satellite rainfall data for long-term applications. The gridded rain gauge dataset was developed by ICPAC to safeguard the decreasing rain gauge information over East Africa.

CHIRPS was chosen based on its close correspondence with rain gauge datasets when compared to other commonly used products over the region [13]. However, CHIRPS, like other satellite-derived products, exhibit systematic errors that vary from place to place. In mountainous areas like East Africa, the retrieval of topographic processes is challenging. These biases need to be reduced to increase the effectiveness of this product over the region. It is understandable that high elevated areas play a significant role in water availability and minimizing these errors would improve the representative of the CHIRPS estimates. Furthermore, as CHIRPS has a high (0.05°) spatial scale, it can be used for future reference for the incoming coarser satellite products. The study aimed at temporally and spatially evaluating how CHIRPS rainfall estimates compared with the gridded rain gauge data in magnitude and distributions after bias correction. For the Bayesian correction, a historical dataset of 22 years of data was used as the calibration dataset, and the ensuing 11 years were used for the validation.

Monthly analysis showed that CHIRPS estimates had systematic errors mainly of underestimations arising in high elevated areas and their surroundings. The approach adequately reduced them in both wet and relatively dry months. The study revealed large biases emerged during the wet months and less so for the relatively drier months. This is understandable as the CHIRPS algorithm has been optimized for drought monitoring, that is, for better accuracy during drier periods. As such, in April, which is the peak rainfall month of the March to May rainfall season, the biases were distributed over East Africa. In May, they were observed near the Lake Victoria region and coastal areas of Kenya where high rainfall is experienced during this month. During the south-east monsoon month of August, they were situated on the border between Kenya and Uganda following the highest rainfall amounts in the June through August rainfall season. However, low linearity improvements of the negative biases were observed during the southeast and northeast monsoon months in June to September and December to February, respectively. These were attributable to indirect rainfall variability related to external factors in high elevated areas. The study revealed that these monsoons influenced orographic processes, reducing further inland impacts.

Furthermore, the Bayesian approach assumes linear relationships between rainfall variability and the mean errors. Overestimations were dominant over Lake Victoria, Southern Tanzania, Mt. Elgon, Eastern Kenya, and the Kilimanjaro region. These biases were attributable to cirrus effects, which in IR based products (like CHIRPS) often results in an overestimation of precipitation. Adequate bias reduction was observed except in a few areas over Southern Tanzania and part of Lake Victoria where misrepresentation due to the sparse rain gauge network affected the correction. It is worth noting that inconsistencies of average errors, due to sparse rain gauge distributions were attributable to CHIRPS large uncorrected errors in those areas. A remarkable difference in the approach to other methods is its ability to leave the satellite estimates close to the uncorrected state when the reference data is inconsistent in time. The overall monthly average reduction indicates that the RMSD reduced between 4% and 60%, and the MAE between 1% and 63%, while the correlations improved by up to 21%, as shown in Table 1.

The yearly analysis showed an increase in the corrected biases which is associated to the reduced rainfall variability at this scale. Similar to monthly findings, the corrected underestimations were more dominant on high elevated areas and were in areas of high rainfall magnitudes that showed seasonality. The study observed that the corrected biases were more during years of high rainfall magnitude. Although large biases were evident during El Niño year of 2006 which also has highest rainfall amount, not all El Niño years showed similar results. This is understandable as El Niño years occur during the October–December rainfall months, but its effect spreads to the March–May rainfall months. As a result, the year that followed was observed to have high rainfall, hence an increased in the biases corrected. An example is shown in the 2003, 2006, and 2009 El Niño years whereby an increased rainfall average amount was observed a year after in 2004, 2007, and 2010. Consequently, a higher percentage of bias were corrected. This observation is vital as the systematic errors are increased with increased rainfall amounts, but are also dependent on the distribution of the rainfall in time. The overall yearly average reduction indicated that the RMSD was reduced by between 17% and 49%, and the MAE was reduced by between 13% and 48%, while the correlations increased by between 9% and 17%, as shown in Table 2.

Further assessments of the performance of the Bayesian approach were carried out using selected pixels (Figure 1) in areas of substantial uncorrected bias. These areas included Mt. Elgon, Southern Tanzania, Lake Victoria, and the Mt. Kilimanjaro areas. Nine pixels from the areas with the largest overestimations were used for the extraction of average rainfall for CHIRPS, and bias-corrected CHIRPS (bc) and compared with the corresponding rain gauge dataset. The percentage change in MAE quantified their relationships. This analysis indicated that CHIRPS considerably overestimated rainfall in those areas of sparse rain gauge distributions. The highest corrected MAE were observed in the years of the lowest MAE. Observations showed that the years 2005, 2007, and 2009 had the lowest annual rainfall in these regions, and consequently, the highest MAE decrease. Although each region received different amounts of rain over the validation period, the errors in corrected estimates showed a similar trend. The lowest MAE before correction corresponded with the highest percentage of corrected errors spatially and temporally. This can be understood as shallow convections in cirrus effects are minimal. Conversely, in the years and areas of deep convections, cirrus clouds increase. However, in areas of relatively dense rain gauges, cirrus related errors are minimized. For example, in the year 2009 over the Mt. Elgon and Lake Victoria areas, the lowest MAEs observed before correction were 49.5 mm/year and 273.5 mm/year, respectively. Consequently, those were the areas of highest corrected MAE, at 38% and 44%, respectively. Similarly, over Southern Tanzania, in the year 2007, the lowest MAE was 175.6 mm/year, and was consistent with the highest corrected MAE of 50%. In Mt. Kilimanjaro, in 2005, the lowest MAE error was 1140.6 mm/year and coincided with the highest percentage improvement of 23%. Therefore, overcorrections were adequately addressed spatially and temporally when rainfall was low in areas of sparse rain gauges, but remain unaddressed in deep convection attributed to monsoons. This effect was minimal in densely distributed rain gauge areas due to the increased representation of rainfall variability.

The Bayesian approach is recommended to adjust rainfall estimates of other algorithms with long-term historical records that are representative of rainfall variability of the area under consideration. The method is suitable for the reduction of systematic errors, especially in high elevated regions. It is, however, sensitive to sparse rain gauge distributions in regions of deep convective systems due to cirrus effects that cause overestimations. Therefore, with a good quality rain gauge network, the approach works better in those areas. It is recommended to have the reference data of a higher or equal spatial scale for higher rainfall spatial representation.

Author Contributions: All three authors contributed in concept building and manuscript preparation. M.W.K. analyzed the data and drafted the manuscript. Z.S. and J.C.B.H. provided conceptual advice and contributed to the overall writing and approval of the final manuscript.

Acknowledgments: This work was made possible through the funding provided by the Netherlands Fellowship Programs (NFP), and we greatly appreciate their support. The University provided funds for open access.

Conflicts of Interest: The authors declare no conflict of interest.

References

1. Kidd, C.; Bauer, P.; Turk, J.; Huffman, G.J.; Joyce, R.; Hsu, K.L.; Braithwaite, D. Intercomparison of high-resolution precipitation products over northwest Europe. *J. Hydrometeorol.* **2012**, *13*, 67–83. [CrossRef]

2. Villarini, G.; Krajewski, W.F. Empirically-based modeling of spatial sampling uncertainties associated with rainfall measurements by rain gauges. *Adv. Water Resour.* **2008**, *31*, 1015–1023. [CrossRef]

3. Dibike, Y.B.; Coulibaly, P. Validation of hydrological models for climate scenario simulation: The case of Saguenay watershed in Quebec. *Hydrol. Process.* **2007**, *21*, 3123–3135. [CrossRef]

4. Sorooshian, S.; Hsu, K.L.; Gao, X.; Gupta, H.V.; Imam, B.; Braithwaite, D. Evaluation of persiann system satellite-based estimates of tropical rainfall. *Bull. Am. Meteorol. Soc.* **2000**, *81*, 2035–2046. [CrossRef]

5. Feidas, H.; Lagouvardos, K.; Kotroni, V.; Cartalis, C. Application of three satellite techniques in support of precipitation forecasts of a NWP model. *Int. J. Remote Sens.* **2005**, *26*, 5393–5417. [CrossRef]

6. Krajewski, W.F.; Ciach, G.J.; McCollum, J.R.; Bacotiu, C. Initial validation of the global precipitation climatology project monthly rainfall over the United States. *J. Appl. Meteorol.* **2000**, *39*, 1071–1086. [CrossRef]

7. Adler, R.F.; Huffman, G.J.; Chang, A.; Ferraro, R.; Xie, P.; Janowiak, J.; Arkin, P. The version 2 global precipitation climatology project (GPCP) monthly precipitation analysis. *J. Hydrometeorol.* **2003**, *4*, 1147–1167. [CrossRef]

8. Arkin, P.A.; Xie, P.P. The global precipitation climatology project—1st algorithm intercomparison project. *Bull. Am. Meteorol. Soc.* **1994**, *75*, 401–419. [CrossRef]

9. Maidment, R.I.; Grimes, D.I.F.; Allan, R.P.; Greatrex, H.; Rojas, O.; Leo, O. Evaluation of satellite-based and model re-analysis rainfall estimates for Uganda. *Meteorol. Appl.* **2013**, *20*, 308–317. [CrossRef]

10. Tote, C.; Patricio, D.; Boogaard, H.; van der Wijngaart, R.; Tarnavsky, E.; Funk, C. Evaluation of satellite rainfall estimates for drought and flood monitoring in Mozambique. *Remote Sens.* **2015**, *7*, 1758–1776. [CrossRef]

11. AghaKouchak, A.; Mehran, A.; Norouzi, H.; Behrangi, A. Systematic and random error components in satellite precipitation data sets. *Geophys. Res. Lett.* **2012**, *39*. [CrossRef]

12. Abera, W.; Brocca, L.; Rigon, R. Comparative evaluation of different satellite rainfall estimation products and bias correction in the Upper Blue Nile (UBN) basin. *Atmos. Res.* **2016**, *178*, 471–483. [CrossRef]

13. Kimani, M.W.; Hoedjes, J.C.B.; Su, Z.B. An assessment of satellite-derived rainfall products relative to ground observations over East Africa. *Remote Sens.* **2017**, *9*, 430. [CrossRef]

14. Maurer, E.P.; Pierce, D.W. Bias correction can modify climate model simulated precipitation changes without adverse effect on the ensemble mean. *Hydrol. Earth Syst. Sci.* **2014**, *18*, 915–925. [CrossRef]

15. Mateus, P.; Borma, L.S.; da Silva, R.D.; Nico, G.; Catalao, J. Assessment of two techniques to merge ground-based and trmm rainfall measurements: A case study about Brazilian Amazon rainforest. *GISci. Remote Sens.* **2016**, *53*, 689–706. [CrossRef]

16. Vila, D.A.; de Goncalves, L.G.G.; Toll, D.L.; Rozante, J.R. Statistical evaluation of combined daily gauge observations and rainfall satellite estimates over continental South America. *J. Hydrometeorol.* **2009**, *10*, 533–543. [CrossRef]

17. Tian, Y.D.; Peters-Lidard, C.D.; Eylander, J.B. Real-time bias reduction for satellite-based precipitation estimates. *J. Hydrometeorol.* **2010**, *11*, 1275–1285. [CrossRef]

18. Stokstad, E. Hydrology—Scarcity of rain, stream gages threatens forecasts. *Science* **1999**, *285*, 1199–1200. [CrossRef]

19. Dinku, T.; Ceccato, P.; Grover-Kopec, E.; Lemma, M.; Connor, S.J.; Ropelewski, C.F. Validation of satellite rainfall products over East Africa's complex topography. *Int. J. Remote Sens.* **2007**, *28*, 1503–1526. [CrossRef]

20. Talagrand, O. Assimilation of observations, an introduction (gtspecial issueltdata assimilation in meteology and oceanography: Theory and practice). *J. Meteorol. Soc. Jpn. Ser. II* **1997**, *75*, 191–209. [CrossRef]

21. Seiz, G.; Foppa, N. National climate observing system of switzerland (GCOS Switzerland). *Adv. Sci. Res.* **2011**, *6*, 95–102. [CrossRef]

22. Funk, C.C.; Peterson, P.J.; Landsfeld, M.F.; Pedreros, D.H.; Verdin, J.P.; Rowland, J.D.; Romero, B.E.; Husak, G.J.; Michaelsen, J.C.; Verdin, A.P. *A Quasi-Global Precipitation Time Series for Drought Monitoring*; U.S. Geological Survey Data Series: Reston, VA 20192, USA, 2014.

23. IGAD. Available online: http://www.Icpac.Net/ (accessed on 6 November 2015).

24. Funk, C.; Nicholson, S.E.; Landsfeld, M.; Klotter, D.; Peterson, P.; Harrison, L. The centennial trends greater horn of Africa precipitation dataset. *Sci. Data* **2015**, *2*, 150050. [CrossRef] [PubMed]

25. ICPAC. Available online: http://chg-wiki.geog.ucsb.edu/wiki/GeoCLIM (accessed on 6 November 2015).

26. SRTM 90m Digital Elevation Database v4.1. Available online: http://www.Cgiar-csi.Org/data/srtm-90m-digital-elevation-database-v4-1 (accessed on 5 July 2018).

27. Carlin, B.P.; Louis, T.A. *Bayes and Empirical Bayes Methods for Data Analysis*; Chapman and Hall: London, UK, 1996; p. 399.

28. Suwendi, A.; Allebach, J.P. Nearest-neighbor and bilinear resampling factor estimation to detect blockiness or blurriness of an image. *J. Electron. Imaging* **2008**, *17*, 023005.

29. Tan, M.L.; Ficklin, D.L.; Dixon, B.; Ibrahim, A.L.; Yusop, Z.; Chaplot, V. Impacts of DEM resolution, source, and resampling technique on SWAT-simulated streamflow. *Appl. Geogr.* **2015**, *63*, 357–368. [CrossRef]

30. Taylor, K.E. Summarizing multiple aspects of model performance in a single diagram. *J. Geophys. Res. Atmos.* **2001**, *106*, 7183–7192. [CrossRef]

31. Nicholson, S. The turkana low-level jet: Mean climatology and association with regional aridity. *Int. J. Climatol.* **2016**, *36*, 2598–2614. [CrossRef]

32. Kinuthia, J.H.; Asnani, G.C. A newly found jet in North Kenya (Turkana Channel). *Mon. Weather Rev.* **1982**, *110*, 1722–1728. [CrossRef]

33. Trejo, F.J.P.; Barbosa, H.A.; Penaloza-Murillo, M.A.; Moreno, M.A.; Farias, A. Intercomparison of improved satellite rainfall estimation with chirps gridded product and rain gauge data over Venezuela. *Atmosfera* **2016**, *29*, 323–342. [CrossRef]

34. Sassen, K.; Wang, Z.; Liu, D. Cirrus clouds and deep convection in the tropics: Insights from CALIPSO and CloudSat. *J. Geophys. Res. Atmos.* **2009**, *114*. [CrossRef]

35. Kinuthia, J.H. Horizontal and vertical structure of the lake Turkana jet. *J. Appl. Meteorol.* **1992**, *31*, 1248–1274. [CrossRef]

36. Indeje, M.; Semazzi, F.H.M.; Ogallo, L.J. Enso signals in East African rainfall seasons. *Int. J. Climatol.* **2000**, *20*, 19–46. [CrossRef]

37. Young, M.P.; Williams, C.J.R.; Chiu, J.C.; Maidment, R.I.; Chen, S.H. Investigation of discrepancies in satellite rainfall estimates over Ethiopia. *J. Hydrometeorol.* **2014**, *15*, 2347–2369. [CrossRef]

38. Paredes-Trejo, F.J.; Barbosa, H.A.; Kumar, T.V.L. Validating chirps-based satellite precipitation estimates in northeast Brazil. *J. Arid Environ.* **2017**, *139*, 26–40. [CrossRef]

remote sensing

Article

Confronting Soil Moisture Dynamics from the ORCHIDEE Land Surface Model With the ESA-CCI Product: Perspectives for Data Assimilation

Nina Raoult [1,*], Bertrand Delorme [1,†], Catherine Ottlé [1], Philippe Peylin [1], Vladislav Bastrikov [1], Pascal Maugis [1] and Jan Polcher [2]

[1]   Laboratoire des Sciences du Climat et de l'Environnement, IPSL, 91190 Saint-Aubin, France; bdelorme@stanford.edu (B.D.); cottle@lsce.ipsl.fr (C.O.); peylin@lsce.ipsl.fr (P.P.); vladislav.bastrikov@lsce.ipsl.fr (V.B.); pascal.maugis@lsce.ipsl.fr (P.M.)
[2]   Laboratoire de Météorologie Dynamique (LMD, CNRS), IPSL, Ecole Polytechnique, 91128 Palaiseau, France; jan.polcher@lmd.jussieu.fr
*   Correspondence: nina.raoult@lsce.ipsl.fr
†   Current address: Department of Earth System Science, Stanford University, Stanford, CA 94305, USA.

Received: 18 July 2018; Accepted: 6 November 2018; Published: 10 November 2018

Abstract: Soil moisture plays a key role in water, carbon and energy exchanges between the land surface and the atmosphere. Therefore, a better representation of this variable in the Land-Surface Models (LSMs) used in climate modelling could significantly reduce the uncertainties associated with future climate predictions. In this study, the ESA-CCI soil moisture (SM) combined product (v4.2) has been confronted to the simulated top-first layers/cms of the ORCHIDEE LSM (the continental part of the IPSL Earth System Model) for the years 2008-2016, to evaluate its potential to improve the model using data assimilation techniques. The ESA-CCI data are first rescaled to match the climatology of the model and the signal representative depth is selected. Results are found to be relatively consistent over the first 20 cm of the model. Strong correlations found between the model and the ESA-CCI product show that ORCHIDEE can adequately reproduce the observed SM dynamics. As well as considering two different atmospheric forcings to drive the model, we consider two different model parameterizations related to the soil resistance to evaporation. The correlation metric is shown to be more sensitive to the choice of meteorological forcing than to the choice of model parameterization. Therefore, the metric is not optimal in highlighting structural deficiencies in the model. In contrast, the temporal autocorrelation metric is shown to be more sensitive to this model parameterization, making the metric a potential candidate for future data assimilation experiments.

Keywords: surface soil moisture; land-surface model; satellite data; representative depth; temporal autocorrelation

---

1. Introduction

Land-surface models (LSMs) are crucial components of climate models, representing the matter and energy transfers at the continental interface with the atmosphere. They therefore play a key role on the climate model simulations, past, present, and future predictions. Within LSMs, surface soil moisture (SM) heavily influences the hydrological interactions between soil, vegetation, and atmosphere. Because of the coupling between water and carbon cycles at the leaf and soil levels, it is a major constraint for the assimilation of carbon by the vegetation through photosynthesis [1], whose net global value is still largely debated [2–4]. Moreover, SM governs the partition between runoff and infiltration and is thus a key element of the global water cycle. Consequently, correctly representing this variable in LSMs could highly improve both short- and long-term forecasts [5–9].

Despite considerable improvements in recent years, estimated water and carbon fluxes in LSMs are still affected with biases and large uncertainties. Three different sources of uncertainty prevail in LSMs: the external forcing, model structure and its parameterization. More specifically, simulated SM exhibit large sensitivities to meteorological forcing data and to land-surface parameterizations [10]. Since SM variability is largely driven by precipitation, the meteorological forcing data used to run the LSM could have a greater weight on the comparison scores than the parameterization of the model itself [11].

Simulated SM depends on the model version and configuration used. Since it is impossible to accurately model and represent all the land-surface processes in an LSM due to lack of knowledge and computing limitations, choices are made to determine which processes are included and how they are described in the model. These choices contribute to the structure of the model. For example, how one chooses to discretize the upper soil layers will affect the soil moisture dynamics [12], similarly the inclusion of a soil resistance term for water transfer will greatly affect evapotranspiration. It is important to separate the different sources of errors to understand whether the deficiencies in the model process are due to the forcing used or the model structure and configuration. To improve the representation of a variable such as SM in LSMs, the model can be confronted with the observations. These observations can be measured in situ, i.e., locally, or remotely, for example, from space instruments. Since soil moisture is difficult to observe at large scales due to its high spatial and temporal variability, satellite data can provide much needed consistent and comparable global datasets. Extra care needs to be taken with SM data since it is not directly observed but rather estimated using backscattering signals and brightness temperatures through different inversion algorithms.

Over the last decade, the quality and quantity of the Earth Observations has vastly improved and has been used to evaluate the performance of LSMs over different climatology and hydrological environments. For example, GRACE (Gravity Recovery and Climate Experiment) data has been used to evaluate land water storage in several LSMs (e.g., ORCHIDEE; [13]). GRACE has been used to evaluate the CLM (Community Land Model) over Africa, specifically to assess a soil resistance parameterization in the model [14]. Ref. [15] evaluate the CLM and GLDAS models over Africa using GRACE and other remote sensing data to highlight discrepancies between the models' evapotranspiration outputs. Other remote sensing products used to evaluate the representation of soil moisture in LSMs include SMOS (Soil Moisture and Ocean Salinity, [16]), SMAP (Soil Moisture Active Passive, [17]) and ASCAT (Advanced Scatterometer, [18]). These have been used to evaluate LSMs over a wide range of climatologies e.g., [19] consider ASCAT and SMOS over Europe and Northern Africa, and [20] use SMAP assess storm event runoff over America. Satellite retrievals have also allowed global comparison (e.g., [21]). Of course, there are several other remote-sensed products used to evaluate LSMs, such as biomass, albedo and NDVI (normalized difference vegetation index); however, these are beyond the scope of this study.

As well as evaluating LSMs, satellite retrieval can be used to improve models through data assimilation (DA). Several recent studies have used data DA to improve the representation of SM in LSMs, by either optimizing initial conditions or the model parameters e.g., [22–27], and therefore improve LSMs ability to forecast events such as droughts and floods e.g., [23,24,26,28]. SM data can also be used to improve other components of the LSM such as the carbon cycle [29]. Furthermore, DA can be used to reveal structural errors in the model [30,31]. However, before attempting to improve the model through DA, a thorough comparison of the observations with the model output is needed. Such a comparison allows us to assess whether the physical properties in both the observations and model output are comparable, and to select which model configuration and forcing to use at a later stage. This allows to evaluate whether the model is sophisticated enough and whether the observations can improve the model through assimilation or not. It then helps determining the most appropriate metrics to use in an objective function to be optimized which will depend on the aspect of the model under focus. In addition, most DA methods rely on Gaussian statistics. Therefore, a proper assessment of the satellite and model errors is needed, with potential bias correction, before any assimilation step.

The ESA-CCI multi-instruments dataset provides the SM observations used in this study. Thanks to its large spatio-temporal coverage, the ESA-CCI product offers a good opportunity to benchmark LSMs and improve their parametrization with DA techniques [32]. Combining several products from a broad range of instruments allows to get a large spatio-temporal coverage, which is suitable for global LSMs evaluations, and provides additional information at these larger scales compared to in situ observations in a global assimilation process. Moreover, active and passive instruments have been found to complement each other as they are affected differently by vegetation density (active instruments perform better over moderate vegetation areas whereas passive instruments give more reliable estimates over arid regions) and radio frequency interference contamination which alters L-band instruments measures [21]. The SMOPS soil moisture product [33] is another example of a blended soil moisture data set; however, this product does not cover as many years as the ESA-CCI product.

There are some known errors linked to the satellite data themselves which need to be considered (see [32]). For example, the observations can be spatially inhomogeneous due to the satellite's retrieval heavy dependence on surface and sub-surface properties such as soil texture and vegetation. Similarly, the satellite data suffer from temporal inhomogeneity because of instrument drift and lifespan issues. The sampling of the observations is not continuous and in the case of the ESA-CCI SM product where each independent dataset has been rescaled to be merged together, combining different instruments from different periods adds temporal inhomogeneity to the final dataset and thus complications for model evaluations.

The quality of the ESA-CCI SM product has already been evaluated on a global scale by several studies with encouraging results. The merging of active and passive products has been shown to increase the number of observations and hence coverage, while keeping the same relative dynamics and minimally changing the accuracy [23,32,34,35]. By comparing the blended product to reanalysis and in situ measurements, [36] found that the product was able to capture the annual cycle of SM and its short-term variability. The quality of the product has been shown to increase with time due to the addition of new satellites and methods use to merge them [32]. A comprehensive list of studies using the ESA-CCI SM product can be found in [32].

In this study, the ESA-CCI SM product is compared with the process-based global LSM ORCHIDEE (ORganizing Carbon and Hydrology In Dynamic EcosystEm, [37]) at the global scale. ORCHIDEE has been recently endowed with a new 11 layers hydrological model which allows to refine the representation of soil water transfers. This discretization scheme offers a unique opportunity to match as closely as possible the representative depth of the satellite instrument products. While the paper focuses on a comparison of the soil moisture dynamics simulated by ORCHIDEE land-surface model with ESA-CCI soil moisture product, the analysis is done with a future data assimilation experiment in mind. Even though no data assimilation is performed in this work, this paper serves as an important preliminary study assessing the following more specific objectives:

- How should the satellite product be processed (bias corrected) for a meaningful comparison with the ORCHIDEE LSM?
- What are the impacts of the selected model representative soil depth and the meteorological forcing on the comparison between the satellite and model SSM?
- What are the strengths and weaknesses of a few (widely used) temporal comparison metrics? What is the potential of these metrics for future model structure/parameter optimization?

2. Methods and Data

2.1. The ORCHIDEE Land-Surface Model

The ORCHIDEE LSM [37] allows the representation as explicit as possible of processes governing water, carbon, and energy budgets of the terrestrial biosphere. It has a high spatial resolution flexibility (based on the meteorological forcing spatial resolution) and a temporal resolution of 30 min. ORCHIDEE conceptually includes (i) a Soil Vegetation Atmosphere Transfer (SVAT) scheme, which calculates energy, water and carbon fluxes between soil, plant and atmosphere; (ii) a module describing carbon flow within ecosystems (plants and soils) and (iii) a sub-model describing the long-term dynamics of vegetation which is not activated in this study. We use the main version of ORCHIDEE (Trunk, revision 4783) that corresponds to the version used in the IPSL earth system model for the ongoing CMIP6 (Coupled Model Intercomparison Project Phase 6) climate simulations. This version has recently been improved with a multi-layer hydrological model which should be ideally suited to confront remotely sensed and simulated soil moisture products, thanks to its high vertical resolution [38–40]. In this new scheme, hydrology is resolved using the Richards equation over the first 2 m of soil with an 11-layer discretization. Increasing grid spacing is used throughout the soil column, allowing a finer discretization at the top of the column (Table 1). The Richards equation describes the diffusion of water according to the soil properties (prescribed following a global texture map). The SM at different levels is thus computed at each time step as influenced by rain infiltration, evaporation, transpiration, and drainage. For the bottom boundary condition, three options are possible: 1- free drainage (under the hypothesis that hydraulic head is constant beneath the last layer), 2- no drainage and 3- a mixed option between the two previous ones. Free drainage is used as default. Concerning the upper boundary condition, bare soil evaporation in ORCHIDEE is the minimum between potential evaporation and the hydrological flux allowed by diffusion. Then, soil water infiltration is obtained by solving the Green-Ampt equation which represents the evolution of the wetting front through the different soil layers.

Table 1. Table showing the depths of the 11-layer discretization based on work by [39] and adapted from [41].

Layer	Layer Thickness (m)	Integrated Depth (m)
1	0.001	0.001
2	0.003	0.004
3	0.006	0.010
4	0.012	0.022
5	0.023	0.045
6	0.047	0.092
7	0.092	0.186
8	0.188	0.374
9	0.375	0.750
10	0.750	1.500
11	0.500	2.000

In ORCHIDEE, vegetation is represented by 15 plant functional types (PFTs). To define soil properties, these PFTs are aggregated into 3 groups classified as bare soil, low vegetation (grasses and crops) and high vegetation (forests). This allows to divide each grid box into 3 tiles for which an independent hydric budget is calculated, by using the 11-layer physically based hydrology scheme. In the analysis presented here only the grid box weighted average SM profile (based on the fraction of vegetation) is considered.

2.2. ESA-CCI SM Product

The ESA-CCI SM project is part of the ESA (European Space Agency) Program on Global Monitoring of Essential Climate Variables (ECV), better known as the Climate Change Initiative (CCI) and initiated in 2010. A first version of the ESA-CCI SM product was released in June 2012 by the Vienna University of Technology [42]. In this study, we use the new version of the global ESA-CCI SM product (v4.2), released in January 2018 and freely available at http://esa-soilmoisture-cci.org/, that extends the data record from November 1978 to December 2016.

ESA-CCI SM is based on a series of active and passive microwave satellite sensors. Three datasets are provided: a merged passive-only (produced from multi-frequency radiometers), a merged active-only (produced from C-band scatterometers) and a combined active and passive SM product. In this new version, the exploited passive instruments are Scanning Multichannel Microwave Radiometer (SMMR), Special Sensor Microwave/Imager (SM/I), Tropical Rainfall Measuring Mission Microwave Imager (TMI), Advanced Microwave Scanning Radiometer for Earth Observation System (ASMR-E) and, added more recently, AMSR-2 and WindSat. The algorithm developed by the Vrije Universiteit Amsterdam in collaboration with the National Aeronautics and Space Administration (VUA-NASA) has been used to retrieve SM from remote-sensed measurements [43]. The active instruments exploited in this new version are the C-band scatterometers on board of the European Remote Sensing (ERS-1&2) and the Advanced SCATterometer (ASCAT). The change detection algorithm developed by the Vienna University of Technology (TU-Wien) has been applied here to retrieve SM from remote-sensed measurements [18,44]).

To provide a global coverage, the active and passive retrievals were combined. Since data provided by the retrievals have different units, to merge them, the SM values need to be rescaled into a common climatology. Therefore, data were scaled by GLDAS-Noah surface soil moisture product (v1, soil depth 10 cm) by a direct CDF-matching of the collocated observations [45]. The final combined ESA-CCI SM dataset is based on the spatial distribution of the temporal correlation between active-only and passive-only datasets. Over regions with correlation coefficients greater than 0.65, a simple average between active and passive measurements is used. Outside these regions, the merged passive-only dataset is used over areas with sparse vegetation density and the merged active-only dataset is used over areas with moderate vegetation density.

The ESA-CCI SM product provides daily SM, with a spatial resolution of 0.25 degrees, together with quality flags indicating the possible presence of dense vegetation, snow, or a temperature below 0 °C. For our study, we used only the final combined dataset which is provided in volumetric units (m^3/m^3) and furnishes the largest spatio-temporal coverage of the three ESA-CCI SM datasets available.

2.3. Simulation Set Up

2.3.1. ORCHIDEE Configurations

In this study, the CMIP6 version of the ORCHIDEE model is run globally at a resolution of 0.5°. Using the half-hourly moisture content output, daily values of soil moisture are calculated using the weighted average of the 3 hydric budgets for soil columns associated with bare soil, forests, and grasslands PFTs. The distribution of PFTs and soil textures are determined according to some ancillary files created for CMIP6.

There is an option to account for a soil resistance term in ORCHIDEE which is used for the calculation of soil evaporation rates. Namely, it affects bare soil evaporation E_g which is calculated using the following equation:

$$E_g = \min(E_{pot}/(1 + r_{soil}/r_a), Q) \tag{1}$$

where E_{pot} denotes the potential evaporation, r_a denotes the aerodynamic resistance and Q is the supply of water by diffusion in the soil. When $r_{soil} = 0$, the resistance term disappears and E_g is simply the minimum of the potential evaporation and the supply. The soil resistance term itself is calculated following the formulation of [46]

$$r_{soil} = \exp(8.2 - 4.3 \cdot W_1) \tag{2}$$

where W_1 is the wetness of the first 4 layers of the soil (2.2 cm) found by dividing the soil moisture in these layers by the corresponding moisture at saturation. Most of our analysis will be conducted over these first 4 layers of the model output. The coefficients in this equation have been determined in [46]'s study and are therefore specific to their model. These values may need to be reconsidered or indeed calibrated for the ORCHIDEE model later.

This soil resistance term is included by default in our simulations, with the exception of one comparative run where r_{soil} is set to zero.

2.3.2. Meteorological Forcing

We use two different meteorological forcings to drive ORCHIDEE in this study. The first is the CERASAT forcing data, used by default in our simulations. This is a 3-hourly climate reanalysis dataset spanning 8 years between 2008 and 2016 generated by ECMWF within the ERACLIM2 project. It has been produced specifically for the satellite era [47]. The second, used for a comparative run, is CRUNCEP, a 6-hourly 0.5° global forcing product (1901–2015) which is a combination of the annually updated CRU-TS monthly climate dataset [48] and NCEP climate reanalysis [49]. Rainfall, cloudiness, relative humidity, and temperature are taken from the CRU while the other fields (pressure, longwave radiation, windspeed) were directly derived from NCEP. The forcings are interpolated to the half-hourly time step of the model.

In addition, Multi-Source Weighted-Ensemble Precipitation (MSWEP; [50]) data is used during the analysis step. The MSWEP data have a 0.25° spatial resolution and a 3-hourly temporal resolution. This forcing is independent from the two datasets used to drive the ORCHIDEE model, as well as the forcing data used to generate the GLDAS reanalysis, used in scaling the ESA-CCI product.

2.3.3. Land Cover

ORCHIDEE uses land-cover maps to prescribe the distribution of vegetation used in the simulation. By default, we use the yearly land-cover maps created for the CMIP6 release of the model. These maps have been created through aggregation of the latest ESA-CCI land-cover map for 2010 (v2.0.7b) at 300 m resolution from 38 ESA land-cover classes into 15 ORCHIDEE PFTs at 0.5° resolution [51,52]. For producing historical PFT maps (1860–2010) the resulting map has been further merged with the LUH2 dataset (Land-Use Harmonization, http://luh.umd.edu/). The overall protocol of the ORCHIDEE PFT maps production is outlined here: https://orchidas.lsce.ipsl.fr/dev/lccci/.

For our comparative run, an older version of the land-cover maps created for the CMIP5 LSM generation is used. These maps differ from the CMIP6 maps by the initial ESA-CCI land-cover map used (the epoch map for 2010 has been used, v1.6.1), and the protocol for merging the ESA data with LUH2 dataset has been slightly improved since that time.

2.3.4. Model Spin Up

The ORCHIDEE simulations include a spin up phase, that brought the prognostic variables including vegetation state and soil moisture content into equilibrium. The protocol follows that from the TRENDY model intercomparison (see http://dgvm.ceh.ac.uk/node/21/) with several hundred years of spin up, recycling the first 10 years of each meteorological forcing. Please note that such protocol includes a transient phase with increasing atmospheric CO_2 concentration and changes in land cover, as it is primarily designed for the spin up of long-term soil carbon stock variables (not necessary for this paper). The spin up is followed by a historical simulation over the period of the ESA-CCI SM product.

2.3.5. Model Data Comparison

Both the model output and the observations used in the comparison were upscaled to the same spatial and temporal resolution (i.e., daily and global using a resolution of 0.5°), making it is possible to match up the time series at each grid point. The observed values with the quality flag indicating dense vegetation, snow, or freezing temperature were disregarded, as well as the modelled data corresponding to these points. Similarly, points are masked when the modelled surface temperature is less than 0 °C, suggesting snow cover or frozen soil moisture values at that time and location.

Satellite SM observations represent only the top few centimeters of the soil column. From the eleven soil moisture layers simulated by ORCHIDEE (as described in Table 1), the first 4 layers spanning the top 2.2 cm of soil are most closely related to the instruments theoretical global mean sensing depth of 2 cm [32]. We thus choose this depth as the reference and averaged the model SM over the 4 top layers for the comparison. In addition, we also tested the impact of choosing different depths. The first 6 layers span the top 9.2 cm of the soil which is closer the GLDAS-Noah model depth simulation (the GLDAS-Noah model is used in combining the different products used to form ESA-CCI SM data). Additionally, the in situ data used in several studies to benchmark the satellite data is typically 5 cm which is closer to the integrated depth of 4.5 cm (layer 5). This raises the question of how to link observations to model variables and thus motivates the sensitivity analysis on the selected model depth, reported in Section 3.2.

2.3.6. Performed Simulations

The different simulations used in this study are outlined in Table 2. Since CERASAT only covers 2008-2016, our analysis will focus on this period. This is also the period for which the ESA-CCI SM combined product provides close to a complete spatio-temporal coverage [32]. In this paper, we compare our default run (ORC_REF) to ORC_CRU in Section 3.4 to assess the effect of meteorological forcing data on simulated SM values. Also in Section 3.4, ORC_REF is compared to ORC_LC5 to evaluate the effects of different land-cover maps. ORC_REF is finally compared to ORC_NoRs in Section 3.7 to discuss the effect different parameterizations can have on modelled SM.

Table 2. The different simulations used in this study; differences from the reference run are in bold.

Simulation	Climate Forcing	Land Cover	Soil Resistance
ORC_REF	CERASAT	LC6	Y
ORC_CRU	**CRUNCEP**	LC6	Y
ORC_LC5	CERASAT	**LC5**	Y
ORC_NoRs	CERASAT	LC6	**N**

2.4. Data Processing and Statistical Analysis

All the metrics described in this section are calculated over the entire time period n and considered at every location i; $x_{i,t=1...n}$ and $y_{i,t=1...n}$ denote the time series of two datasets at that given location.

2.4.1. Statistically Rescaling and Bias Correcting the Observations

Cumulative density function (CDF)-matching schemes are used in this study to rescale the observed soil moisture y to the modelled soil moisture x (further explained in Section 3.1). Two CDF-matching algorithms are implemented [53]. The first is a linear CDF-matching which matches the mean and standard deviation in the following manner. Let $x_{i,t=1...n}$ and $y_{i,t=1...n}$ denote the time series of the two datasets for a given location i. Similarly, let σ_{x_i} and σ_{y_i} denote the standard deviation of each dataset respectively at location i. Then to obtain rescaled $y_{i,t}^*$ at each grid point, we use the following equation

$$y_{i,t}^* = \frac{\sigma_{x_i}(y_{i,t} - \overline{y_i})}{\sigma_{y_i}} + \overline{x_i}. \tag{3}$$

The second algorithm used a full CDF-matching which, in addition to the mean and standard deviation, matches higher moments of the distribution. The CDFs function of y_i is matched to the CDF y_i such that y_i^* satisfies the following equation:

$$\text{cdf}_{x_i}(y_i^*) = \text{cdf}_{y_i}(y_i). \tag{4}$$

We use the python software (pytesmo package) to do this matching [54].

In another approach to rescale data, it is also normalized in the following manner:

$$\widehat{x}_{i,t} = \frac{x_{i,t} - \min x_i}{\max x_i - \min x_i}. \tag{5}$$

When applied to SM data, this normalization translates values to what is known as the Soil Water/Wetness Index (SWI).

2.4.2. Seasonal and Inter-Annual Variability

To separate seasonal effects from faster components, sub-monthly variations (also shown as monthly anomalies) are calculated. The raw data are divided into two components:

1. a slow-varying component (SC) which represents seasonal variability. This is calculated in two steps. First, the mean annual cycle over the 8 years of data is calculated $(\overline{x_i^a})$. Then, following [55], the signal is averaged using a 35-day moving window. Considering a 35-day period $P = [t - 17, t + 17]$, a minimum of 5 values are needed to calculate the average soil moisture for this period resulting in the slow-varying component:

$$x_{i,t}^{SC} = \overline{\overline{x_i^a}(P)}. \tag{6}$$

   When there are not 5 values within P for a given grid point, the grid point is then ignored for that given period.
2. a fast-varying component (FC), or daily anomaly, given by:

$$x_{i,t}^{FC} = x_{i,t} - x_{i,t}^{SC}. \tag{7}$$

2.4.3. Correlation Scores

We use the Pearson product-moment correlation coefficient, r, to assess the relative agreement of the temporal structures between two datasets:

$$r = \frac{\sum_{t=1}^{n}(x_{i,t} - \overline{x_i})(y_{i,t} - \overline{y_i})}{\sqrt{\sum_{t=1}^{n}(x_{i,t} - \overline{x_i})^2}\sqrt{\sum_{t=1}^{n}(y_{i,t} - \overline{y_i})^2}}. \tag{8}$$

We also consider spatial correlations. These are calculated using same formula but on single vector of locations at each time step.

2.4.4. Root Mean Standard Deviation

To calculate the difference between two datasets, we used the root mean standard deviation (RMSD):

$$\text{RMSD} = \sqrt{\frac{\sum_{t=1}^{n}(x_{i,t} - y_{i,t})^2}{n}}. \tag{9}$$

2.4.5. MSD Decomposition

Following [56], we can decompose the mean standard deviation (MSD = RMSD^2) into three informative components. Let $x_{i,t=1...n}$ and $y_{i,t=1...n}$ denote the two datasets, then

$$\text{MSD} = (\overline{x_i} - \overline{y_i})^2 + (\sigma_{x_i} - \sigma_{y_i})^2 + 2\sigma_{x_i}\sigma_{y_i}(1 - r) \tag{10}$$

$$= \text{bias}^2 + \text{VAR} + \text{LCS} \tag{11}$$

where the bias = $(\overline{x_i} - \overline{y_i})$, VAR = $(\sigma_{x_i} - \sigma_{y_i})^2$ and LCS = $2\sigma_{x_i}\sigma_{y_i}(1 - r)$.

The first term of the decomposition is the squared bias. The second term, VAR, shows the difference in the magnitude of fluctuation between the simulation and measurements. The larger this term, the more the model fails to simulate the magnitude of the fluctuation among the n measurements. The last term of the decomposition is the Lack of Correlation weighted by the Standard deviations (LCS). This term is large when the model fails to simulate the pattern of fluctuation across the n measurements [56].

2.4.6. Autocorrelation and Lag Time

For a dataset $x_{i,t=1...n}$, the lag-k autocorrelation coefficient can be expressed as:

$$r_k = \frac{\sum_{t=1}^{n-k}((x_{i,t} - \overline{x_i}) - (x_{i,t+k} - \overline{x_i}))^2}{\sum_{t=1}^{n}(x_{i,t} - \overline{x_i})^2}. \tag{12}$$

Following [57], we define the characteristic lag time (c_t) as the time at which the autocorrelation coefficient reduces to below $e^{-1} \approx 0.37$. This is also known as the e-folding time. This quantity determines a threshold after which the dataset is no longer considered as significantly autocorrelated. This can thus be seen as a way to quantify the inertia of the studied variable. This metrics is applied to the fast-varying component of the data.

2.4.7. Singular Value Decomposition

Singular Value Decomposition (SVD) offers an adequate tool to analyze the temporal co-variability between two gridded datasets. A detailed description of the method can be found in [58]. The spatio-temporal analysis will consider the total explained covariance of the two fields that is explained by the decomposition into pairs of optimally correlated observation and model patterns, as well as homogeneous correlation maps (depicting the geographic localization of the covarying parts of observation and model respectively) and observation heterogeneous correlation maps (reflecting how well observations are predicted by model patterns). Since the SVD analysis concerns itself with spatial patterns, the SVD algorithm is directly applied to the raw data instead of using the CDF-matched data. This is because the rescaling is applied pixel by pixel and hence would affect the spatial distribution.

3. Results

3.1. Rescaling the ESA-CCI SM Product

Since the ESA-CCI SM product has undergone a rescaling process with the GLDAS-Noah LSM, a direct comparison with the ORCHIDEE model outputs is likely misleading as those models have different soil representation: spatial distribution, textural and hydraulic properties and vertical discretization. In this context, agreement between the ORCHIDEE and ESA-CCI SM products should only be evaluated in terms of relative dynamics unless the observations are first bias corrected or scaled to be consistent with the model climatology. Bias correction is also vital for future data assimilation experiments. Most data assimilation algorithms assume that the uncertainties in the retrievals and the model output are Gaussian distributed, bias correcting the retrievals ensures this condition is satisfied.

In Figure 1 we consider how different scaling algorithms introduced in Section 2.4.1 affect the ESA-CCI SM values. Since the rescaling used in the creation of ESA-CCI product was applied individually at every pixel, we also do the same. Three different scaling are considered. The first rescaling translates both the ESA-CCI and the modelled values to soil water index (SWI) values. Calculated using the minimal and maximal SM values as proxy for the residual/minimum and volumetric field capacity respectively, the values obtained are between 0 and 1. Since the residual value and volumetric field capacity vary between land-surface models, through this rescaling, the ESA-CCI values remain within the GLDAS-Noah climatology. This scaling is more of a type of normalization than a type of bias correction. The second type of rescaling, referred to as linear CDF-matching, matches the mean and standard deviation of the ESA-CCI SM values to that of the modelled values (as described in Section 3.1). The third type of rescaling considered here, is a more complex full CDF-matching which is a refinement of the linear approach, but also matching higher moments of the distributions such as skewness and kurtosis. This latter algorithm for CDF-matching is a non-linear transformation, and as such we can expect small changes in the relative magnitude of the temporal dynamics [35]. Since a full CDF-matching was used in the creation of the ESA-CCI product, it follows that this is the best suited method for rescaling the data to the ORCHIDEE climatology. It is also the method using in several SM assimilation studies e.g., [59–61].

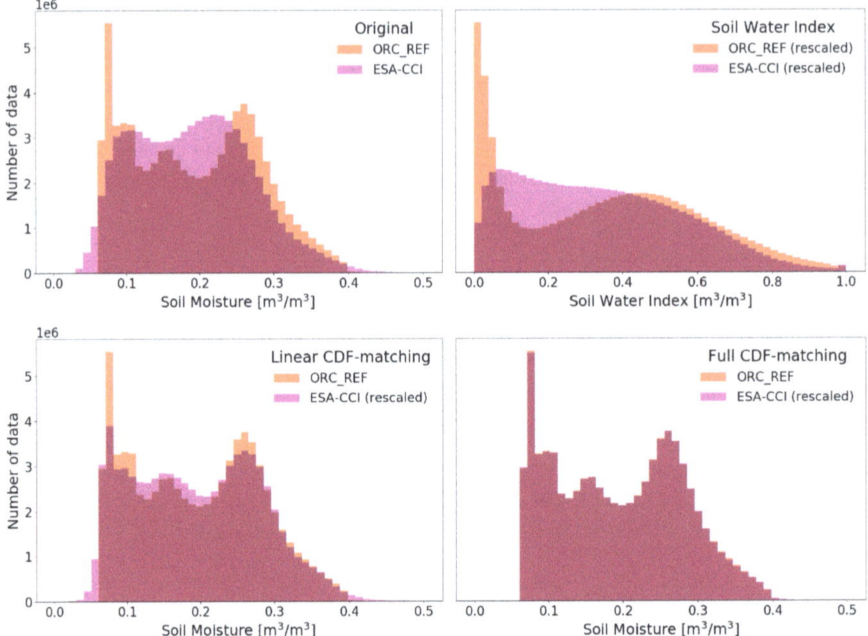

Figure 1. Global distribution of daily SM within ORC_REF and the ESA-CC1 product for 2008–2016. The histogram found prior to any rescaling is shown in the top left-hand corner. Histograms for the SWI values of ESA-CCI and the model are shown in the top right-hand corner. The ESA-CCI values rescaling via a CDF-matching to the ORCHIDEE model are shown on the bottom row. The data used in each panel is shown by the title of the legend.

The histogram for the ESA-CCI soil moisture values prior to any rescaling is binomial with the second peak (around 0.22 m^3/m^3) much larger than the peaks seen for the modelled soil moisture values (Figure 1). When translated to SWI values, this second peak is lost as the histogram is flattened. After the rescaling using CDF-matching, this peak is split in two to match the three narrower peaks observed in the modelled data. Using the linear CDF-matching, the total global RMSD over the whole 8-year period drops from 0.059 m^3/m^3 to 0.039 m^3/m^3, which equates to a 34% discrepancy reduction. Logically, the SM density distribution overlapping increased from 68.3% to 87.9%, associated with the creation of a third intermediate peak. This similarity between the distributions at this stage is promising. It assures us that the observed and modelled SM values are already relatively consistent and can be compared and used in the future to further optimize model parameters and bring potential improvements. When applying the full CDF-matching, the distributions are nearly identical.

The MSD decomposition shown in Figure 2 illustrates how the full CDF-matching performs at different latitudes focusing on one year (2016). Prior to this rescaling, we can see that bias makes up most of the error. On average it makes up over 72% of the MSD. Errors are particularly high in the northern latitudes for the winter months (January–March). This is because of snow and frozen areas found at these latitudes. In ESA-CCI SM product, points with high levels of uncertainty are flagged and removed. Similarly, using the model output, points where surface temperature is lower than negative one are also removed. Therefore few data points remain to calculate the average MSD in this area during this season as can be seen in Table A1. Otherwise, errors are relatively consistent at each of the latitude and over each three-month period prior to rescaling.

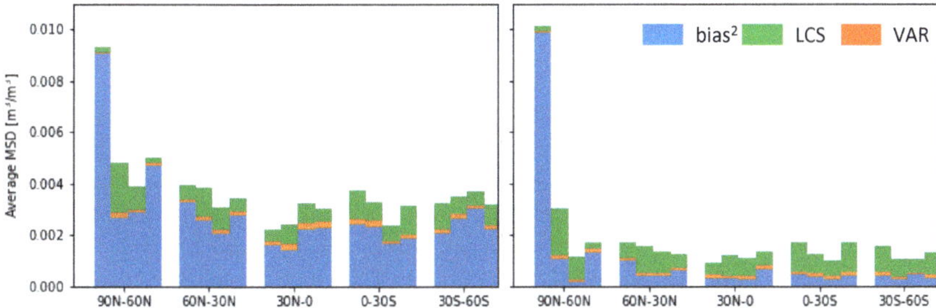

Figure 2. Example of the averaged MSD decomposition for the year 2016. The MSD between the daily SM values of the ESA-CCI SM product and ORC_REF model at each pixel is averaged by latitude and then decomposed. The MSD calculated using the ESA-CCI SM values before the full CDF-matching can be seen on the left and after the full CDF-matching on the right. Each bar covers a three-month period (January–March, April–June, July–September, October–December) and the number of points used to calculate each bar is described in Table A1.

In each case, the discrepancy is reduced after the full CDF-matching. Each decomposition still displays a bias component, it on average makes up 37% the bar. This is because the rescaling is done using data from multiple years whereas the figure only displays one single year, and therefore the remnant bias is the bias with respect to the yearly average. Generally, for each latitude-month (one bar), most of the bias is removed, VAR decreases slightly and the LCS portion of the MSD remain the same size. This shows that overall correlation amplitudes are relatively insensitive to this choice of rescaling.

Spatial correlations were also considered between the model and ESA-CCI SM values both prior to and after rescaling. Before rescaling, mean spatial correlations globally range according to a seasonal cycle between 0.66 and 0.86. After rescaling, these values follow the same pattern but drop between 0.50 and 0.79 for the SWI algorithm and increase to between 0.82–0.95 for both versions of CDF-matching.

For the remainder of this paper, unless stated otherwise, the rescaled ESA-CCI SM values obtained through the full CDF-matching are the ones considered.

3.2. An Analysis of Model Depth Selection

We investigate in this section the effect of integrating the model soil moisture over different depths. The reference depth is 2.2 cm, corresponding to the 4 top layers (see Section 2.3). We first consider the RMSD with the reference depth that makes the most theoretical sense (Figure 3a). The lowest errors are observed in the Sahara and other desert areas. RMSD is relatively low in these dry areas, with errors averaging 0.017, since these places will have low SM values. The largest values of RMSD of about 0.15 tend to be found in the northern latitudes and along the Himalayan mountain range. These are snowy regions and, as discussed previously, these may be areas where most of the model and ESA-CCI SM values are flagged. There are also a few areas with higher RMSD values near the masked areas of dense vegetation (i.e., Brazil and central Africa). The rest of the map shows a relatively constant error close to 0.04 m^3/m^3.

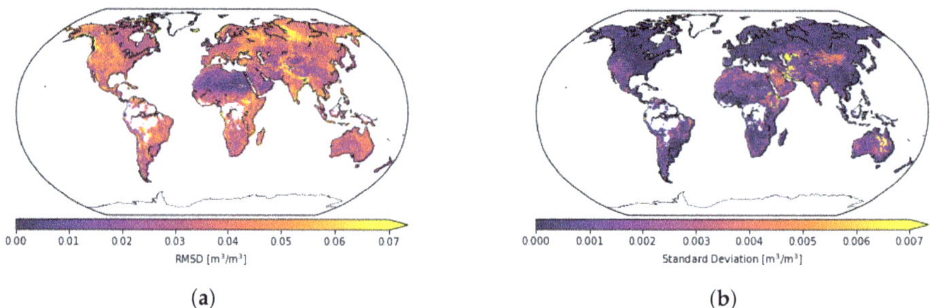

(**a**)   (**b**)

Figure 3. An analysis of daily soil moisture RMSD between ORC_REF and the rescaled ESA-CCI SM at different depths of the ORCHIDEE model averaged over the 2008–2016 period. The left-hand figure considers the RMSD at the reference depth of 2.2 cm (top 4 layers of the model). The right-hand figure shows the standard deviation of the RMSD values calculated at the different integrated depths in ORCHIDEE between the top 2 layers and down to the top 7 layers (18.6 cm). Note the difference of scale in parts (**a**,**b**). Areas with dense vegetation have been masked (e.g., Brazil and central Africa). (**a**) Surface soil moisture RMSD (top 4 layers); (**b**) Vertical standard deviation of RMSD over layers 2–7.

When considering the standard deviation globally in Figure 3b, we can see that there is little to no variation in RMSD between layers. Indeed, for 50% of the map, the standard deviation between layers is less than 0.001 m^3/m^3. This proportion increases to over 90% of the map for standard deviations less than 0.003 m^3/m^3. The highest scores of around 0.01 (less than 0.12% of the points globally) can be found in some of the dry areas. These RMSD values remain low considering the values of soil moisture are of magnitude $\mathcal{O}(0.1)$ m^3/m^3. The consistency in RMSD values over these first few layers of the model may be partly due to the use of daily averages. It is possible that we may see variations at different depths if we could get instantaneous values for the different rainfall events. However, since the ESA-CCI SM product currently only provides daily values, we cannot reduce our time step.

With the first few layers virtually the same, even though the errors are slightly lower calculated over the top 5 layers, we choose to keep using the top 4 layers in our analysis since this depth makes the most theoretical sense [32,62]. Therefore, for the remainder of this study, a depth of 2 cm (top 4 layers in the model) is used.

3.3. Mean Temporal Correlations between Model and ESA-CCI SM Product

Due to the rescaling process undergone to create ESA-CCI SM product, as well as the rescaling procedure implemented, considering the RMSD between the observations and modelled time series is not the most informative metric. Temporal correlation, on the other hand, evaluates the agreement between the ORCHIDEE and ESA-CCI SM product in terms of relative dynamics and therefore is less sensitive to the choice of rescaling. Overall, we can see strong correlations between the model output and the ESA-CCI SM values (Figure 4). Highest correlations are found in the tropics and southern latitudes when the average correlation score is 0.7. Temporal correlations are highest over areas where remote sensing methods are known to be efficient. Poorer results are found over desert, mountainous and frozen areas. In desertic areas, this may be explained by forcing errors, i.e., the precipitation used to drive the model. Rain is spatially heterogeneous over deserts with few events, these can be hard to observed, and hence can be a large source of uncertainty over such regions. Furthermore, the satellite representative depth over deserts tends to be larger in dry conditions and the larger surface ground heating during the day may also increase the errors on the retrievals [43]. The poor results in mountainous and frozen areas can be partly explained by the difficulty to model the radiative transfer in those conditions (i.e., separate roughness and soil moisture effects), and therefore to inverse SM from remote-sensed data. This is particularly noticeable in the high latitudes where the average correlation reduces to 0.15 compared to the global mean of 0.49. In lines with previous studies, best scores are obtained in areas with strong annual cycles and higher density data [32].

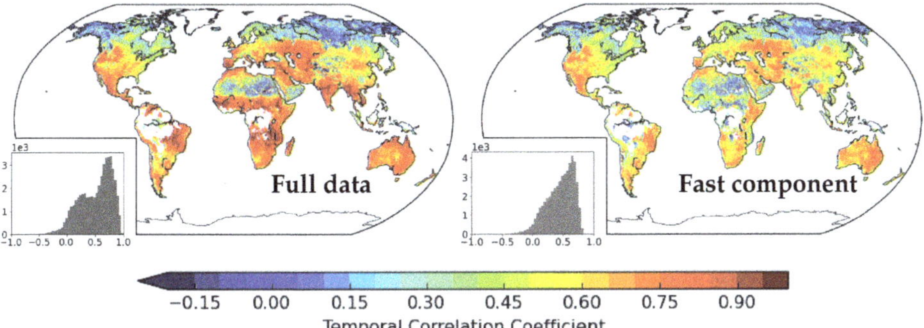

Figure 4. Temporal correlation of the full data (**left**) and the fast component (**right**) between the daily ESA-CCI SM values and ORC_REF SM output integrated over a depth of 2 cm. Histograms show the distribution of the correlation values for each map. The slow seasonal component is not shown as correlations are uniformly one globally.

For the purposes of data assimilation, we are interested in the sub-monthly variations. From Figure 4 we see that the sub-monthly variations are less correlated than the full data. The histograms show that for the full data, there is a peak large at 0.76. For the anomalies, the peak is lower found at 0.63. The full data reaches close to values of one whereas the fast component does not exceed 0.9. The differences observed between the full data and the fast component highlight the fact that factors such as the spatio-temporal variability of precipitation strongly affect the temporal correlations. Though not shown here, the correlations found here are nearly identical to correlations calculated at the other depths between layers 2 and 7. This is also the case when considering the raw and different rescalings of the ESA-CCI SM data. The full CDF matching used to rescale the observations can have a slight effect on the temporal dynamics. However, since the strong correlations are also observed when the raw ESA-CCI values are used, we can be confident that the strong correlations are due to the agreement between the model and satellite data instead of the rescaling procedure itself.

3.4. The Effect of Meteorological and Land Cover Forcing Data

The choice of forcing data heavily influences the simulation of SM in LSMs. In this section, we first compare our default ORCHIDEE run ORC_REF to one with a different meteorological forcing, CRUNCEP, introduced in Section 2.3. From Figure 5 we can see that the ORC_REF run outperforms the ORC_CRU for the full data and the fast component. This is particularly notable in arid or desert regions (i.e., Sahara, Asia, and Australia). ORC_REF shows stronger correlations than ORC_CRU for the fast component. This shows that the ORCHIDEE run using the CERASAT meteorological forcing not only captures the same seasonal patterns as the satellite data but also more of the sub-monthly behavior of SM.

CERASAT is a new dataset of the ECMWF based on recent advances in data assimilation and uses satellite data in its creation. Another reason for the CERASAT forcing data outperforming CRUNCEP may be due to the difference in temporal resolution between both products. CERSAT is 3-hourly whereas CRUNCEP is 6-hourly. This difference in frequency may affect how the precipitation is partitioned for each different time step and hence the response of SM in the model. The noticeable differences between the correlations for each run show that the choice of meteorological forcing is critical for the accurate simulation of SM. Many studies have in fact used SM to correct precipitation input [63–67]. Conversely, to correct the SM parameters in ORCHIDEE, it is important to find a metric that is less sensitive to the meteorological forcing.

(**a**)

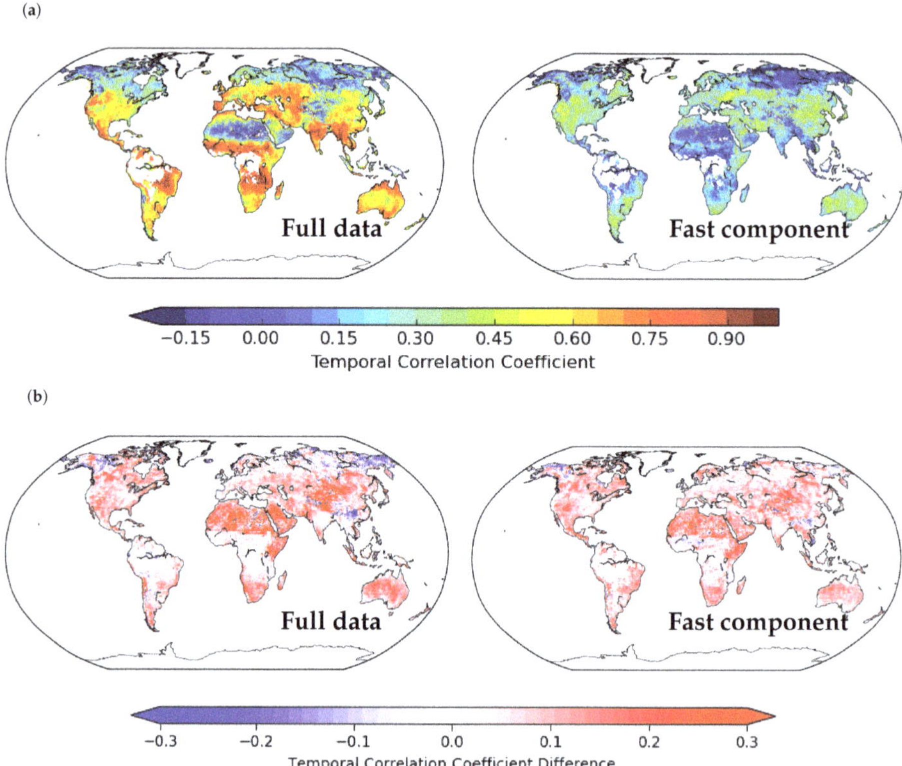

(**b**)

Figure 5. Temporal correlations between the daily ESA-CCI SM observations and the ORC_CRU output shown on the top row, and the difference in correlation scores found using different forcing to run the model is shown on the bottom (i.e., $r(\text{ORC\_REF}, \text{ESA-CCI}) - r(\text{ORC\_CRU}, \text{ESA-CCI})$). Both the full data (**left**) and fast components (**right**) are considered. (**a**) Correlations "ORC_CRU vs. ESA-CCI"; (**b**) Correlations "ORC_REF vs. ESA-CCI" minus Correlations "ORC_CRU vs. ESA-CCI".

Forcing data is not limited to the meteorological data used to drive the model. It encompasses all external data used for a given model run. Another dataset used as forcing data in ORCHIDEE is the choice of land-cover map, which determines the distribution of the global vegetation. In this study, we only tested two slightly different land-cover maps, and the correlations between the ESA-CCI SM product and ORCHIDEE SM output remained very similar. This suggest that, in this case, the choice of land cover is not the primary source of variability.

3.5. Measuring the Covariance of Rainfall and Soil Moisture

The chosen meteorological forcing data prescribes (among others) the precipitation in the model. Since precipitation, both the amount and intensity, has been shown to be a major driver to surface soil moisture dynamics [68], this dependency can be assessed through their covariance. Following [69], we use the SVD method described in Section 2.4.7 to identify and study coupled rainfall-SM covarying spatio-temporal patterns. To ensure that the comparison does not favor the model output and that therefore any differences found originate from the SM only [69], independent rainfall data is needed. For this experiment, it was provided by MSWEP data and the period 2008–2014 was considered because it overlaps all datasets.

We focus our analysis over Australia and the Iberian Peninsula. Australia was selected since the ESA-CCI SM product has shown a good performance over this country in terms of correlation with reanalysis and in situ datasets [35,36]. Also, passive, active, and combined products have all been used over this region [70] resulting in the fractions of days with valid observations to be very high [35]. Therefore, we can expect a reliable quality of the product over this region. The Iberian Peninsula was selected to provide direct comparison with [69] who used satellite SM values from the ESA's Soil Moisture and Ocean Salinity product (SMOS-BEC).

Considered in Table 3 are the first three covarying patterns representing approximately 95% and 99% of the covariance of SM and P for Australia and the Iberian Peninsula, respectively. The explained covariance is of the same proportion whether the ESA-CCI product is considered, or whether the different model runs are considered. For the Iberian Peninsula, most of the covariance is explained by PC1. When the SM time-expansion coefficients are plotted, (see Figure 6) we can identify a large annual cycle in PC1. PC1 therefore represents the seasonal co-variability of rainfall and SM. The usually shallow trough observed in 2012 corresponds to the severe drought that Spain experienced with precipitation less than 30% than its normal amount between December and February. PC2 and PC3 represent the higher frequency (synoptic) variability of SM linked to rainfall events.

For Australia on the other hand, PC1 still explains most of the covariance; however, PC2 and PC3 also explain a small part. From Figure 7, PC1 appears to be explaining the monsoon events occurring in north-east Australia. The pronounced trough seen in late 2010/early 2011 correspond to the Australia floods of 2010–2011. These floods were due to unusually heavy and prolonged rain caused by the seasonal wet monsoon coinciding with an unusually strong La Niña events [71]. The spatial representation of the PC1 for the ESA-CCI SM suggest that the monsoon comes down too low in both versions of the model runs. PC2 and PC3 described the rainfall patterns found in the south and southeast.

By considering the spatial patterns in Figures 6 and 7, we can see that PC2 and PC3 for the ESA-CCI SM and the models are reversed for both Australia and the Iberian Peninsula. The PCs are ordered by the amount of explained covariance. For Australia, the satellite product suggests that the rainfall in the south explains more of the covariance than the rainfall in the south-east, whereas the models suggest the opposite. For the Iberian Peninsula, the ESA-CCI SM product highlights the Mediterranean coastal rainfall explaining more than the rainfall over the Pyrenees. However, when we consider the values of explained covariance in Table 3 we can see that the difference between the two PCs is negligible.

The homogeneous explained variance fraction in Table 3 refers to the SM variance explained by the soil moisture itself. The soil moisture PC1 for the Iberian Peninsula and Australia explains a slightly higher proportion of the total SM variance when using the model runs that the ESA-CCI product; 81% compared to 76% for the Iberian Peninsula and 30% compared to 23% for Australia. This slight discrepancy between the satellite observations and models is consistent with [69]'s findings and similarly suggests that there are structures that are not rainfall-driven captured by the satellite product that are missing in the model.

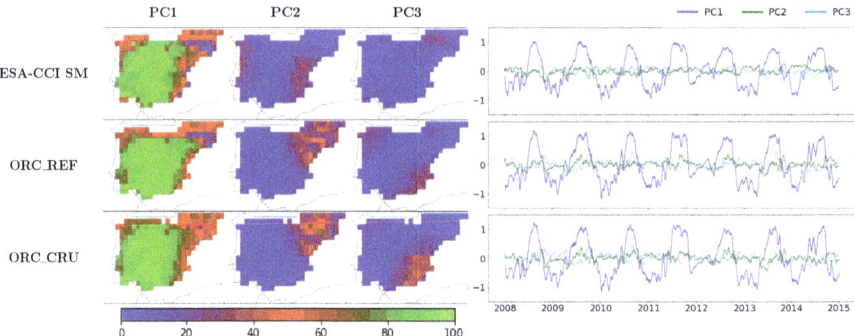

Figure 6. Visualization of the SVD results for the Iberian Peninsula depicted in Table 3. On the left-hand side, spatial patterns of the explained homogeneous variance by SM patterns for the ESA-CCI product and two different runs of the ORCHIDEE model over the Iberian Peninsula; the first run with the CERASAT forcing data (ORC_REF) and the second with CRUNCEP (ORC_CRU). On the right-hand side, the expansion coefficients for the first three pairs of covarying patterns (i.e., the temporal projection of these covarying patterns). From top to bottom the SVD analysis corresponds to MSWEP/ESA-CCI, MSWEP/ORC_REF and MSWEP/ORC_CRU.

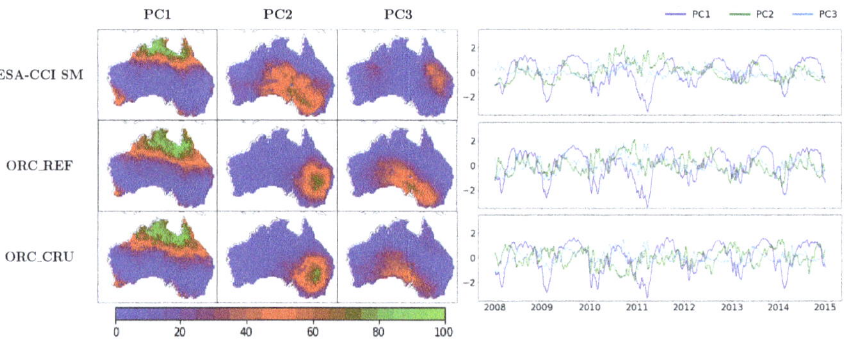

Figure 7. Same as Figure 6 but for Australia.

The heterogeneous explained variance fraction in Table 3 refers to the SM variance explained by the rainfall. For both the Iberian Peninsula and Australia, the PCs show a slightly weaker link between rainfall and SM for the satellite data than for the model. The sub-seasonal variability of precipitation extracted by the SVD analysis (shown by PC2 and PC3) seems to explain a negligible part of the SM variability. This is because the seasonal variations of the precipitation are better represented than the higher frequencies due to the precipitation's high spatio-temporal variability.

The best thing that can explain the variance in the SM patterns is SM itself. Therefore, the values for homogeneous explained are the highest possible. The fact that, for Australia, the precipitation can explain the variance of the SM pattern by the same magnitude as SM itself, clearly shows that precipitation is the main driver of SM in this region. For the Iberian Peninsula, the precipitation explains nearly the same amount of the SM pattern variance as it does for Australia, but this is only a third of the homogeneous values.

Table 3. Characteristics of the first three pairs of covarying patterns of precipitation (P) and SM over Australia and the Iberian Peninsula for 2008–2014. Explained (co-)variance are shown in percentage. Found in bold are the total explained (co-)variances and mean correlation scores.

Data	Covarying Pattern	Explained Covariance	Explained Variance of SM Patterns		Correlations between Covarying Patterns of SM and P	
			by Precip. (Heterogeneous)	by SM Itself (Homogeneous)	Spatial	Temporal (Expansion Coeff.)
Australia						
ESA-CC1	PC1	85.2	12.73	21.71	0.92	0.76
	PC2	5.15	4.05	23.6	0.84	0.42
	PC3	4.55	2.34	9.87	0.92	0.5
		94.9	**19.12**	**55.18**	**0.89**	**0.56**
ORC_REF	PC1	81.97	20.61	29.46	0.90	0.83
	PC2	6.77	4.01	13.82	0.95	0.56
	PC3	5.03	5.50	17.97	0.92	0.57
		93.77	**30.12**	**61.25**	**0.92**	**0.65**
ORC_CRU	PC1	86.69	19.91	29.76	0.87	0.80
	PC2	5.08	3.01	15.37	0.92	0.46
	PC3	3.50	3.50	14.27	0.86	0.52
		95.27	**26.42**	**59.40**	**0.88**	**0.59**
The Iberian Peninsula						
ESA-CC1	PC1	96.78	15.89	75.81	0.56	0.46
	PC2	1.65	1.39	8.89	0.89	0.4
	PC3	1.04	0.82	5.25	0.87	0.41
		99.47	**18.1**	**89.95**	**0.77**	**0.42**
ORC_REF	PC1	95.12	23.68	81.96	0.52	0.54
	PC2	2.52	3.11	12.4	0.95	0.52
	PC3	1.79	2.29	8.36	0.96	0.54
		99.43	**29.08**	**102.72**	**0.81**	**0.53**
ORC_CRU	PC1	95.71	18.99	80.24	0.68	0.49
	PC2	2.12	1.71	11.92	0.89	0.42
	PC3	01.58	1.51	10.15	0.94	0.44
		99.41	**22.21**	**102.31**	**0.84**	**0.45**

The spatial correlation values are stronger for Australia than for the Iberian Peninsula. One of the reason we observe higher spatial correlations in our analysis may be due to topological and land-use differences between Australia and the Iberian peninsula. Australia has more homogeneous and natural landscape, whereas the Iberian Peninsula has more irrigation, lakes, mountainous areas in a smaller area.

As observed in [69], we find that the slow-varying PC1 for the Iberian Peninsula is less spatially correlated than the fast-varying PC2 & 3. Since the PC1s represent the seasonality for the Iberian Peninsula, the variability represented is determined not only by rainfall, but also by other processes such as evaporation and drainage. These processes are less important for the fast-varying PCs. In contrast, for Australia, we find that spatial correlations of each of the PCs show the same magnitude in each of the datasets. This may be due to the fact each of the PC1s for Australia is representative of different rainfall types. Overall these spatial correlations are stronger than those observed in [69]. In [69], the SVD analysis was conducted using SMOS, a pure remote sensing estimate, whereas we are using a merged product where a model has been used. It is possible that the usage of the GLDAS-Noah model in the merging process has smoothed out a large part of the spatial variability associated with landscape features or measurement noise.

3.6. Temporal Autocorrelation

The strong correlations found between the ESA-CCI SM product and the ORCHIDEE output are very promising. In this section, we further investigate the temporal autocorrelation, a similar metric that should also be insensitive to the choice of rescaling. It represents the correlation of a variable with its own future and past values [72]. Intuitively, since SM temporal autocorrelation should be more controlled by processes such as evaporation and drainage, it should be less sensitive to the choice of meteorological forcing and more sensitive to the choice of parameterization used in the model.

Temporal autocorrelation, and thus characteristic lag time, is a relatively simple metric but with potentially complex interpretations which can integrate notions of dry-down. Temporal autocorrelation, in cases when the interval between precipitation events is large enough, can sometimes be interpreted as a useful indicator of SM inertia (i.e., persistence and depletion time), less influenced than correlation metrics by the climate forcing; it should allow to assess more precisely the model's structure. In past studies, people rarely used such analysis, probably because it often leads to large disparities between modelled and observed products (see for example [12]). Such disagreements could be explained by the gap between the observed representative depth and the simulation depth. It is indeed well-known that depth has a major influence on the inertia of SM [69]. However, we have shown in Section 3.2 that signal representative depth in ORCHIDEE for daily SM value is relatively consistent.

The characteristic lag time c_t defined in Section 2.4.6, is a measure of temporal decorrelation rate. More specifically, it measures soil moisture memory which can be partly related to the time taken by the soil to 'forget' an anomaly [73]. c_t of the ESA-CCI SM product varies greatly depending on different regions (Figure 8a). Areas with the longest c_t, between 15 and 30 days, tend to be regions with a temperate climate. Shorter c_t times occur in areas with a drier climate, for example 1.4 days over the Sahara, where there is less rainfall. Figure 8b allows us to compare c_t for the model run and the ESA-CCI product. Overall there is higher SM stability in the model than in the satellite product (red areas in Figure 8b), though parts of Australia, South America and Asia are modelled as more reactive (blue areas in Figure 8b). The model predicts much higher characteristic lag-time values around the tropics than the ESA-CCI SM, mainly around the masked dense vegetation; these values can differ up to 15–20 days. On average; however, the differences between the model and the product tend to be of the order 4–5 days.

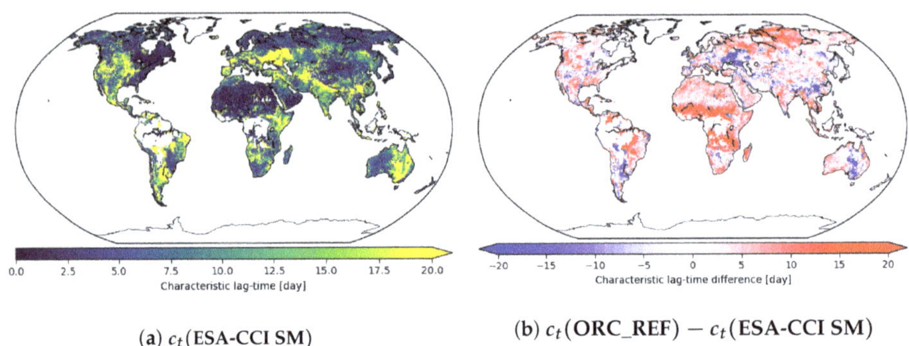

(a) c_t(ESA-CCI SM)   (b) c_t(ORC_REF) − c_t(ESA-CCI SM)

Figure 8. Spatial distribution of the characteristic lag time of the daily anomalies of the ESA-CCI product (**left**) and the difference between the characteristic lag time of the ORCHIDEE anomalies and the ESA-CCI anomalies (**right**). Both figures cover the 2008–2016 period.

3.7. Potential Impact of Parameterization on SM Dynamics

In this section, we consider the effect of removing the r_{soil} parameterization from the model (see Section 2.3), as was chosen for a configuration of ORCHIDEE in the IPSL-ESM for the CMIP6 exercise. This is not exhaustive but will help to illustrate the impact that changing a parameterization can have on SM dynamics. The soil resistance term in the model will control how long the soil moisture is kept in the first few centimeters of the soil. Water is removed from these centimeters either by evapotranspiration into the atmosphere, by vertical infiltration into the deeper layers of the soil or by surface-runoff to rivers. When comparing ORC_REF to ORC_NoRs (an ORCHIDEE run without soil resistance, Figure 9), since both versions of the model runs use the same PFT and soil maps, and same topography, the evapotranspiration will be the flux quantity that differs the most between the different model runs.

Evapotranspiration comprises soil evaporation and plant transpiration. In arid regions, since soil evaporation is dominant, it will have a larger control on SM than in vegetated areas where transpiration will play a key role. When soil resistance is not included in the model (i.e., ORC_NoRs), evaporation rates are higher and therefore the SM in the model will dry out faster. This is why for ORC_NoRs, arid regions, where evaporation dominates the total evapotranspiration, show the greatest difference in characteristic lag time between the run and the ESA-CCI values. The differences in the characteristic lag-time values for ORC_NoRs and ORC_REF are of the same magnitude as the differences between ORC_NoRs and the ESA-CCI values. This shows that adding or removing a parameterization can largely affect SM output and thus structural model changes need to be considered before parameter estimation experiments.

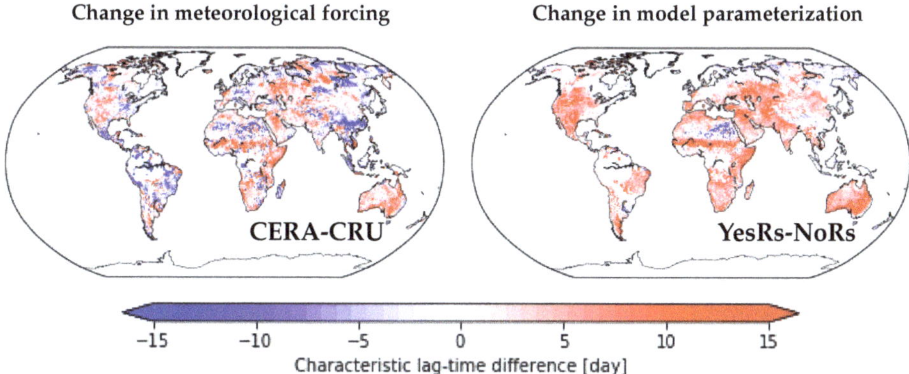

Figure 9. Difference in the characteristic lag-times found between the model and ESA-CCI values for different model runs. On the left-hand side, the model runs differ in the meteorological forcing used to drive the model - specifically CERASAT (ORC_REF) and CRUNCEP (ORC_CRU). On the right-hand side, the model runs differ in the parameterization used - namely the inclusion (ORC_REF) and exclusion (ORC_NoRs) of a soil resistance term.

When considering the difference between the autocorrelation lag time of the model and the ESA-CCI product for the different meteorological forcings, similar regional patterns are identified. However, when taking the difference between both model runs(i.e., ORC_REF and ORC_CRU), the intensity is seen to differ (Figure 9). The forcing induced difference is also comparable in magnitude to the difference between two model runs with different parameterizations (with change in model parameterization being slightly stronger; the average global absolute difference is of 5 days compared to 4 days). Without the soil resistance parameterization, the c_t values are much lower nearly globally. Changing the meteorological forcing or the model parameterization in this example affects different regions. For example, the characteristic lag time of south China is more sensitive to the meteorological forcing than the soil resistance term. In contrast, the Australia c_t value is more affected by the parametrization change than the change in meteorological forcing. This is an example where the autocorrelation metric is sensitive to the specifically selected parameterization. Since the characteristic lag time was shown to differ between the different configurations of the model and the satellite observations, it is potentially a useful metric to identify and correct systematic errors and parameters in ORCHIDEE.

In contrast, if instead we consider the correlation metric, as shown in Figure 10, we observed that changing the meteorological forcing has a greater effect on the correlation scores than changing the model parameterization. The global mean absolute difference in correlation score is 0.11 for the meteorological forcing compared to 0.07 the parameterization change. This increases to 0.25 over the Sahara compared to 0.08. In this example, the correlation metric is more sensitive to the meteorological forcing used and it may be harder to use this metric to pick out the internal deficiencies of the model but would be more appropriate if one wanted to diagnose or correct input precipitation data.

The choice to include or exclude soil resistance will also have a direct effect on the carbon cycle. As such, to conclude this section, we consider how different vegetation (based on maximum leaf area index values LAI_{MAX}) and soil classes perform with respect to the ESA-CCI product and the different metrics discussed (see Table A2 in Appendix A). The strongest correlations are found at points where $LAI_{MAX} \geq 3$ or where points have a fine soil texture. Areas with finer soil textures tend to have the most fertile soils and hence vegetation with the largest LAI. These regions, found in the temperate and tropical zones, have precipitation regimes with frequent rainfall events. The RMSD is relatively consistent over the different vegetation and soil texture classes considered, though slightly lower in areas with sparse vegetation ($LAI_{MAX} < 0.5$) and coarse soils. These areas include desert areas where soil moisture values will be very low. The average RMSD is slightly higher for the run without the soil resistance parameterization. The characteristic lag time increases with increasing LAI_{MAX}, and these values are higher when using soil resistance in the model compared to running the model without soil resistance. When $LAI_{MAX} < 0.5$, these pixels will be close to having bare soil. With increasing LAI_{MAX} values, the vegetation will become more complex (for example with larger roots) as will the processes involved in control the soil moisture. Characteristic lag time is also highest for fine soil textures which have a higher water-holding capacity and lower hydraulic conductivity. The average characteristic lag-times for the model differs the most in areas with sparse vegetation or areas with fine soils, and more for the ORC_REF (with soil resistance) than for the ORC_NoRs run (no soil resistance).

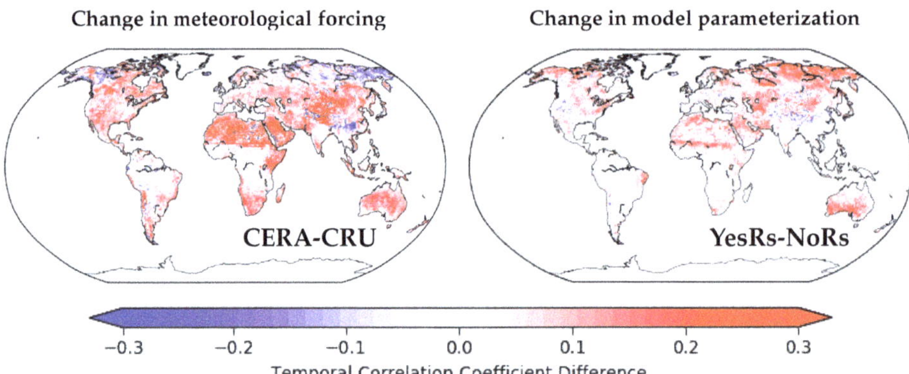

Figure 10. Difference in the correlations found between the model and ESA-CCI values for different model runs. On the left-hand side, the model runs differ in the meteorological forcing used to drive the model - specifically CERASAT (ORC_REF) and CRUNCEP (ORC_CRU). The figure shows $r(\text{ORC\_REF}, \text{ESA-CCI}) - r(\text{ORC\_CRU}, \text{ESA-CCI})$. On the right-hand side, the model runs differ in the parameterization used - namely the inclusion (ORC_REF) and exclusion (ORC_NoRs) of a soil resistance term. The figure shows $r(\text{ORC\_REF}, \text{ESA-CCI}) - r(\text{ORC\_NoRs}, \text{ESA-CCI})$.

4. Discussion

Throughout this paper, we have been assessing the potential of the ESA-CCI SM product to be used to evaluate and further improve the ORCHIDEE LSM. A first step in doing so, was to rescale the observations to the climatology of the ORCHIDEE model. This is necessary due to the steps undergone to create the ESA-CCI SM product: i.e., active and passive satellite retrievals were combined using the climatology of the GLDAS-Noah model reanalysis. Since this model has its own soil moisture climatology with different dynamical ranges [74], it is important to rescale data to the ORCHIDEE's climatology. In addition to the rescaling undertaken in creating the ESA-CCI product, there are also biases due the satellite signal processing. SM is estimated through inversion procedures based on radiative transfer modelling. This procedure uses additional assumptions and data such as vegetation distribution maps. For data assimilation studies, proper treatment of systematic errors is critical for the success of the experiment [75]. Since Gaussian error statistics is one of the key assumptions of most data assimilation systems, bias correction is a vital first step. However, it is worth noting that any form of bias correction will not only correct systematic biases and the biases due to the climatology mismatch but will also remove biases linked to the ORCHIDEE model itself. In an ideal case, using DA techniques, we would correct this latter source of bias. Since it is impossible to distinguish between sources of biases, we accept that, through the rescaling, we lose some of the information contained in the observations that could have potentially, corrected parts of the model. This choice is common practice when using satellite products even though it has been shown that the current bias-correction methods may remove up to 50% of information from the original satellite observations [76,77].

Since the microwave instruments used to generate the ESA-CCI product, are only able to monitor a thin surface layer, it was important to assess which depth has to be considered in the model. Fortunately, the first two meters of the soil column in ORCHIDEE is discretized by an 11-layered hydrology scheme, allowing us to select the corresponding depth. When unable to directly match the sensing depth, or when one wants to constrain model SM further down the soil column, a potential approach to the issue of representativity is to use an exponential filter [44]. This method translates the SM observed in the top few centimeters by the satellites to the SM representative of deeper layers following certain exponential assumptions. Such exponential filter has been used in several previous DA studies e.g., [55,78,79]. Since we can directly match the sensing depth in this study, we simply use the first four layers of the ORCHIDEE soil column discretization to perform the intercomparisons.

Comparison scores were found to be very similar at different depths of ORCHIDEE soil column. Since this trend was observed globally, this is the case when both the passive satellites dominate the product (i.e., over deserts) and when the active retrievals dominate the combined product (i.e., over high latitudes). The merging process may have contributed to the smoothing of the depth profile, as might the fact that the product only provides daily values. The first 4 layers of the model's soil column have been evaluated in this study. Although little difference was observed between the different integrated depths, if the data assimilation experiment is conducted at this depth, it will be important to consider its effect of the improvements over the rest of the 11 layers. The soil hydro-thermal parameters are currently homogeneous over the soil column but improving the surface soil moisture may highlight the need for different parameters further down the soil column or lead to conclusions that parts of the model such as the bare soil evaporation or root dynamics need to be modified.

An SVD analysis showed that most of the SM variability could be explained by the precipitation patterns. Since the simulation of SM in the model is driven by precipitation and hence the meteorological forcing data, it is impossible to find metrics unaffected by the forcing errors. Two main metrics were considered: correlation and autocorrelation. Both allow to focus on relative temporal dynamics for the chosen rescaling and they were also found to be consistent over the first few layers of the ORCHIDEE discretization scheme.

The strong correlations between the daily model outputs and the ESA-CCI values were identified. The mean correlation globally for the full data had a value of 0.49, this average increased to 0.7 when considering the tropics and southern latitudes. For the fast component, the mean global correlation was 0.45, and 0.51 over the tropics and south latitudes. Such strong correlations give confidence in both the model and satellite product. However, the correlation metric was shown to be sensitive to the meteorological forcing used, with variations of 10% globally, and therefore not necessarily informative about the model structure and parameter errors. Indeed, we saw when considering the soil resistance parameterization, changing the meteorological forcing had a greater effect on the correlation scores than this structural change. For example, over the Sahara, changing the meteorological forcing saw an average change of 21% in correlation scores compared to a change of 5% found when changing the soil resistance parameterization.

Temporal autocorrelation, which characterizes quickness and extension of SM variations, highlighted different behaviors in the model *vs* ESA-CCI product. In some areas, the ESA-CCI values are found to be more autocorrelated than the model, and in the others, the opposite is seen. This metric is sensitive to the parameterization of the model that control variables linked to the simulation of SM, such as evaporation, infiltration and drainage. Given the large uncertainties that are still associated with these processes [10,74] using SM temporal autocorrelation to optimize the main parameters involved in these parameterizations, through DA techniques, is promising.

Temporal autocorrelation can be used to measure soil moisture memory and can be related in some cases to notions of dry-out. The characteristic lag-time metric used in this study is one of a number to the measure of these processes. Although it is informative, it has a few limitations. Firstly, since this metric is based on anomalies, to calculate these anomalies, the mean state is removed from the full data. The mean state was averaged only over 8 years of data which is not long enough to properly extract the climatology for regions with high year to year variation in seasonal precipitation. Hence the calculation of the mean state introduces sampling error into the data [73]. Secondly, autocorrelation-based-metrics ignore the sign of the anomaly which can provide physically meaningful information. Other metrics such as mean persistence time can overcome some of these limitations and hence could be interesting metrics to consider in the future [73,80].

We have shown that the choice of parameterization used in the model is vital in the simulation of SM. In our example, removing soil resistance in the model was seen to decrease soil inertia and therefore create greater discrepancies with the ESA-CCI SM values. Of course, other key variables related to water fluxes should be considered for DA such as soil porosity and the other textural properties and hydraulic parameters (saturated hydraulic conductivity and variation with depth according to soil compaction and vegetation roots) as well as plant water uptake parameters such as root profile. The water and energy balance formulations in ORCHIDEE are representative of the equations found in other LSMs (e.g., JULES [81,82], JSBACH [83]) used in wider Earth system models. Therefore, improvements and structural changes highlighted in a future DA study could potentially also be applicable to these other models.

Overall the ESA-CCI product has great potential to identify missing processes and structural issues in the ORCHIDEE model, and potentially to correct the internal model parameters. However, the effectiveness of any data assimilation frameworks relies on the quality of satellite soil moisture observations and assimilation technique used. We will need to quantify the errors in the satellite retrievals to ensure the success of the assimilation, as well as account for forcing errors in the DA scheme.

Improving soil moisture will influence other modelled products such as evapotranspiration and infiltration. It will be important to consider the impact improving soil moisture has on such variables. Incorporating other products in the assimilation process can help ensure the quality of these outputs do not degrade when correcting soil moisture values. In addition, due to SM's role in driving the terrestrial carbon balance (through photosynthesis, ecosystem dynamics, and soil respiration), SM data has been highlighted as one of the remotely sensed Earth observation products relevant for terrestrial carbon cycle data assimilation [29]. As such, SM data can also be used as one of multiple data streams to improve the carbon cycle. Two approaches exist when assimilating multiple data streams. The first is simultaneously, as shown in [84] where remotely sensed soil moisture data along with in situ atmospheric CO_2 are used to constrain global terrestrial carbon cycle. The second approach is stepwise, by taking each data stream one at a time. This approach is used in [85], where three different data streams are assimilated to constrain the ORCHIDEE LSM's carbon cycle. A follow up study will consider the addition of SM into the carbon cycle DA system.

Throughout this study, we have focused on off-line runs which use meteorological data to drive the model. Another avenue for further work would be to consider coupled surface-atmosphere runs, which would allow to analyze the role of soil moisture on the atmospheric boundary layer properties and on the generation of precipitation, especially in semi-arid areas where these links are debated [5,9].

5. Conclusions

Even though there have been many studies evaluating the usefulness of the ESA-CCI SM dataset, this study is unique in considering the product on a global scale using a cutting-edge terrestrial CMIP6 model. This global comparison has shown that both the model and the product are at a stage where useful information can be derived and assimilated. The strong temporal correlations seen between the ESA-CCI product and the model show that the dynamics in the model are developed enough to capture the main trends identified by the combined satellite product. This is especially the case when the CERASAT meteorological forcing data was used.

Temporal autocorrelation was identified as a useful metric to investigate the discrepancies in the inertia of SM. Since it measures the response time of the model and in some cases is more intrinsically linked to the internal dynamics of the model, it gives us an opportunity to evaluate the structure of the model more precisely. The focus of upcoming DA studies could be to correct the internal parameters of the model by minimizing characteristic lag-time discrepancies, or similar metrics of soil moisture memory.

Author Contributions: Conceptualization, C.O. and P.P.; Data curation, N.R. and B.D.; Formal analysis, N.R. and B.D.; Funding, C.O. and P.P.; Investigation, N.R., B.D. and V.B.; Methodology: N.R., B.D., C.O. and P.P.; Software, N.R., B.D. and V.B.; Validation: N.R.; Visualization: N.R.; Supervision, C.O. and P.P.; Writing—original draft, N.R., B.D., C.O. and P.P.; Writing—review & editing, N.R., C.O., P.P., P.M. and J.P.

Funding: This research was partly funded by H202 MULTIPLY grant number 687320 and the French national program TOSCA-CNES.

Acknowledgments: We would like to acknowledge the ORCHIDEE team for their provision of the latest ORCHIDEE version. Special thanks go to Agnès Ducharne for her documentation of the ORCHIDEE hydrological modules and development of the soil resistance parameterisation. We would also like to thank the anonymous referees for their thoughtful and constructive reviews.

Conflicts of Interest: The authors declare no conflict of interest. The founding sponsors had no role in the design of the study; in the collection, analyses, or interpretation of data; in the writing of the manuscript, and in the decision to publish the results.

Appendix A

Table A1. Number of points used in the MSD decomposition shown in Figure 2.

	90 N–60 N	60 N–30 N	30 N–0	0–30 S	30 S–30 N
Jan–Mar	34	10,662	9398	6543	1874
Apr–Jun	9212	19,385	9578	6531	1866
Jul–Sep	9597	19,456	9557	6501	1820
Oct–Dec	3310	17,003	9545	6528	1865

Table A2. Properties for different classes of maximum LAI and soil textures. All values correspond the mean calculated over all the points in each class. Each metric is calculated between the model (both ORC_REF and ORC_NoRs runs) and the rescaled ESA-CCI SM product (in each case, CDF-matched to the corresponding ORCHIDEE run). The distribution of the different classes is displayed in Figure A1.

Index	Range Covered	% of Points		r		RMSD		c_t (Mod)		Δc_t (Mod-Obs)	
		REF	NoRs	REF	NoRs	REF	NoRs	REF	NoRs	REF	NoRs
1	$LAI_{MAX} < 0.5$	29.0	25.0	0.47	0.37	0.029	0.033	9.605	6.744	4.859	1.969
2	$0.5 \leq LAI_{MAX} < 2$	28.0	27.0	0.46	0.38	0.038	0.049	12.14	8.017	2.768	-0.760
3	$2 \leq LAI_{MAX} < 3$	23.0	25.0	0.51	0.47	0.037	0.046	11.841	9.711	1.744	-0.244
4	$3 \leq LAI_{MAX}$	18.0	21.0	0.67	0.68	0.038	0.043	14.943	12.631	1.796	-0.166
1	Coarse	25.0	25.0	0.50	0.46	0.028	0.029	8.217	5.719	2.756	0.254
2	Medium	62.0	62.0	0.49	0.44	0.037	0.049	12.392	9.734	2.768	0.111
3	Fine	11.0	11.0	0.68	0.65	0.038	0.042	17.380	13.497	4.207	0.350

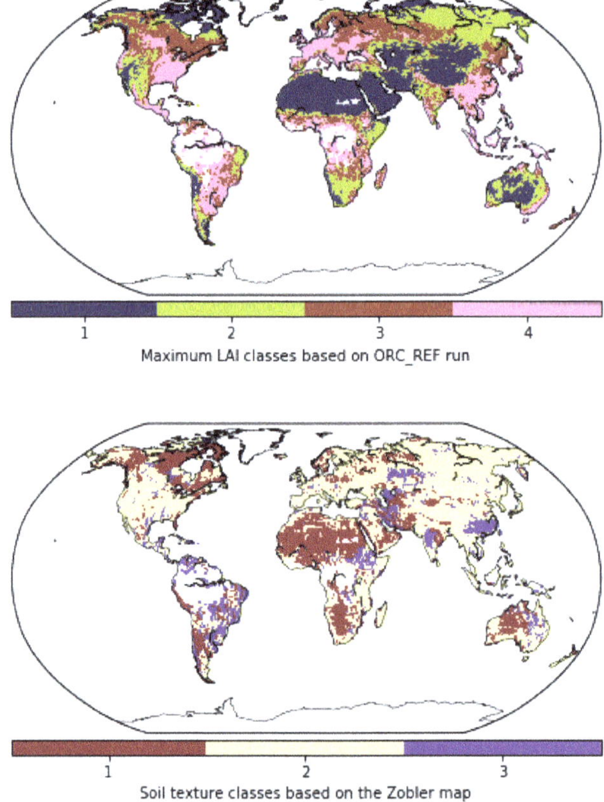

Figure A1. Global grid-points partitioned down into different LAI_{MAX} classes and soil classes. The distribution of LAI_{MAX} point differs slightly when considering the ORC_NoRs run (not shown).

To further understand characteristic lag time, we consider the global points divided into different categories: namely by LAI class (as a proxy for PFT classes) and by soil type. To be able to still consider temporal correlations, each pixel is consider only once by categorising the points by maximum LAI

(LAI$_{MAX}$). For the soil textures, the 5 soil textures from the Zobler map are reduced to only three, which are then called the Coarse (holding class 1), Medium (gathering classes 2, 3, 4) and Fine (class 5) soil textures.

References

1. Nemani, R.R.; Keeling, C.D.; Hashimoto, H.; Jolly, W.M.; Piper, S.C.; Tucker, C.J.; Myneni, R.B.; Running, S.W. Climate-driven increases in global terrestrial net primary production from 1982 to 1999. *Science* **2003**, *300*, 1560–1563. [CrossRef] [PubMed]
2. Booth, B.B.; Jones, C.D.; Collins, M.; Totterdell, I.J.; Cox, P.M.; Sitch, S.; Huntingford, C.; Betts, R.A.; Harris, G.R.; Lloyd, J. High sensitivity of future global warming to land carbon cycle processes. *Environ. Res. Lett.* **2012**, *7*, 024002. [CrossRef]
3. Le Quéré, C.; Moriarty, R.; Andrew, R.M.; Peters, G.P.; Ciais, P.; Friedlingstein, P.; Jones, S.D.; Sitch, S.; Tans, P.; Arneth, A.; et al. Global carbon budget 2014. *Earth Syst. Sci. Data* **2015**, *7*, 47–85. [CrossRef]
4. Anav, A.; Friedlingstein, P.; Beer, C.; Ciais, P.; Harper, A.; Jones, C.; Murray-Tortarolo, G.; Papale, D.; Parazoo, N.C.; Peylin, P.; et al. Spatiotemporal patterns of terrestrial gross primary production: A review. *Rev. Geophys.* **2015**, *53*, 785–818. [CrossRef]
5. Koster, R.D.; Dirmeyer, P.A.; Guo, Z.; Bonan, G.; Chan, E.; Cox, P.; Gordon, C.; Kanae, S.; Kowalczyk, E.; Lawrence, D.; et al. Regions of strong coupling between soil moisture and precipitation. *Science* **2004**, *305*, 1138–1140. [CrossRef] [PubMed]
6. Chen, C.; Haerter, J.O.; Hagemann, S.; Piani, C. On the contribution of statistical bias correction to the uncertainty in the projected hydrological cycle. *Geophys. Res. Lett.* **2011**, *38*. [CrossRef]
7. Brocca, L.; Moramarco, T.; Melone, F.; Wagner, W.; Hasenauer, S.; Hahn, S. Assimilation of surface-and root-zone ASCAT soil moisture products into rainfall–runoff modeling. *IEEE Trans. Geosci. Remote Sens.* **2012**, *50*, 2542–2555. [CrossRef]
8. Miralles, D.; Van Den Berg, M.; Teuling, A.; De Jeu, R. Soil moisture-temperature coupling: A multiscale observational analysis. *Geophys. Res. Lett.* **2012**, *39*. [CrossRef]
9. Taylor, C.M.; De Jeu, R.A.; Guichard, F.; Harris, P.P.; Dorigo, W.A. Afternoon rain more likely over drier soils. *Nature* **2012**, *489*, 423. [CrossRef] [PubMed]
10. Pitman, A.; Henderson-Sellers, A.; Desborough, C.; Yang, Z.L.; Abramopoulos, F.; Boone, A.; Dickinson, R.; Gedney, N.; Koster, R.; Kowalczyk, E.; et al. Key results and implications from phase 1 (c) of the project for intercomparison of land-surface parametrization schemes. *Clim. Dyn.* **1999**, *15*, 673–684. [CrossRef]
11. Guimberteau, M.; Zhu, D.; Maignan, F.; Huang, Y.; Chao, Y.; Dantec-Nédélec, S.; Ottlé, C.; Jornet-Puig, A.; Bastos, A.; Laurent, P.; et al. ORCHIDEE-MICT (v8, 4.1), a land surface model for the high latitudes: Model description and validation. *Geosci. Model Dev.* **2018**, *11*, 121. [CrossRef]
12. Rebel, K.; De Jeu, R.; Ciais, P.; Viovy, N.; Piao, S.; Kiely, G.; Dolman, A. A global analysis of soil moisture derived from satellite observations and a land surface model. *Hydrol. Earth Syst. Sci.* **2012**, *16*, 833–847. [CrossRef]
13. Ngo-Duc, T.; Laval, K.; Ramillien, G.; Polcher, J.; Cazenave, A. Validation of the land water storage simulated by Organising Carbon and Hydrology in Dynamic Ecosystems (ORCHIDEE) with Gravity Recovery and Climate Experiment (GRACE) data. *Water Resour. Res.* **2007**, *43*. [CrossRef]
14. Swenson, S.; Lawrence, D. Assessing a dry surface layer-based soil resistance parameterization for the Community Land Model using GRACE and FLUXNET-MTE data. *J. Geophys. Res. Atmos.* **2014**, *119*, 10299–10312. [CrossRef]
15. Ahmed, M.; Sultan, M.; Yan, E.; Wahr, J. Assessing and improving land surface model outputs over Africa using GRACE, field, and remote sensing data. *Surv. Geophys.* **2016**, *37*, 529–556. [CrossRef]
16. Kerr, Y.H.; Waldteufel, P.; Wigneron, J.P.; Delwart, S.; Cabot, F.; Boutin, J.; Escorihuela, M.J.; Font, J.; Reul, N.; Gruhier, C.; et al. The SMOS mission: New tool for monitoring key elements ofthe global water cycle. *Proc. IEEE* **2010**, *98*, 666–687. [CrossRef]
17. Entekhabi, D.; Njoku, E.G.; O'Neill, P.E.; Kellogg, K.H.; Crow, W.T.; Edelstein, W.N.; Entin, J.K.; Goodman, S.D.; Jackson, T.J.; Johnson, J.; et al. The soil moisture active passive (SMAP) mission. *Proc. IEEE* **2010**, *98*, 704–716. [CrossRef]

18. Bartalis, Z.; Wagner, W.; Naeimi, V.; Hasenauer, S.; Scipal, K.; Bonekamp, H.; Figa, J.; Anderson, C. Initial soil moisture retrievals from the METOP-A Advanced Scatterometer (ASCAT). *Geophys. Res. Lett.* **2007**, *34*, 5–9. [CrossRef]

19. Fascetti, F.; Pierdicca, N.; Pulvirenti, L.; Crapolicchio, R.; Muñoz-Sabater, J. A comparison of ASCAT and SMOS soil moisture retrievals over Europe and Northern Africa from 2010 to 2013. *Int. J. Appl. Earth Obs. Geoinf.* **2016**, *45*, 135–142. [CrossRef]

20. Crow, W.; Chen, F.; Reichle, R.; Xia, Y.; Liu, Q. Exploiting Soil Moisture, Precipitation, and Streamflow Observations to Evaluate Soil Moisture/Runoff Coupling in Land Surface Models. *Geophys. Res. Lett.* **2018**, *45*, 4869–4878. [CrossRef] [PubMed]

21. Al-Yaari, A.; Wigneron, J.P.; Ducharne, A.; Kerr, Y.; De Rosnay, P.; De Jeu, R.; Govind, A.; Al Bitar, A.; Albergel, C.; Munoz-Sabater, J.; et al. Global-scale evaluation of two satellite-based passive microwave soil moisture datasets (SMOS and AMSR-E) with respect to Land Data Assimilation System estimates. *Remote Sens. Environ.* **2014**, *149*, 181–195. [CrossRef]

22. Francois, C.; Quesney, A.; Ottlé, C. Sequential assimilation of ERS-1 SAR data into a coupled land surface–hydrological model using an extended Kalman filter. *J. Hydrometeorol.* **2003**, *4*, 473–487. [CrossRef]

23. Liu, Y.Y.; Parinussa, R.M.; Dorigo, W.A.; De Jeu, R.A.M.; Wagner, W.; Van Dijk, A.I.J.M.; McCabe, M.F.; Evans, J.P. Developing an improved soil moisture dataset by blending passive and active microwave satellite-based retrievals. *Hydrol. Earth Syst. Sci.* **2011**, *15*, 425–436. [CrossRef]

24. De Lannoy, G.J.; Reichle, R.H. Assimilation of SMOS brightness temperatures or soil moisture retrievals into a land surface model. *Hydrol. Earth Syst. Sci.* **2016**, *20*, 4895. [CrossRef]

25. Yang, K.; Zhu, L.; Chen, Y.; Zhao, L.; Qin, J.; Lu, H.; Tang, W.; Han, M.; Ding, B.; Fang, N. Land surface model calibration through microwave data assimilation for improving soil moisture simulations. *J. Hydrol.* **2016**, *533*, 266–276. [CrossRef]

26. Kolassa, J.; Reichle, R.; Draper, C. Merging active and passive microwave observations in soil moisture data assimilation. *Remote Sens. Environ.* **2017**, *191*, 117–130. [CrossRef]

27. Pinnington, E.; Quaife, T.; Black, E. Impact of remotely sensed soil moisture and precipitation on soil moisture prediction in a data assimilation system with the JULES land surface model. *Hydrol. Earth Syst. Sci.* **2018**, *22*, 2575–2588. [CrossRef]

28. Bolten, J.D.; Crow, W.T.; Zhan, X.; Jackson, T.J.; Reynolds, C.A. Evaluating the utility of remotely sensed soil moisture retrievals for operational agricultural drought monitoring. *IEEE J. Sel. Top. Appl. Earth Obs. Remote Sens.* **2010**, *3*, 57–66. [CrossRef]

29. Scholze, M.; Buchwitz, M.; Dorigo, W.; Guanter, L.; Quegan, S. Reviews and syntheses: Systematic Earth observations for use in terrestrial carbon cycle data assimilation systems. *Biogeosciences* **2017**, *14*, 3401. [CrossRef]

30. Vrugt, A.; Diks, G.H.; Gupta, V.; Bouten, W.; Verstraten, M. Improved treatment of uncertainty in hydrologic modeling: Combining the strengths of global optimization and data assimilation. *Water Resour. Res.* **2005**, *41*. [CrossRef]

31. Ajami, K.; Qingyun, D.; Soroosh, S. An integrated hydrologic Bayesian multimodel combination framework: Confronting input, parameter, and model structural uncertainty in hydrologic prediction. *Water Resour. Res.* **2007**, *43*. [CrossRef]

32. Dorigo, W.; Wagner, W.; Albergel, C.; Albrecht, F.; Balsamo, G.; Brocca, L.; Chung, D.; Ertl, M.; Forkel, M.; Gruber, A.; et al. ESA CCI Soil Moisture for improved Earth system understanding: State-of-the art and future directions. *Remote Sens. Environ.* **2017**, *203*, 185–215. [CrossRef]

33. Yin, J.; Zhan, X.; Zheng, Y.; Liu, J.; Fang, L.; Hain, C.R. Enhancing model skill by assimilating SMOPS blended soil moisture product into Noah land surface model. *J. Hydrometeorol.* **2015**, *16*, 917–931. [CrossRef]

34. Liu, Y.Y.; Dorigo, W.A.; Parinussa, R.M.; De Jeu, R.A.M.; Wagner, W.; McCabe, M.F.; Evans, J.P.; Van Dijk, A.I.J.M. Trend-preserving blending of passive and active microwave soil moisture retrievals. *Remote Sens. Environ.* **2012**, *123*, 280–297. [CrossRef]

35. Dorigo, W.A.; Gruber, A.; De Jeu, R.A.M.; Wagner, W.; Stacke, T.; Loew, A.; Albergel, C.; Brocca, L.; Chung, D.; Parinussa, R.M.; et al. Evaluation of the ESA CCI soil moisture product using ground-based observations. *Remote Sens. Environ.* **2015**, *162*, 380–395. [CrossRef]

36. Albergel, C.; Dorigo, W.; Reichle, R.; Balsamo, G.; De Rosnay, P.; Muñoz-Sabater, J.; Isaksen, L.; De Jeu, R.; Wagner, W. Skill and global trend analysis of soil moisture from reanalyses and microwave remote sensing. *J. Hydrometeorol.* **2013**, *14*, 1259–1277. [CrossRef]

37. Krinner, G.; Viovy, N.; de Noblet-Ducoudré, N.; Ogée, J.; Polcher, J.; Friedlingstein, P.; Ciais, P.; Sitch, S.; Prentice, I.C. A dynamic global vegetation model for studies of the coupled atmosphere-biosphere system. *Glob. Biogeochem. Cycles* **2005**, *19*, 1–33. [CrossRef]

38. De Rosnay, P.; Bruen, M.; Polcher, J. Sensitivity of surface fluxes to the number of layers in the soil model used in GCMs. *Geophys. Res. Lett.* **2000**, *27*, 3329–3332. [CrossRef]

39. De Rosnay, P.; Polcher, J.; Bruen, M.; Laval, K. Impact of a physically based soil water flow and soil-plant interaction representation for modeling large-scale land surface processes. *J. Geophys. Res. Atmos.* **2002**, *107*. [CrossRef]

40. D'Orgeval, T.; Polcher, J.; De Rosnay, P. Sensitivity of the West African hydrological cycle in ORCHIDEE to infiltration processes. *Hydrol. Earth Syst. Sci.* **2008**, *12*, 1387–1401. [CrossRef]

41. Ducharne, A.; Ottlé, C.; Maignan, F.; Vuichard, N.; Ghattas, J.; Wang, F.; Peylin, P.; Polcher, J.; Guimberteau, M.; Maugis, P.; et al. *The Hydrol Module of ORCHIDEE: Scientific Documentation [Rev 3977] and On, Work in Progress, Towards CMIP6v1*; Technical Report; Institut Pierre Simon Laplace: Paris, France, 2017.

42. Wagner, W.; Dorigo, W.; de Jeu, R.; Fernandez, D.; Benveniste, J.; Haas, E.; Ertl, M. Fusion of active and passive microwave observations to create an essential climate variable data record on soil moisture. *ISPRS Ann. Photogramm. Remote Sens. Spat. Inf. Sci. (ISPRS Ann.)* **2012**, *7*, 315–321. [CrossRef]

43. Owe, M.; De Jeu, R.; Holmes, T. Multisensor historical climatology of satellite-derived global land surface moisture. *J. Geophys. Res. Earth Surf.* **2008**, *113*, 1–17. [CrossRef]

44. Wagner, W.; Lemoine, G.; Rott, H. A method for estimating soil moisture from ERS scatterometer and soil data. *Remote Sens. Environ.* **1999**, *70*, 191–207. [CrossRef]

45. Rodell, M.; Houser, P.R.; Jambor, U.; Gottschalck, J.; Mitchell, K.; Meng, C.J.; Arsenault, K.; Cosgrove, B.; Radakovich, J.; Bosilovich, M.; et al. The Global Land Data Assimilation System. *Bull. Am. Meteorol. Soci.* **2004**, *85*, 381–394. [CrossRef]

46. Sellers, P.J.; Heiser, M.D.; Hall, F.G. Relations between surface conductance and spectral vegetation indices at intermediate (100 m^2 to 15 km^2) length scales. *J. Geophys. Res. Atmos.* **1992**, *97*, 19033–19059. [CrossRef]

47. CERA-SAT ECMWF Archived Dataset. Available online: https://www.ecmwf.int/en/forecasts/datasets/archive-datasets/reanalysis-datasets/cera-sat (accessed on 29 May 2018).

48. New, M.; Hulme, M.; Jones, P. Representing twentieth-century space–time climate variability. Part II: Development of 1901–96 monthly grids of terrestrial surface climate. *J. Clim.* **2000**, *13*, 2217–2238. [CrossRef]

49. Kalnay, E.; Kanamitsu, M.; Kistler, R.; Collins, W.; Deaven, D.; Gandin, L.; Iredell, M.; Saha, S.; White, G.; Woollen, J.; et al. The NCEP/NCAR 40-year reanalysis project. *Bull. Am. Meteorol. Soci.* **1996**, *77*, 437–471. [CrossRef]

50. Beck, H.E.; van Dijk, A.I.J.M.; Levizzani, V.; Schellekens, J.; Miralles, D.G.; Martens, B.; de Roo, A. MSWEP: 3-hourly 0.25° global gridded precipitation (1979–2015) by merging gauge, satellite, and reanalysis data. *Hydrol. Earth Syst. Sci.* **2017**, *21*, 589–615. [CrossRef]

51. Poulter, B.; MacBean, N.; Hartley, A.; Khlystova, I.; Arino, O.; Betts, R.; Bontemps, S.; Boettcher, M.; Brockmann, C.; Defourny, P.; et al. Plant functional type classification for earth system models: Results from the European Space Agency's Land Cover Climate Change Initiative. *Geosci. Model Dev.* **2015**, *8*, 2315–2328. [CrossRef]

52. Bontemps, S.; Boettcher, M.; Brockmann, C.; Kirches, G.; Lamarche, C.; Radoux, J.; Santoro, M.; Van Bogaert, E.; Wegmüller, U.; Herold, M.; et al. Multi-Year Global Land Cover Mapping at 300 M and Characterization for Climate Modelling: Achievements of the Land Cover Component of the Esa Climate Change Initiative. *Int. Arch. Photogramm. Remote Sens. Spat. Inf. Sci.* **2015**, 323–328. [CrossRef]

53. Scipal, K.; Drusch, M.; Wagner, W. Assimilation of a ERS scatterometer derived soil moisture index in the ECMWF numerical weather prediction system. *Adv. Water Resour.* **2008**, *31*, 1101–1112. [CrossRef]

54. Paulik, C.; Plocon, A.; Hahn, S.; Mistelbauer, T.; Schmitzer, M.; Alegrub88; iteubner.; Reimer, C. *TUW-GEO/Pytesmo: V0.6.10*; TU Wein: Vienna, Austria, 2018. [CrossRef]

55. Albergel, C.; Rüdiger, C.; Pellarin, T.; Calvet, J.C.; Fritz, N.; Froissard, F.; Suquia, D.; Petitpa, A.; Piguet, B.; Martin, E. From near-surface to root-zone soil moisture using an exponential filter: An assessment of the method based on in-situ observations and model simulations. *Hydrol. Earth Syst. Sci. Discuss.* **2008**, *12*, 1323–1337. [CrossRef]

56. Kobayashi, K.; Salam, M.U. Comparing simulated and measured values using mean squared deviation and its components. *Agron. J.* **2000**, *92*, 345–352. [CrossRef]

57. Delworth, T.L.; Manabe, S. The Influence of Potential Evaporation on the Variabilities of Simulated Soil Wetness and Climate. *Am. Meteorol. Soc.* **1988**. [CrossRef]

58. Bretherton, C.S.; Smith, C.; Wallace, J.M. An Intercomparison of Methods for Finding Coupled Patterns in Climate Data. *Am. Meteorol. Soc.* **1992**. [CrossRef]

59. Reichle, R.H.; Koster, R.D. Bias reduction in short records of satellite soil moisture. *Geophys. Res. Lett.* **2004**, *31*. [CrossRef]

60. Martens, B.; Miralles, D.G.; Lievens, H.; van der Schalie, R.; De Jeu, R.A.M.; Fernández-Prieto, D.; Beck, H.E.; Dorigo, W.A.; Verhoest, N.E.C. GLEAM v3: Satellite-based land evaporation and root-zone soil moisture. *Geosci. Model Dev.* **2017**, *10*, 1903–1925. [CrossRef]

61. Albergel, C.; Munier, S.; Leroux, D.J.; Dewaele, H.; Fairbairn, D.; Barbu, A.L.; Gelati, E.; Dorigo, W.; Faroux, S.; Meurey, C.; et al. Sequential assimilation of satellite-derived vegetation and soil moisture products using SURFEX_v8.0: LDAS-Monde assessment over the Euro-Mediterranean area. *Geosci. Model Dev.* **2017**, *10*, 3889–3912. [CrossRef]

62. Entekhabi, D.; Reichle, R.H.; Koster, R.D.; Crow, W.T. Performance metrics for soil moisture retrievals and application requirements. *J. Hydrometeorol.* **2010**, *11*, 832–840. [CrossRef]

63. Brocca, L.; Melone, F.; Moramarco, T.; Wagner, W.; Naeimi, V.; Bartalis, Z.; Hasenauer, S. Improving runoff prediction through the assimilation of the ASCAT soil moisture product. *Hydrol. Earth Syst. Sci.* **2010**, *14*, 1881–1893. [CrossRef]

64. Crow, W.; van Den Berg, M.; Huffman, G.; Pellarin, T. Correcting rainfall using satellite-based surface soil moisture retrievals: The Soil Moisture Analysis Rainfall Tool (SMART). *Water Resour. Res.* **2011**, *47*. [CrossRef]

65. Román-Cascón, C.; Pellarin, T.; Gibon, F.; Brocca, L.; Cosme, E.; Crow, W.; Fernández-Prieto, D.; Kerr, Y.H.; Massari, C. Correcting satellite-based precipitation products through SMOS soil moisture data assimilation in two land-surface models of different complexity: API and SURFEX. *Remote Sens. Environ.* **2017**, *200*, 295–310. [CrossRef]

66. Massari, C.; Brocca, L.; Moramarco, T.; Tramblay, Y.; Lescot, J.F.D. Potential of soil moisture observations in flood modelling: Estimating initial conditions and correcting rainfall. *Adv. Water Resour.* **2014**, *74*, 44–53. [CrossRef]

67. Massari, C.; Camici, S.; Ciabatta, L.; Brocca, L. Exploiting Satellite-Based Surface Soil Moisture for Flood Forecasting in the Mediterranean Area: State Update Versus Rainfall Correction. *Remote Sens.* **2018**, *10*, 292. [CrossRef]

68. Loew, A.; Stacke, T.; Dorigo, W.; De Jeu, R.; Hagemann, S. Potential and limitations of multidecadal satellite soil moisture observations for selected climate model evaluation studies. *Hydrol. Earth Syst. Sci.* **2013**, *17*, 3523–3542. [CrossRef]

69. Polcher, J.; Piles, M.; Gelati, E.; Barella-Ortiz, A.; Tello, M. Comparing surface-soil moisture from the SMOS mission and the ORCHIDEE land-surface model over the Iberian Peninsula. *Remote Sens. Environ.* **2016**, *174*, 69–81. [CrossRef]

70. Kidd, R.; Haas, E. *ESA Climate Change Initiative Phase II Soil Moisture—USER GUIDE*; EODC: Vienna, Austria, 2015; p. 56.

71. Boening, C.; Willis, J.K.; Landerer, F.W.; Nerem, R.S.; Fasullo, J. The 2011 La Niña: So strong, the oceans fell. *Geophys. Res. Lett.* **2012**, *39*. [CrossRef]

72. Wilks, D. *Statistical Methods in the Atmospheric Sciences*; Academic Press: Cambridge, MA, USA, 1995; pp. 51–54.

73. McColl, K.A.; Alemohammad, S.H.; Akbar, R.; Konings, A.G.; Yueh, S.; Entekhabi, D. The global distribution and dynamics of surface soil moisture. *Nat. Geosci.* **2017**, *10*, 100. [CrossRef]

74. Koster, R.D.; Milly, P. The interplay between transpiration and runoff formulations in land surface schemes used with atmospheric models. *J. Clim.* **1997**, *10*, 1578–1591. [CrossRef]

75. Dee, D.P.; Da Silva, A.M. Data assimilation in the presence of forecast bias. *Q. J. R. Meteorol. Soc.* **1998**, *124*, 269–295. [CrossRef]

76. Nearing, G.S.; Gupta, H.V.; Crow, W.T.; Gong, W. An approach to quantifying the efficiency of a Bayesian filter. *Water Resour. Res.* **2013**, *49*, 2164–2173. [CrossRef]

77. Kumar, S.; Peters-Lidard, C.; Santanello, J.; Reichle, R.; Draper, C.; Koster, R.; Nearing, G.; Jasinski, M. Evaluating the utility of satellite soil moisture retrievals over irrigated areas and the ability of land data assimilation methods to correct for unmodeled processes. *Hydrol. Earth Syst. Sci.* **2015**, *19*, 4463–4478. [CrossRef]

78. Massari, C.; Brocca, L.; Tarpanelli, A.; Moramarco, T. Data Assimilation of Satellite Soil Moisture into Rainfall-Runoff Modelling: A Complex Recipe? *Remote Sens.* **2015**, *7*, 11403–11433. [CrossRef]

79. Alvarez-Garreton, C.; Ryu, D.; Western, A.W.; Crow, W.T.; Su, C.H.; Robertson, D.R. Dual assimilation of satellite soil moisture to improve streamflow prediction in data-scarce catchments. *Water Resour. Res.* **2016**, *52*, 5357–5375. [CrossRef]

80. Ghannam, K.; Nakai, T.; Paschalis, A.; Oishi, C.A.; Kotani, A.; Igarashi, Y.; Kumagai, T.; Katul, G.G. Persistence and memory timescales in root-zone soil moisture dynamics. *Water Resour. Res.* **2016**, *52*, 1427–1445. [CrossRef]

81. Best, M.; Pryor, M.; Clark, D.; Rooney, G.; Essery, R.; Ménard, C.; Edwards, J.; Hendry, M.; Porson, A.; Gedney, N.; et al. The Joint UK Land Environment Simulator (JULES), model description—Part 1: Energy and water fluxes. *Geosci. Model Dev.* **2011**, *4*, 677–699. [CrossRef]

82. Clark, D.; Mercado, L.; Sitch, S.; Jones, C.; Gedney, N.; Best, M.; Pryor, M.; Rooney, G.; Essery, R.; Blyth, E.; et al. The Joint UK Land Environment Simulator (JULES), model description—Part 2: Carbon fluxes and vegetation dynamics. *Geosci. Model Dev.* **2011**, *4*, 701–722. [CrossRef]

83. Raddatz, T.; Reick, C.; Knorr, W.; Kattge, J.; Roeckner, E.; Schnur, R.; Schnitzler, K.G.; Wetzel, P.; Jungclaus, J. Will the tropical land biosphere dominate the climate–carbon cycle feedback during the twenty-first century? *Clim. Dyn.* **2007**, *29*, 565–574. [CrossRef]

84. Scholze, M.; Kaminski, T.; Knorr, W.; Blessing, S.; Vossbeck, M.; Grant, J.; Scipal, K. Simultaneous assimilation of SMOS soil moisture and atmospheric CO2 in-situ observations to constrain the global terrestrial carbon cycle. *Remote Sens. Environ.* **2016**, *180*, 334–345. [CrossRef]

85. Peylin, P.; Bacour, C.; MacBean, N.; Leonard, S.; Rayner, P.; Kuppel, S.; Koffi, E.; Kane, A.; Maignan, F.; Chevallier, F.; et al. A new stepwise carbon cycle data assimilation system using multiple data streams to constrain the simulated land surface carbon cycle. *Geosci. Model Dev.* **2016**, *9*, 3321. [CrossRef]

 *remote sensing*

Review

Satellite and In Situ Observations for Advancing Global Earth Surface Modelling: A Review

Gianpaolo Balsamo [1,*,†], Anna Agustì-Parareda [1], Clément Albergel [2], Gabriele Arduini [1], Anton Beljaars [1], Jean Bidlot [1], Eleanor Blyth [3] Nicolas Bousserez [1], Souhail Boussetta [1], Andy Brown [1], Roberto Buizza [1,4], Carlo Buontempo [1], Frédéric Chevallier [5], Margarita Choulga [1], Hannah Cloke [6], Meghan F. Cronin [7], Mohamed Dahoui [1], Patricia De Rosnay [1], Paul A. Dirmeyer [8], Matthias Drusch [9], Emanuel Dutra [10], Michael B. Ek [11], Pierre Gentine [12], Helene Hewitt [13], Sarah P. E. Keeley [1], Yann Kerr [14], Sujay Kumar [15], Cristina Lupu [1], Jean-François Mahfouf [2], Joe McNorton [1], Susanne Mecklenburg [9], Kristian Mogensen [1], Joaquín Muñoz-Sabater [1], Rene Orth [16], Florence Rabier [1], Rolf Reichle [15], Ben Ruston [17], Florian Pappenberger [1], Irina Sandu [1], Sonia I. Seneviratne [18], Steffen Tietsche [1], Isabel F. Trigo [19], Remko Uijlenhoet [20], Nils Wedi [1], R. Iestyn Woolway [6] and Xubin Zeng [21]

[1]   European Centre for Medium-range Weather Forecasts (ECMWF), Reading RG2 9AX, UK;
      Anna.Agusti-Panareda@ecmwf.int (A.A.-P.); Gabriele.Arduini@ecmwf.int (G.A.); Anton.Beljaars@ecmwf.int
      (A.B.); Jean.Bidlot@ecmwf.int (J.B.); Nicolas.Bousserez@ecmwf.int (N.B.); Souhail.Boussetta@ecmwf.int
      (S.B.); Andy.Brown@ecmwf.int (A.B.); Carlo.Buontempo@ecmwf.int (C.B.);
      Margarita.Choulga@ecmwf.int (M.C.); Mohamed.Dahoui@ecmwf.int (M.D.);
      Patricia.Rosnay@ecmwf.int (P.D.R.); Sarah.Keeley@ecmwf.int (S.P.E.K.); Cristina.Lupu@ecmwf.int (C.L.);
      Joe.McNorton@ecmwf.int (J.M.); Kristian.Mogensen@ecmwf.int (K.M.); Joaquin.Munoz@ecmwf.int (J.M.-S.);
      Florence.Rabier@ecmwf.int (F.R.); Florian.Pappenberger@ecmwf.int (F.P.); Irina.Sandu@ecmwf.int (I.S.);
      Steffen.Tietsche@ecmwf.int (S.T.); Nils.Wedi@ecmwf.int (N.W.)
[2]   Météo-France, Centre National de Recherches Météorologique, 31000 Toulouse, France;
      clement.albergel@meteo.fr (C.A.); jean-francois.mahfouf@meteo.fr (J.-F.M.)
[3]   Centre for Ecology & Hydrology, Wallingford OX10 8BB, UK; emb@ceh.ac.uk
[4]   Scuola Superiore Sant'Anna, 56127 Pisa, Italy; roberto.buizza@santannapisa.it
[5]   Laboratoire des Sciences du Climat et de l'Environnement, Institut Pierre-Simon-Laplace, Commissariat à
      l'énergie atomique et aux énergies alternatives, LSCE/IPSL/CEA, 91190 Gif sur Yvette, France;
      frederic.chevallier@lsce.ipsl.fr
[6]   Meteorology Depart., University of Reading, Reading RG6 7BE, UK; h.l.cloke@reading.ac.uk (H.C.);
      r.i.woolway@reading.ac.uk (R.I.W.)
[7]   National Oceanic and Atmospheric Administration, Pacific Marine Environmental Laboratory, Seattle,
      WA 98115, USA; meghan.f.cronin@noaa.gov
[8]   Center for Ocean-Land-Atmosphere Studies, George Mason University, Fairfax, VA 22030, USA;
      pdirmeye@gmu.edu
[9]   European Space Agency (ESA), European Space Research and Technology Centre (ESTEC),
      2201 AZ Noordwijk, The Netherlands; matthias.drusch@esa.int (M.D.); Susanne.Mecklenburg@esa.int (S.M.)
[10] Instituto Dom Luiz, University of Lisbon, 1749-016 Lisbon, Portugal; endutra@fc.ul.pt
[11] National Center for Atmospheric Research, Boulder, CO 80305, USA; ek@ucar.edu
[12] Department of Earth and Environmental Engineering, Columbia University, New York, NY 10027, USA;
      pg2328@columbia.edu
[13] UK MetOffice, Exeter EX1 3PB, UK; helene.hewitt@metoffice.gov.uk
[14] Centre National d'Etudes Spatiales, CESBIO, 31401 Toulouse, France; yann.kerr@cesbio.cnes.fr
[15] National Aeronautics and Space Administration (NASA), Goddard Space Flight Center (GSFC), Greenbelt,
      MD 20771, USA; sujay.v.kumar@nasa.gov (S.K.); rolf.reichle@nasa.gov (R.R.)
[16] Max Planck Institute for Biogeochemistry, 07745 Jena, Germany; rene.orth@bgc-jena.mpg.de
[17] Naval Research Laboratory (NRL), Monterey, CA 93943, USA; Ben.Ruston@nrlmry.navy.mil
[18] Eidgenössische Technische Hochschule (ETH), 8092 Zürich, Switzerland; sonia.seneviratne@ethz.ch
[19] Instituto Português do Mar e da Amosfera (IPMA), 1749-077 Lisbon, Portugal; isabel.trigo@ipma.pt
[20] Department of Environmental Sciences, Wageningen University and Research, 6708 PB Wageningen,
      The Netherlands; remko.uijlenhoet@wur.nl

21   Department Hydrology and Atmospheric Sciences, University of Arizona, Tucson, AZ 85721, USA;
      xubin@atmo.arizona.edu
*   Correspondence: gianpaolo.balsamo@ecmwf.int; Tel.: +44-759-5331517
†   Current address: European Centre for Medium-Range Weather Forecasts (ECMWF), Shinfield Park, Reading
      RG2 9AX, UK.

Received: 15 October 2018; Accepted: 5 December 2018; Published: 14 December 2018

Abstract: In this paper, we review the use of satellite-based remote sensing in combination with in situ data to inform Earth surface modelling. This involves verification and optimization methods that can handle both random and systematic errors and result in effective model improvement for both surface monitoring and prediction applications. The reasons for diverse remote sensing data and products include (i) their complementary areal and temporal coverage, (ii) their diverse and covariant information content, and (iii) their ability to complement in situ observations, which are often sparse and only locally representative. To improve our understanding of the complex behavior of the Earth system at the surface and sub-surface, we need large volumes of data from high-resolution modelling and remote sensing, since the Earth surface exhibits a high degree of heterogeneity and discontinuities in space and time. The spatial and temporal variability of the biosphere, hydrosphere, cryosphere and anthroposphere calls for an increased use of Earth observation (EO) data attaining volumes previously considered prohibitive. We review data availability and discuss recent examples where satellite remote sensing is used to infer observable surface quantities directly or indirectly, with particular emphasis on key parameters necessary for weather and climate prediction. Coordinated high-resolution remote-sensing and modelling/assimilation capabilities for the Earth surface are required to support an international application-focused effort.

Keywords: earth-observations; earth system modelling; direct and inverse methods

---

1. Introduction

The era of information technology and advances in high-performance computing are opening up possibilities to realistically simulate the evolution of the Earth system at scales not thought possible before. In hydrology, this has been labeled hyper-resolution, referring to a finer spatial discretisation than what was previously attempted at global scale [1–3] and it has triggered a fruitful debate on the importance of information-driven advances [4,5]. The debate is partly about the resolution and model complexity needed to resolve and represent regional-scale water-cycle processes [6–8]. In weather and climate studies, calls for higher resolution have also been made [9,10], as currently unresolved cloud processes are one of the obstacles for improved climate predictions. Attaining these finely resolved scales with the entire Earth System including the land, atmosphere, oceans and ice surfaces involves major technological advances in scalability [11] due to computing energy barriers, time-to-solution requirements, and hardware/software limitations. In this review, we emphasize the essential role of Earth Observations data from polar orbiting and geostationary satellites as a key driver for advancing models, and we present a number of examples of how satellite data combined with in situ observations have successfully driven model improvement. A global kilometric representation of the Earth surface that can feed into the next generation of weather and climate models will have to be driven by satellite observations [12,13]. The expected increased use of subseasonal-to-seasonal (S2S) forecasts [14] makes it even more important to improve the representation of the land and ocean surface in current models, given the essential role of land–ocean–atmosphere interactions in S2S predictability. Such an ambitious goal can be achieved only via the synergistic development of models and model-data fusion techniques that make use of the wealth of satellite observations, ingested into the modelling chain, to genuinely characterize the details of land physiography, coastal and inland water and all surface heterogeneity that impacts water and energy fluxes. The drive to achieve physical realism in Earth surface modelling is common to hydrological and environmental applications [15] and resolution and model complexity are key ingredients. Not all surface variables can be observed routinely and constraining most sub-surface properties remains challenging. Therefore, the representation of uncertainties is essential in forecasting applications [16].

Information on the parameters necessary to describe the thermal, radiative, and hydrological properties of the surface must also be extracted, and uncertainties evaluated, so that the models can predict the main reservoirs and fluxes for water, energy and mass exchanges. The availability of stored energy and water within the Earth surface is further modulated by land use (such as irrigation, agricultural practice, forest management) and sub-surface properties which drive the high spatial heterogeneity. That is why there is a growing realization that human activities, impacts and feedback need to be represented in models [17]. The projected increase in the occurrence of extreme weather events (heatwaves, but also droughts and floods in some regions [18]) in connection with global warming in the 21st century, combined with the proven fact of observed warming trends since pre-industrial time [19], are both of global-scale and local relevance. The latest IPCC (Intergovernmental Panel on Climate Change) assessment [18] indicates that further warming, even for only 0.5 °C of global warming, would further increase the occurrence or intensity of warm spells on global scale, of heavy precipitation in several regions, and of droughts in some regions; heavy precipitation associated with tropical cyclones is also projected to increase with increased global warming. Similarly, the fast-paced change observed for the cryosphere including the significant trends in both snow and sea-ice cover [20–22] means that we need an improved account of surface thermodynamic processes at the relevant scales in interaction with the diurnal cycle and synoptic variability. Observing and simulating the response of land biophysical variables and the cryosphere to extreme events is a major scientific challenge in Numerical Weather Prediction (NWP) and environmental applications, which is relevant also in the context of climate change adaptation.

The modeling of terrestrial variables can be improved through the integration of observations. Integrating observations into models covers several aspects: (1) the dynamic integration of observations into models through data assimilation techniques, (2) the use of observations for model validation and evolution and (3) the mapping of the model parameters used to characterize the representation of land properties within the model (e.g., soil properties, land cover). Although the assimilation from satellite data is becoming widely adopted in Earth system modelling also at the surface [23], limited use tends to be made of the wealth of Earth observations (EO) data available for model validation, evolution and parameterisation. There is a thus a great opportunity to improve upon current development strategies. For example, it is well known that the use of indirect observations such as the near surface temperatures to constrain model soil moisture [24] can lead to compensating errors. Remote sensing can also greatly support model development for the cold processes associated with seasonal snow cover. Snow is a key component of the Earth system due to the unique surface "shielding" properties and its presence over large parts of the Earth's surface. The radiative (albedo), thermal (insulation) and mechanic (roughness) characteristics of a snow covered surface differ greatly from those of a snow-free surface. While the particular microphysical characteristics of snow are challenging to represent in models, they provide unique radiative signatures that can be explored by EO data. Remote sensing observations are particularly useful in this context because they are now available for all parts of the globe. However, most snow remote sensing products greatly benefit from in situ observations [25]. Many satellite-derived products relevant to the hydrological and vegetation cycles are already available and the use of combined observing platforms is increasingly adopted in order to enable the move towards the kilometer-scale and support global Earth surface modelling.

The focus of this review is to cover relevant successful examples, inspired by the National Aeronautics and Space Administration (NASA) decadal survey [26] and by the World Meteorological Organisation (WMO) Global Climate Observing System (GCOS) essential climate variable observing capabilities [27,28], without aiming to be exhaustive. We present recent use of Earth Observations data pertinent to constraining the surface energy, water and carbon cycles in models. We discuss their relevance for the next generation of surface models for Earth system monitoring and forecasting applications, and we present options for closer collaboration and more rapid information uptake between data providers and model developers. In Section 2, satellite and in situ observations of the earth surface are listed with the relevant instruments information. Section 3 presents the Earth surface model development examples over land and ocean. Section 4 presents some of the pathways in which EO data guide enhanced global-scale local relevant monitoring and forecasting applications. Section 5 discusses how to increase EO data uptake in earth system modelling moving towards global km-scale, including international coordination activities. Section 6 presented a summary and the outlook.

2. EO Satellite and In Situ Observations for Earth Surface

This section reviews the satellites and in situ platforms that have dedicated sensors for Earth surface monitoring and that have been used to inform model development. Both surface dedicated and surface sensitive instruments are considered.

2.1. SMOS Soil Moisture Ocean Salinity Mission

The Soil Moisture and Ocean Salinity SMOS mission [29–31] was launched on 2 November 2009. This is an interferometric synthetic aperture radiometer mission that can measure the Earth's natural emission at a relatively low microwave frequency of 1.4 GHz in full-polarization and covering multiple incidence angles from 0 to 60 degrees with angles in the 40–45 range accessible all across the swath. SMOS has 69 small static antennae that, thanks to interferometry, are equivalent to a filled antenna of 8 m, achieving a spatial resolution on the ground ranging from 27 to 55 km. The satellite follows a sun-synchronous polar orbit with 6:00 a.m. (ascending half-orbit) and 6:00 p.m. (descending half-orbit) Equator overpass times. The mission provides global brightness temperature and retrieves soil moisture fields in near real time [32,33]. The land products are also generated globally at 15 km resolution with a latency of 24 h. These products are surface soil moisture, vegetation opacity, and surface dielectric constant [30]. Level-3 and Level-4 are also produced [34] and cover a large range of products [35,36], e.g., root zone soil moisture and drought index, L-Band Vegetation Optical Depth (VOD, [37–39]), thin sea ice [40], global rainfall estimates [41], to name a few.

2.2. SMAP Soil Moisture Active Passive Mission

The Soil Moisture Active Passive SMAP mission [42], launched on 31 January 2015 carries both an L-band (1.4 GHz) passive microwave radiometer and a radar (although the latter is no longer functioning). Using a 6-m diameter spinning antenna, the observatory provides near-global coverage every two to three days. The primary focus of the SMAP mission is on soil moisture retrieval and freeze/thaw detection, but SMAP observations have also been used for ocean salinity [43] and wind retrievals [44]. To provide estimates of root zone soil moisture and complete spatio-temporal coverage, SMAP radiometer observations are routinely assimilated into the NASA Catchment land surface model to generate the SMAP Level-4 Soil Moisture (L4SM) data product [45]. The L4SM product is generated globally at 9 km resolution every three hours and distributed to the public with a latency of 2.5 days from the time of observation.

2.3. TERRA/AQUA—MODIS

Terra and Aqua (Latin for land and water) are NASA Earth Science satellite missions equipped with multi-sensors aimed at studying the Earth's water cycle including evaporation from the oceans, water vapor in the atmosphere, clouds, precipitation, soil moisture, sea ice, land ice, and snow cover on the land and ice. Terra was launched in December 1999 and initiated data transmission in February 2000, with its five instruments, ASTER, CERES, MISR, MOPITT, and MODIS (see acronyms list for details, page 55). Aqua was launched on 4 May 2002, and had six Earth-observing instruments on board, AIRS, AMSU, AMSR-E, CERES, HSB, and MODIS, which are collecting a variety of global data sets. Aqua was originally developed for a six-year design life but has now far exceeded that original goal. About half of the sensors continue transmitting high-quality, in particular from AIRS, CERES, and MODIS, with reduced functionality from AMSU-A channels. AMSR-E became inactive since 2011, with HSB, collected approximately nine months of high quality data but failed in February 2003. For both satellite platforms MODIS (Moderate-resolution Imaging Spectroradiometer) is a key instrument with its 36 spectral bands ranging in wavelength from 0.4 μm to 14.4 μm and at varying spatial resolution (two bands at 250 m, five bands at 500 m and 29 bands at 1 km). Together, the instruments image the entire Earth every one to two days. A large amount of MODIS-based Earth's surface products are currently used in weather and climate modelling, as detailed in Section 2.

2.4. LANDSAT and Its Legacy

The Landsat program initiated with the launch of the first satellite in July 1972 and the latest Landsat-8 launched in February 2013 is the longest-serving EO platform of high resolution mapping satellites. Landsat-9 foreseen to launch in 2020 will continue this legacy that has allowed to monitor key surface changes over time, including water-bodies changes, glaciers retreat, large fires, urbanization and agricultural land expansions. Landsat 8 ensures the continued acquisition and availability of Landsat data utilizing a two-sensor payload, Operational Land Imager (OLI) and Thermal Infrared Sensor (TIRS). These provide imagery from 15 m to 100 m resolution and up to 700 scenes per day. Landsat data volume is prohibitive for most research applications at global scale and often used in mapping applications to produce more accurate medium resolution (e.g., at 1 km) as mentioned in Section 4.

2.5. SEASAT and Its Legacy

The Seasat was the first ocean dedicated satellite launched in June 1978 although it operated only for 106 days. Earth-orbiting satellites such as TOPEX/Poseidon, and scatterometers on NASA Scatterometer (NSCAT), QuickSCAT, all launched in the 1990s and Jason-1, launched in 2001, Jason-2, launched in June 2008, and Jason-3 launched 17 January 2016 followed the initial mission and continued its legacy. The MetOp and its successor MetOp-SG programmes are the current and future legacy to enhance and expand the European EO capability.

2.6. Copernicus Sentinels

The Copernicus Sentinels program developed by the European Space Agency consists of a family of missions designed for the operational monitoring of the Earth system with continuity up to 2030 and beyond (Figure 1 and Table 1). The Sentinels' concept is based on two satellites per mission necessary to guarantee a good revisit and global coverage and to provide more robust datasets in support of the Copernicus Services.

On-board sensors include both radar and multi-spectral imagers for land, ocean and atmospheric monitoring:

Sentinel-1 is a polar-orbiting, all-weather, day-and-night radar imaging mission for land and ocean services. Sentinel-1A was launched on 3 April 2014 and Sentinel-1B on 25 April 2016.

Sentinel-2 is a polar-orbiting, multi-spectral high-resolution imaging mission for land monitoring to provide, for example, imagery of vegetation, soil and water cover, inland waterways and coastal areas. Sentinel-2 can also deliver information for emergency services. Sentinel-2A was launched on 23 June 2015 and Sentinel-2B followed on 7 March 2017.

Sentinel-3 is a polar-orbiting multi-instrument mission to measure sea-surface topography, sea- and land-surface temperature, ocean colour and land colour with high-end accuracy and reliability. The mission will support ocean forecasting systems, as well as environmental and climate monitoring. Sentinel-3A was launched on 16 February 2016 and Sentinel-3B has been launched on 25 April 2018.

Sentinel-5 Precursor—Sentinel-5P—polar-orbiting mission is dedicated to trace gases and aerosols with a focus on air quality and climate. It has been developed to reduce the data gaps between the Envisat satellite—in particular the Sciamachy instrument—and the launch of Sentinel-5. Sentinel-5P is orbiting since 13 October 2017.

Sentinel-4 is devoted to atmospheric monitoring that will be embarked on a Meteosat Third Generation-Sounder (MTG-S) satellite in geostationary orbit and it will provide European and North African coverage.

Sentinel-5 will monitor the atmosphere from polar orbit aboard a MetOp Second Generation satellite.

Sentinel-6 will be a polar-orbiting mission carrying a radar altimeter to measure global sea-surface height, primarily for operational oceanography and for climate studies.

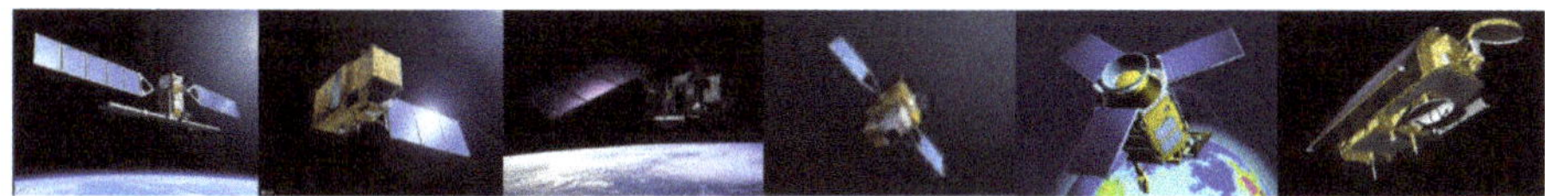

Figure 1. Copernicus Sentinels 1 to 6 (left to right).

Table 1. Copernicus Sentinels launch dates and relevance in modelling application (aggregate use refers to composites in space and time, while direct use refers to single observation).

Sentinel 1	Launch Date	ESM Relevance
Sentinel 1A	3 April 2014	aggregate
Sentinel 1B	25 April 2016	aggregate
Sentinel 2A	23 June 2015	aggregate
Sentinel 2B	7 March 2017	aggregate
Sentinel 3A	16 February 2016	direct
Sentinel 3B	25 April 2018	direct
Sentinel 5P	13 October 2017	direct
Sentinel 6	2020 (expected)	direct

Future Sentinels missions include polar-orbiting satellites dedicated to support an anthropogenic CO_2 monitoring capacity. In the longer term, thermal Infrared imagers with focus drought monitoring and on polar regions are being planned, and a hyper-spectral instrument will enable more precise agricultural monitoring.

Special focus here is given to the Sentinel-3 mission [46], jointly operated by the European Space Agency (ESA) and the European Organisation for the Exploitation of Meteorological Satellites (EUMETSAT), and which tackles precisely the scales of interest of our study and has operational requirements. The objectives of Sentinel-3 are to measure sea-surface topography, sea- and land-surface temperature and ocean- and land-surface color in support of ocean forecasting systems, and for environmental and climate monitoring.

The Sentinel-3 satellites series ensures global, frequent and near-real time ocean, ice and land monitoring, with the provision of observation data in routine, long term (up to 20 years of operations) and continuous fashion, with a consistent quality and a high level of reliability and availability. The Sentinel-3 mission addresses these requirements by implementing and operating the following instruments, building on experience and heritage from the ERS and ENVISAT missions:

A dual frequency, delay-Doppler Synthetic Aperture Radar Altimeter (SRAL) instrument supported by a dual frequency passive microwave radiometer (MWR) for wet-tropospheric correction, and a Precise Orbit Determination package. This combined package provides measurements of sea-surface height and topography measurements over sea ice, ice sheets, rivers and lakes.

A highly sensitive Ocean and Land Colour Imager (OLCI) delivering multi-channel wide-swath optical measurements for ocean and land surfaces. With 21 bands (compared to the 15 on Envisat's MERIS) and a design optimised to minimise sun-glint, OLCI marks a new generation of measurements over the ocean and land attaining a resolution of 300 m over all surfaces.

A dual-view Sea and Land Surface Temperature Radiometer (SLSTR) delivering accurate surface ocean, land, and ice temperature, with an accuracy better than 0.3 K. SLSTR measures in nine spectral channels and two additional bands optimised for fire monitoring. SLSTR has a spatial resolution in the visible and shortwave infrared channels of 500 m and 1 km in the thermal infrared channels. The swath of OCLI and nadir SLSTR fully overlap.

Sentinel-3A and 3B were launched in February 2016 and April 2018, respectively. Full performance will be achieved once both Sentinel-3A and 3B will be in routine operations at the end of 2018, with a revisit time of less than two days for OLCI and less than one day for SLSTR at the equator. The satellite orbit provides a 27-day repeat for the topography package, with a 4-day sub-cycle (defined as the minimum number of days after which the ground track of the satellite nearly repeats itself within a small offset).

The Sentinel-3 ground segment systematically acquires, processes and distributes a set of pre-defined core data products to the users (see https://earth.esa.int/web/sentinel/missions/sentinel-3/data-products).

An example of the sensor capability is given in Figure 2 for land surface temperature, vegetation state (chlorophyll index) and snow cover.

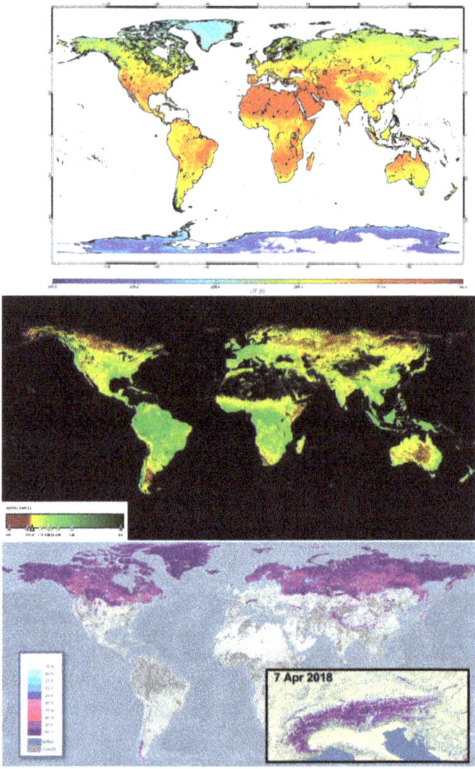

Figure 2. Sentinel-3 derived kilometre-scale products for Earth System Modelling. (**top**) SLSTR Land Surface Temperature monthly composite for September 2016 (Credit: D. Ghent, University of Leicester) (**middle**) OLCI Terrestrial Chlorophyll Index April–May 2017 (Credit: University of Southampton–J. Dash/Brockman Consult (S3-MPC)) (**bottom panel**) SLSTR Snow Extent Product, 5–10 April 2018 (Credit: ENVEO). Copyright: All figures contain modified Copernicus Sentinel data (2018), processed by ESA, CC BY-SA 3.0 IGO.

2.7. GOES, METEOSAT and Other Geostationary Satellites

Geosynchronous satellites have the great advantage of orbiting with the Earth's rotation and therefore of providing observations of the same location at high temporal frequency. Such capability is particularly relevant for the estimation of variables rapidly changing with time over land and ocean surfaces. EUMETSAT currently operates four satellites of the Meteosat Second Generation (MSG) series over Europe and Africa (Meteosat-9, Meteosat-10 and Meteosat-11) and over the Indian Ocean (Meteosat-8). The redundancy over the former (nominal $0°$E) disk allows a full disk service together with a rapid scan service over Europe, keeping one satellite as backup. MSG series observations are available for more than 14 years, since 2004. If combined with Meteosat First Generation satellites, these can be extended back in time to the early 1980s, despite the poorer spatial and temporal samplings, as well as lower spectral resolutions of that series.

The National Oceanic and Atmospheric Administration, NOAA's operating strategy calls for two GOES satellites to be active at all times, one satellite to observe the Atlantic Ocean and the eastern half of the USA, and the other to observe the Pacific Ocean and the western part of the country. GOES-16 (or GOES-East) positioned at $75°$W longitude and GOES-15 (or GOES-West) positioned at $135°$W longitude are currently, the two operational meteorological satellites in geostationary orbit over the equator operated by NOAA. GOES-16 replaced GOES-13 on 8 January 2018 and GOES-15 replaced GOES-11 on 6 December 2011. Additionally, GOES-13 located at $60°$W supports Central and South America to prevent data outages during the GOES-16 rapid scan operations, while GOES-14 is maintained as on-orbit spare to replace either, GOES-15 or GOES-16, in the event of failure.

Russia's new-generation weather satellite Elektro-L No.1 (GOMS-2) operates at $76°$E over the Indian Ocean. India also operates geostationary satellites called INSAT-3D which carry instruments for meteorology. China currently maintains three Fengyun geostationary satellites (FY-2E at $86.5°$E, FY-2F at $112°$E, and FY-2G at $105°$E) in operations, while Japan maintains MTSAT-2 (very similar to GOES satellites prior to GOES-16) located over the mid Pacific at $145°$E (although inactive since 10 March 2017) and the Himawari-8 (with instruments similar to GOES-16/17) at $140°$E. Table 2 provides an overview of geostationary satellite series launched until this day and their main coverage area.

Table 2. Geostationary satellites operated positions as of 2018. Operational position can vary over time.

Geostationary Satellite	Agency	Operating Longitude	Area Coverage
Meteosat First Generation series (up to Meteosat-7)	EUMETSAT	$0°$E; $57°$E–$63°$E; $50°$W	Africa, Europe, partly South America; Indian Ocean Coverage
MSG series (Meteosat-8 to Meteosat-11)	EUMETSAT	$0°$E; $9.5°$E; $45°$E	Africa, Europe, partly South America ; Indian Ocean Coverage
GOES-East/West	NOAA	$75°$W; $135°$W	East Satellite: North and South America, Atlantic Ocean; West Satellite: North and South America, Pacific Ocean
GOMS-2	Roshydromet	$76°$E	Eurasia, Indian Ocean
INSAT-3D	ISRO	$82°$E	Asia, Indian Ocean
FY-2	CMA	$86.5°$E; $105°$E; $112°$E	Asia, Indian Ocean, Australia
MTSAT-2	JMA	$145°$E	Asia, Indian and Western Pacific Ocean, Australia (inactive since 10 March 2017)
Himawari	JMA	$140°$E	Asia, Indian and Western Pacific Ocean, Australia

Accurate derivations of land surface temperature and land surface emissivity from satellite measurements are difficult because the two variables are closely coupled. At the very surface, the land surface temperature is characterised by a strong diurnal cycle that is driven by solar radiation and is dependent on the exact sensing depth and thermal properties of soil and vegetation. Geostationary observations with temporal samplings ranging between 30-min (earlier satellites) and 10-min are particularly suitable for monitoring variables with pronounced diurnal cycles, with the exception of the polar regions where revisit times of polar orbiters are generally short. The Spinning Enhanced Visible and Infrared Imager (SEVIRI) onboard MSG takes top-of-atmosphere measurements within 12 bands, every 15-min and with spatial sampling of at least 3 km at nadir. SEVIRI-based retrievals of Land Surface Temperature (LST) can therefore capture the high spatial and temporal variability (see Figure 3, as example), which in turn are closely linked to surface conditions, such as vegetation cover and state, or soil moisture availability. LSTs (or equivalently Sea Surface Temperatures) are usually derived from top-of-atmosphere thermal infrared measurements in the atmospheric window, by correcting for the atmospheric absorption and emission and surface emissivity. Such surface temperature estimates are fairly close to actual observations [47], though they are restricted to clear-sky conditions and to the disk field of view. Multiple geostationary satellites can be combined to extend coverage, maintaining a good temporal sampling.

The use of satellite LSTs to understand and improve the modelling of land-surface processes is still largely under-utilised [13,48]. The LST spatial and temporal variability reflects changes in the surface energy balance. The amplitude and phase of the LST diurnal cycle are strongly linked to net radiation and turbulent energy flux partitioning at the surface [49], and to surface properties such as surface wind, the surface type and vegetation state, soil moisture availability, amongst others.

2.8. Ground-Based Networks

A great wealth of Earth system observations are available to constrain land surface state. These observations continue to grow in volume and diversity but gaps still exist for certain key parameters (e.g., soil moisture, snow depth and river discharge). Despite their limited coverage and limited expansion, in situ observations continue to be the backbone of the observing system used by NWP land-surface applications. Some of the key in situ parameters used in NWP are: surface pressure, 2 m temperature, 2 m humidity, snow depth, soil temperature and wind (see Figure 4). They are mainly provided by the Synoptic Operations (SYNOP) and METeorological Aerodrome Report (METAR) networks and by radiosondes. These observations are used to constrain the atmospheric model, land-surface and snow analyses together. Large spatial gaps and representativity issues [50] continue to affect soil moisture and snow observations. In addition, most variables are retrieved in the atmosphere so that the constraint on the land surface is only indirect. Indeed, for instance, the soil moisture updates reflect changes in the surface energy partitioning, which affects the atmosphere (e.g., air temperature and humidity). Any incorrect physical representation between the surface (soil moisture) and the atmosphere (fluxes) would lead to incorrect soil moisture inversion.

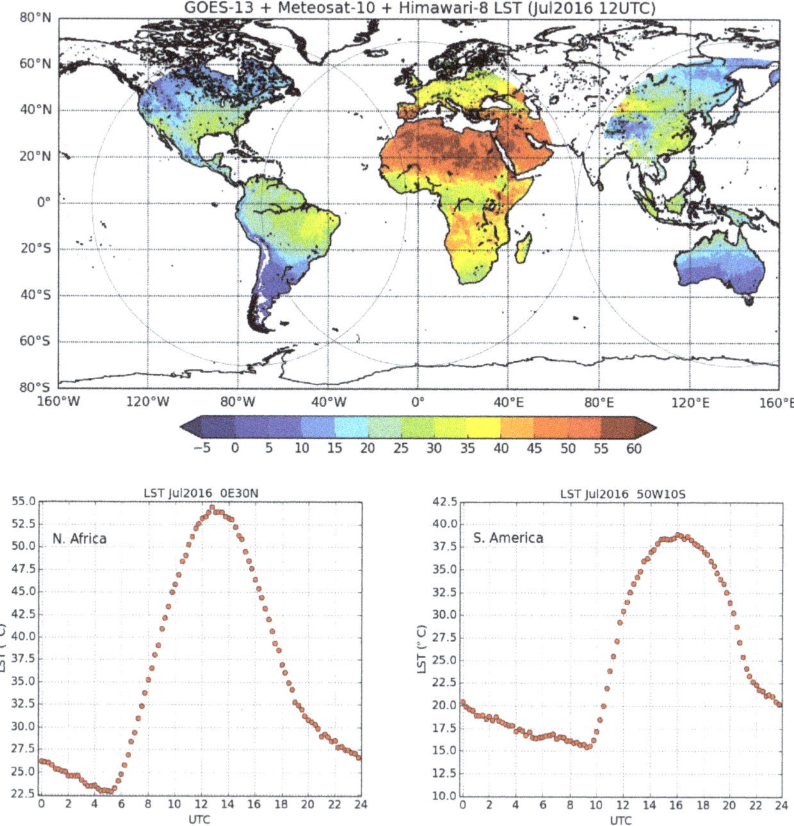

Figure 3. Median Land Surface Temperature (LST, °C) for all 12 UTC LST estimates in July 2016, derived from GOES-13, MSG/SEVIRI and Himawari-8 observations (upper panel; the circles mark the respective 70° disks). The median LST diurnal cycle for July 2016 is shown in the lower panels, for two pixels (desert landscape in Northern Africa, lower-left, and Savannah in South America, lower-right).

In the past five years, the number of observations available from the SYNOP (Synoptic observations) and METAR (Meteorological Report at the Airports) has increased (Figure 5) thanks to the inclusion of automatic data reports not previously considered.

Long-term, large scales networks of ground measurements are also of paramount importance. For instance, detailed ground based soil moisture information is provided by in situ soil moisture databases such as the International Soil Moisture Network (ISMN) [51]. Ground-based measurements of soil moisture and temperature from ISMN [52–55] have been used to evaluate the European Centre for Medium-Range Weather Forecasts (ECMWF) operational analyses, forecasts and reanalyses ability to represent soil moisture and temperature. Verification studies using in situ networks contribute to a better understanding of the deficiencies in models and point to processes that need a better representation.

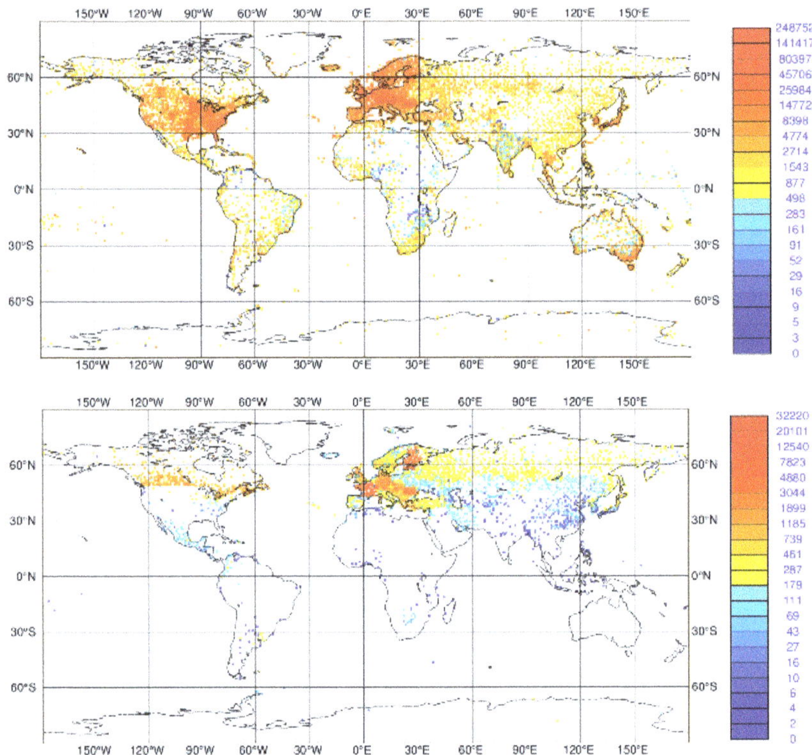

Figure 4. Number of available observations per unit area of $1° \times 1°$ from the SYNOP and METAR networks in January, February and March 2018 for 2 m temperature (**upper**) and snow depth (**lower**).

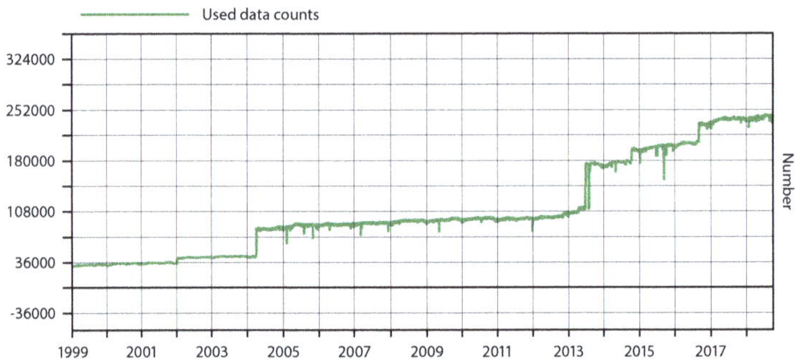

Figure 5. Daily global data counts from SYNOP and METAR observations. The sudden increase of data counts in 2013 is related to the inclusion of existing METAR observations not previously sourced.

Particular attention must also be devoted to rainfall measurements, for which numerous datasets exist combining in situ rain gauge observations from the SYNOP and SYNOP-like national networks with satellite and ground-based remote sensing to obtained gridded products, as recently compared in a study [56]. The combination of multi-sources observations with models to ensure meteorological consistency is advocated as a way to increase accuracy. However, precipitation estimation for hydrological applications remains a challenge and motivates the continuous effort of the International Precipitation Working Group (IPWG), a permanent Working Group of the Coordination Group for

Meteorological Satellites, active since 2001 [57]. Despite the effort of data collection of long term precipitation observations e.g., [58], a precipitation dataset that can support climate change studies and can be readily considered in reanalysis, remains largely unavailable. Earth system models also rely increasingly on non-atmospheric ground-based observation networks including information on river discharge from river gauging stations such as the databases held by the WMO Global Runoff Data Centre (GRDC) and groundwater levels from wells. Such networks are notably at risk from decline in network coverage and restricted national-scale data access [59]. More comprehensive coverage of the in situ networks availability is included in the status report of the global observing system for climate by GCOS in 2015 [60].

2.9. Ocean-Based Networks

The ocean in situ observations coordinated under the Global Ocean Observing System (GOOS, http://www.goosocean.org) is a sustained observing system, unified by the Framework for Ocean Observing [61], designed to provide access to timely ocean observations. In many cases, data are telemetered to data centers in near real-time for assimilation into NWP. Higher quality data are also available in delay mode and can be used for validation and evaluation of processes. With its unique status within the United Nations system, GOOS has been able to gather a network of independently-managed and independently-funded observing elements. The network includes Argo floats [62–64], surface drifters [65], OceanSITES moorings (http://www.oceansites.org), and ships (http://www.jcommops.org/sot). GOOS surface Essential Ocean Variables include sea state, ocean surface stress, sea ice, sea surface height, sea surface temperature, surface currents, sea surface salinity, and ocean surface heat flux, with the turbulent heat flux typically measured as a bulk flux from state variables.

3. Earth Surface Modelling Advances and Links with EO Datasets

In this section, we describe some of the progress that has been made over the last few years in modelling that can be directly linked to EO data availability. For instance, advances in different land-surfaces components introduced in the ECMWF operational forecasts can be linked to availability of key observational datasets. It is acknowledged that surface related developments need to be discussed in the wider context of Earth system modelling. Common areas of development in surface parameterisation going towards environmental applications are presented by clustering the following groups: land-surface reservoirs, land-atmosphere fluxes, land-surface properties, inland and open waters.

3.1. Land-Surface Reservoirs

The land surface is an important component of the Earth system and numerical weather prediction and climate models are evolving continuously to include more of its natural complexity. The relevance of an accurate description of the land surface for atmospheric modelling has been widely established in the scientific community for over thirty years [66–69]. The land controls the partitioning of available energy at the surface between sensible and latent heat fluxes, which in turn has a strong impact on the atmospheric heat and moisture budgets, especially on diurnal time scales. The land also determines the partitioning of available water between evaporation, drainage and runoff. The water fluxes and storage terms interact with the land morphology forming lakes and rivers. The land surface influences weather and climate on all time and space scales [70,71], and responds actively to modifications of weather patterns and climate change [72]. The role of the land surface in Earth system models is to provide a consistent description of the water, energy and carbon exchanges (between atmosphere, biosphere, hydrosphere, and cryosphere) at various time scales ranging from hours to decades. Many sensitivity studies have shown that the description of physical processes of continental surfaces can significantly affect the prediction of meteorological variables such as precipitation, wind or temperature in the lower troposphere [73–77]. Evapotranspiration directly affects weather parameters such as temperature, humidity, boundary layer development and clouds [78–82]. Furthermore, a strong feedback between evaporation and precipitation exists which appears to be negative at the convective scale [83–86] and positive at the continental scale [84]. This feedback involves soil water in the root-zone layer, which is one of the most important variables controlling large-scale continental summer temperature extremes [87,88].

The way land surfaces store and regulate water and energy fluxes is firstly controlled by soil moisture in the unsaturated zone, snow, ground water, lakes and open water. Such water bodies are de facto reservoirs of energy and water and have a "memory" much longer than atmospheric components (on the order of a few days). Secondly, the energy and water fluxes are also controlled by land use/management and biosphere [72], which are complex, heterogeneous and difficult to characterize in practise. Modelling upgrades can be clustered in two areas of development:

- Enhanced realism of the representation of water and energy stocks in soil, snow and inland water bodies, via parameterisations and physiography revisions.
- Improved fluxes for land-atmosphere energy and water exchanges, inclusion of natural and anthropogenic carbon emissions, and improved river discharges.

These developments are summarised in the following subsections: soil, snow, vegetation and river hydrology development are illustrated in light of the EO dataset that have guided the model improvements.

3.1.1. Soil

Soil is a porous medium that can store water, energy and carbon and these can be exchanged with the atmosphere and the oceans via transport mechanisms. The observations within the soil layer are typically in situ measurements, while remote sensing data can only penetrate the top few centimeters; therefore, a combination of satellite based and in situ based information is essential in constraining soil models. The amount of water in the soil and its vertical distribution in the column are important for the regulation of heat and water vapor fluxes towards the atmosphere and involves a range of time scales from minutes to years in the coupled land-atmosphere system.

For example, at ECMWF, the Hydrology—Tiled ECMWF Surface Scheme for Exchanges over Land (H-TESSEL, [89])—has included a revised soil hydrology developed and tested for the global scale. These model developments were a response to known weaknesses of the TESSEL hydrology as used in the ERA-Interim reanalysis: specifically the choice of a single global soil texture, which does not characterize different soil moisture regimes, and a Hortonian runoff scheme which produces hardly any surface runoff. Therefore, a revised formulation of the soil hydrological conductivity and diffusivity (spatially variable according to a global soil texture map) and surface runoff (based on the variable infiltration capacity approach) were introduced in the Integrated Forecasting System (IFS) in November 2007, and is used in weather and sub-seasonal-to-seasonal forecasting applications and climate reanalysis products at ECMWF.

The soil hydrology revisions involved the use of selected field sites as well as global atmospheric coupled simulations and data assimilation experiments [89]. Figure 6 illustrates the impact of these hydrology changes on the water budget of a number of European river catchments. H-TESSEL increases the seasonal amplitude of the Terrestrial Water Storage (TWS, Figure 6) change due the increased water holding capacity of the soil resulting from the new hydrological parameters and soil texture. The H-TESSEL scheme compared better than TESSEL with the terrestrial water storage dataset [90], which was derived as the residual of reanalysed atmospheric moisture convergence and observed river catchment runoff and referred to as the Basin Scale Water Budget (BSWB) dataset.

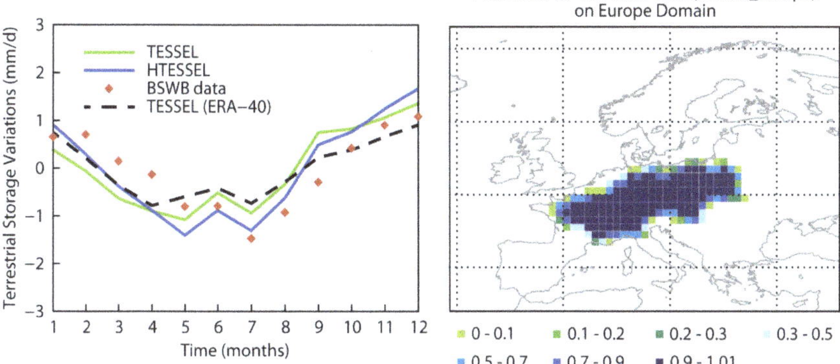

Figure 6. Monthly Terrestrial Water Storage (TWS) changes (left panel) for the Central European catchments Wisla, Oder, Elbe, Weser, Rhine, Seine, Rhone, Po, North-Danube (the coverage is shown in the right panel). The curves are for TESSEL (GSWP-2-driven, green line), H-TESSEL (GSWP-2-driven, blue line), TESSEL in ERA-40 (black dashed line). The red diamonds are the monthly values derived from atmospheric moisture convergence and runoff for the years 1986–1995 in the BSWB dataset [90].

Bare ground evaporation was improved by adopting a lower soil water content threshold for evaporation than the one used for vegetation. The most right panel of Figure 7 shows the difference in Root Mean Square Error (RMSE) of soil moisture between the revised model and the old model. RMSE is computed with respect to in situ measurements of soil moisture over the United States of America [54].

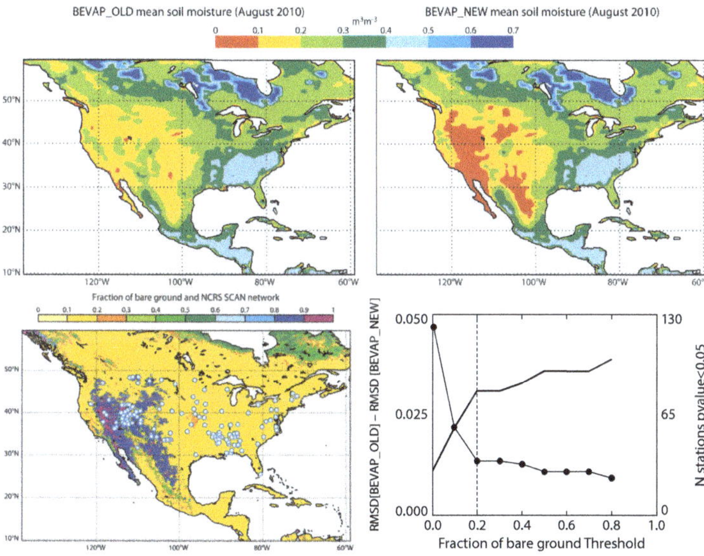

Figure 7. Mean soil moisture in the ECMWF model (August 2010), before (BEVAP OLD, **upper left**) and after (BEVAP NEW, **upper right**) the bare soil evaporation revision, evaluated using the US in situ soil network (**lower left**) to calcuate the difference in soil moisture Root Mean Square Error when using the revised bare soil evaporation threshold instead of the old model version (curve with dots, **lower right**)). The solid curve indicates the number of in situ observations in each of the bare soil fraction classes. For further details, see Figures 1, 2, 4 and 5 in [54].

The new bare ground H-TESSEL evaporation formulation resulted in more realistic soil moisture values when compare to in situ data, particularly over dry areas. Considering ground measurements sites where fraction of bare ground was greater than 0.2, The Root Mean Squared Difference (RMSD) of IFS's soil moisture with the observations decreased from 0.110 m^3 m^{-3} to 0.088 m^3 m^{-3}.

Figure 7 illustrates the impact of the new parameterisation on soil moisture over the USA. The old and new schemes (first and second panel from the left) shows the August soil moisture in the western part of the USA is much lower than with the old scheme. These changes correlate with the bare ground fraction (third panel). This is clearly beneficial, as can be demonstrated by verification based on the Soil Climate Analysis Network (SCAN) over the USA (see Figure 7 for the location of the stations). The positive differences in the right panel of Figure 7 indicate a reduction of the root-mean-square error (RSME) of soil moisture particularly at high bare ground fractions. A better match with Soil Moisture and Ocean salinity (SMOS) satellite observations [53] reinforced the generality of the results obtained from the in situ comparison, and confirmed once again the synergy of satellite and ground-based observations for the model development.

At NASA, the Catchment land surface model [91] has been continuously developed for use in the Goddard Earth Observing System NWP and reanalysis products [92] and the SMAP L4SM product. The latter product in particular depends critically on the skill of the land surface model. Some of the model developments prior to the launch of SMAP were focused on a revised set of soil texture and soil hydraulic parameters [93], as well as on the calibration of the microwave radiative transfer model that converts the modeled soil moisture and temperature values into L-band passive microwave temperatures [94]. Within the L4SM assimilation system, these modeled radiances were then confronted with the SMAP observations, with deviations between the modeled and observed radiances resulting in adjustments to the modeled soil moisture [95].

Since the first release of the L4SM product in 2015, new input parameter datasets for land cover, topography, and vegetation height were derived based on recent, high-quality satellite observations [96,97]. Land cover inputs were updated using the GlobCover2009 product, which is based on satellite observations from the Medium Resolution Imaging Spectrometer. Topographic statistics now rely on observations from the Shuttle Radar Topography Mission. Finally, vegetation height inputs are derived from space-borne Lidar measurements. Additionally, SMAP Level-2 soil moisture retrievals [98] were used to calibrate a particular Catchment model parameter that governs the recharge of soil moisture from the model's root-zone excess reservoir into the surface excess reservoir [99]. Specifically, the replenishment of soil moisture near the surface from below under non-equilibrium conditions was substantially reduced, which brought the model's surface soil moisture better in line with the SMAP Level-2 soil moisture retrievals (not shown).

The benefits of this change are illustrated in Figure 8, which shows that the surface soil moisture dynamics from the new model better agree than the old model with independent in situ measurements at SMAP core validation sites [45].

Figure 8a shows an example of surface soil moisture time series for one such site at Little River, Georgia, USA. Figure 8b quantifies the improvements across 18 sites. Between the old and new model versions, the average correlation increased from 0.64 to 0.67 and the bias decreased from 0.032 to 0.017 m^3 m^{-3}, respectively, while the unbiased RMSE remains unchanged at 0.041 m^3 m^{-3}. The improved Catchment model version will be used in future versions of the L4SM product, thereby resulting in an improved soil moisture data product.

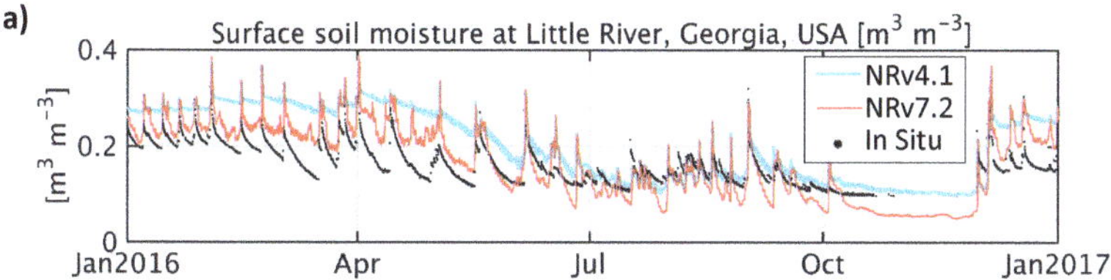

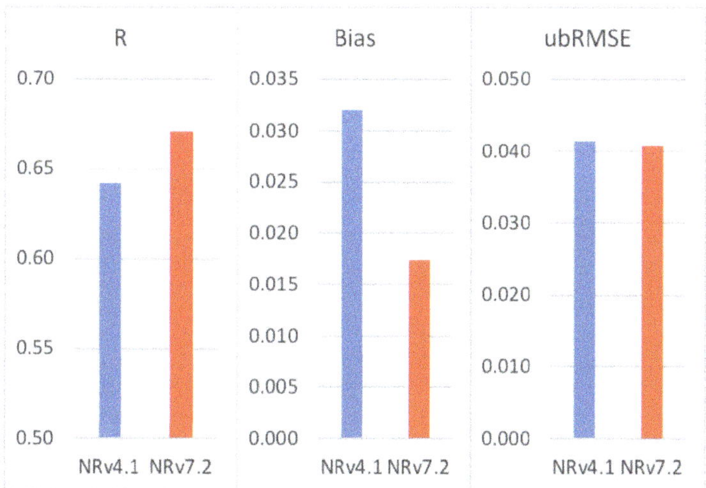

Figure 8. Surface soil moisture at Little Washita, OK, USA for two versions of the NASA Catchment model (**a**). Average surface soil moisture skill (**b**) in terms of time series correlation (R), bias, and unbiased root mean squared error (RMSE) (or standard-deviation of the error). Skill is measured vs. in situ measurements from 18 SMAP core validation sites [45] and then averaged across the sites. The improved model version (NRv7.2) uses updated land cover, topography, and vegetation parameters based on remote sensing observations and was further calibrated using SMAP soil moisture retrievals.

3.1.2. Seasonal Snow Cover

Snow acts as a fast climate switch [100] with implications ranging from weather forecasts to climate change projections. Among the main evidence of the importance of snow in the Earth System, we can enumerate: (i) the snow albedo feedback, (ii) driver of sub-seasonal to seasonal atmospheric predictability, (iii) important role in the recent Arctic amplification and (iv) its impact on future water resources [101]. The snowpack lying on top of the soil affects the evolution of atmospheric temperatures via its high albedo and its thermal insulation capacity that can create a decoupling between the soil and the atmosphere [102–104]. This snow insulating effect causes strong temperature inversions near the surface in winter, which represents a challenge for daily minimum temperature forecasting. Snow also affects the freezing of water in the soil, with an impact on the hydrology in spring, on near-surface temperatures and on the stable boundary layer development [105–107]. Snow cover also acts as a water reservoir, which is released by snowmelt in the spring, influencing runoff, soil moisture, evapotranspiration and thus precipitation and the entire hydrological cycle [108]. Until 2009, the snow pack was parameterised in H-TESSEL following Douville et al. [109]. The snow pack was represented with a single layer of dry snow (i.e., neglecting liquid water) with four snow prognostic variables: mass, albedo, density and temperature. Snow albedo decreased in time at an exponential or linear rate, for melting and normal conditions respectively, and snow density increased with time according to an exponential relaxation.

With the participation of H-TESSEL in the snow model inter-comparison project 2 (SnowMIP2, [110]), and after the soil hydrology revision presented in the previous subsection, several shortcomings of the Douville snow pack representation were identified such as the lack of liquid water representation with freeze/thaw cycles during the melting season, unrealistic evolution of snow density, and unrealistic albedo of shaded snow (i.e., snow under high vegetation). These shortcomings were partially addressed with a full revision of the snow pack parameterisation in 2009 [111] including: (i) a new parameterisation of snow density, (ii) a liquid water reservoir and (iii) revised formulations for the sub-grid snow cover fraction and snow albedo. In offline mode, forced with near-surface observations, the revised scheme reduced the end of season ablation biases from ten to two days in open areas, and from 21 to 13 days in forest areas. The evolution of snow mass, depth and density during one winter season at the Fraser forest and open stations is shown in Figure 9, a research location in the Rocky Mountains, CO, USA. The results show the improvements of the snow scheme revision (NEW) with respect to the version used in ERA-Interim, labelled as control (CTR). The new snow density parameterisation increased the snow thermal insulation, reduced soil freezing, improved the hydrological cycle, and substantially reduced a warm winter bias compared to Siberian screen-level (2m) observations (not shown).

The dramatic change in surface reflectivity when snow is present has been widely explored using EO data, in particular in the visible range, and is mainly limited by cloud presence and contamination. The detection of snow on the surface was one of the first applications of EO data associated with surface characteristics and it is currently one of the surface variables with longer observed datasets using remote sensing. Furthermore, it is the main direct data source in land-surface assimilation in several operational weather forecasting centers [112]. In addition to snow presence, snow mass is an important water reservoir and there have been many studies focusing on the retrieval of snow mass from satellite [113].

Monitoring of snow cover area is well established based on optical data, whereas current Snow Water Equivalent (SWE, i.e., the total water mass in liquid equivalent) products mostly rely on passive microwave measurements [114]. Active microwave measurements at high frequency (Ku-band) are sensitive to volume scattering of the dry snow as well as the wet or dry status of the snow cover [115]. They are therefore also relevant for future mission concepts dedicated to accurate snow water equivalent retrieval from space. Dry snow is, however, largely transparent at lower frequencies (C or L-bands), making it a challenge to observe snow from space [116,117].

Physically-based forward models that rely on a good understanding of the electromagnetic interactions with the snowpack are best suited to obtain consistent retrievals of snow macrophysical and microphysical properties or for data assimilation in NWP systems. The new generation of snow forward models, such as the Snow Microwave Radiative Transfer (SMRT), account for multi-layer snow emission and backscatter [118]. Using satellite data from microwave sensors with appropriate multi-layer forward modeling and retrieval approaches for the snow microwave properties characterisation is of great interest to support physically based multi-layer snow model developments. Recent studies showed that SMOS data [119] and ground-based L-band measurements [120] could give insight on snow density and liquid water in snow using a two stream approach. This is promising for constraining model errors in snow depth that are not associated with snow mass but with its state of compaction.

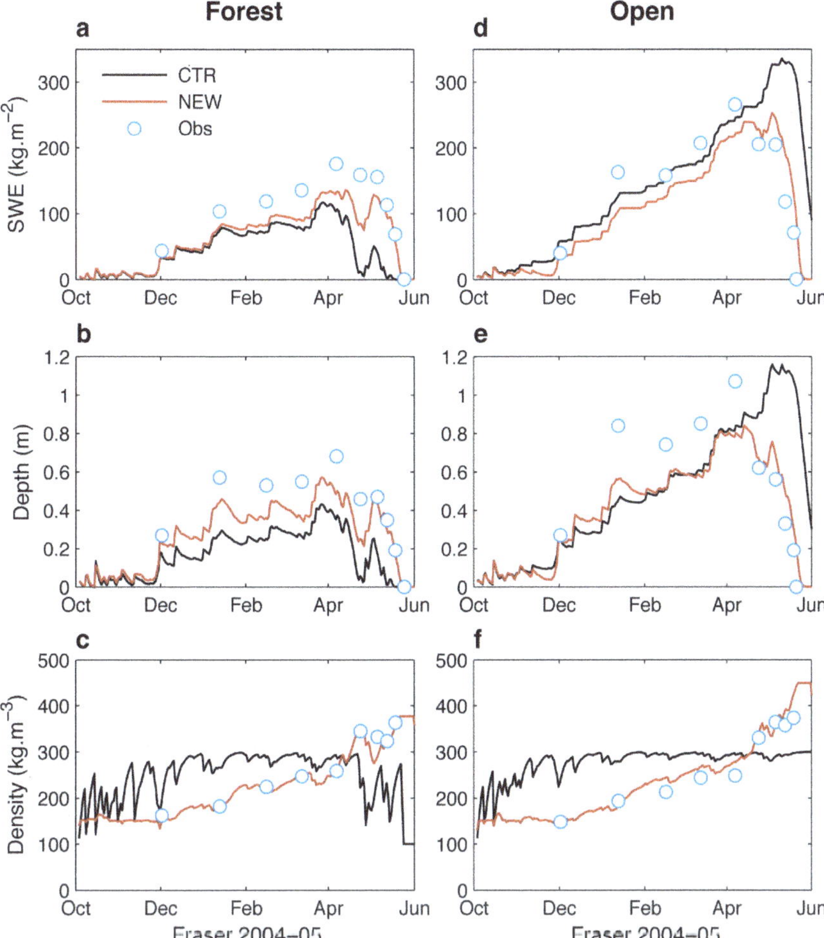

Figure 9. Offline simulation results with the older (CTR ECMWF) snow scheme (black line—before 2009 revision), and NEW (red line—after 2009 revision) for the 2004–2005 winter season at Fraser forest (**a–c**) and open (**d–f**) sites. Snow mass (**a,d**), snow depth (**b,e**) and snow density (**c,f**) are shown, with time on the *x*-axis (from 1 October 2004 to 1 June 2005). Observations are represented by open blue circles.

3.1.3. Permanent Snow and Ice

In glaciers and over ice-sheets such as Antarctica and Greenland, the representation of snow for weather and seasonal forecasting has slightly different challenges. The spatial variability is somewhat reduced and open-snow is the dominant cover. There is also less emphasis on the mass balance as this is often not directly relevant. It is, however, essential to correctly represent the albedo to resolve the snow energy balance with a particular focus on melting. EO snow albedo, along with snow cover, plays a crucial role in model evaluation and development [121] with some important examples of data assimilation [122].

The surface skin temperature provides information particularly over Antarctica where polar orbiting satellites have multiple passages per day. The presence of snow largely affects the diurnal cycle of LST, for which current state-of-the-art models show some limitations. Recent results have also shown the potential for monitoring internal temperature of the ice sheet in Antarctica [123,124] using SMOS data. EOs can provide reliable information of LST over snow and ice in clear-sky conditions [125]. However, its application for model evaluation has been limited so far because of the poor coverage of geostationary satellites in high latitudes and because of the low temporal resolution of polar satellites far from the poles. LST data have been used to evaluate the diurnal cycle of skin temperature in the ECMWF ERA-Interim reanalysis over Antarctica, showing a homogeneous and persistent (throughout the year) warm bias over the Antarctic plateau [126]. Numerical experiments by [127] indicated that this could be associated to the high thermal inertia of deep snowpacks in the bulk (single-layer) snow scheme used in the reanalysis model (see Figure 10), indicating the potential of more sophisticated multi-layer snowpack schemes in future operational weather forecasting models.

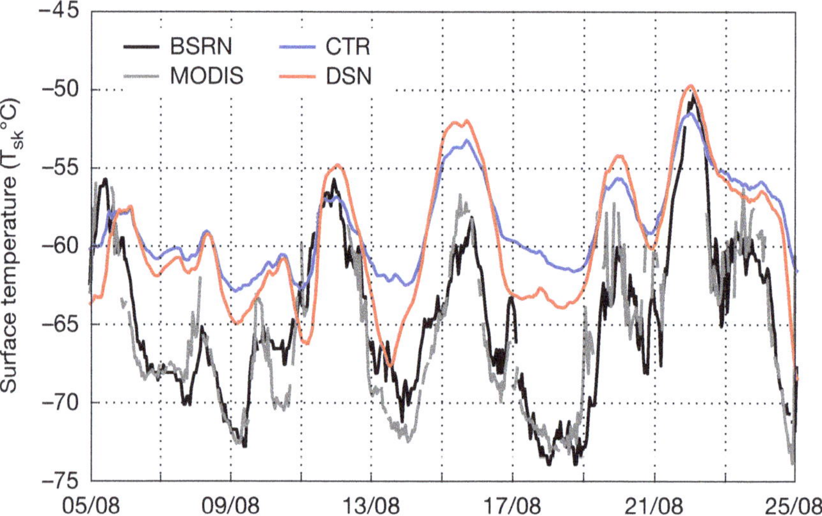

Figure 10. Time-series of the land surface (skin) temperature (LST, Celsius) at the south-pole for in situ observations (BRSN), MODIS satellite data (MODIS), numerical experiment with the bulk (single-layer) snow scheme (CTR) and numerical experiment with the thermal depth of the snow reduced by a factor of 10 (DSN), the latter being a proxy for the potential impact of a multi-layer snow scheme. It is clear that DSN has a larger amplitude LST diurnal cycle than CTR, and a better representation of strong cooling events [127].

3.1.4. Vegetation and Carbon Cycle

The biosphere plays a prominent role in regulating the flux of gases, energy and momentum into the atmosphere, so it is important to properly represent it in models. Parameterisations of the biosphere are simplified representations of the natural processes in which the spatial scales of interest such as the plant, field or watershed remain largely sub-grid in the foreseeable future and can only describe the main feedback mechanisms sometimes marred by sizeable systematic errors. A key characteristic for water vapour and carbon fluxes modeling is the so-called canopy resistance, which is a bulk representation of stomatal resistance, vegetation type and leaf area. The stomata are leaf pores through which the plants absorb carbon dioxide and transpire water vapour.

There are continuing efforts to calculate large-scale evapotranspiration using EO data. Globally available products include MODIS [128], GLEAM [129,130], SEBS [131] and PT-JPL [132], FLUXCOM [133] or WECANN [134]. Since evapotranspiration cannot be observed directly from radiative, observable properties of the surface and atmosphere, typically the models use EO data such as radiation, rainfall and soil moisture to train or derive simple models.

Refs. [135,136] focused on evaluating these products using flux tower and river streamflow under the LandFlux-EVAL initiative. They found significant differences between the products, which limited their use for model evaluation. More recently, Refs. [137,138] have assessed the latest products. They conclude that (i) the models tend to overestimate soil evaporation, (ii) the way the models treat soil moisture stress is important and that (iii) finding ways to evaluate the split of evapotranspiration into its three components (transpiration, soil surface and interception evaporation) is currently not possible but a critical task for the future. For the latter, Ref. [139] showed that the flux attribution is a complex issue.

Despite these caveats, it is imperative that developers use observed data to evaluate their models. To that end, iLAMB (integrated Land Atmosphere Model Benchmarking, [140]) was developed as a shared framework for large-scale model evaluation. It was used by the EartH2Observe project [141] to evaluate a suite of ten models including the evaluation of evaporation against the GLEAM EO product. As other researchers, they found that the comparison had limited value due to limited accuracy of the EO product.

In the ECMWF land surface scheme, the Leaf Area Index (LAI) expresses the phenological phase of vegetation (growing, mature, senescent, dormant). Initially it was kept constant [142] and assigned using a look-up table depending on the vegetation type. Thus, vegetation appeared to be fully developed throughout the year. In November 2010, Ref. [143] modified this and introduced a seasonality of vegetation via a LAI monthly climatology based on the MODIS (collection 5) satellite product by [144]. The new monthly LAI climatology was shown to affect particularly the spring season when the radiative forcing is already strong, but when the vegetation not yet fully developed. The sensitivity generally indicated a warming of the spring as a consequence of the lower LAI and reduced evaporation (and consequently more sensible heat flux). This resulted in a reduction of the systematic 2-m temperature errors in the spring [143].

Recently, the Copernicus Global Land Service (CGLS) products of surface albedo and LAI based on observations from the SPOT-VEGETATION and PROBA-V sensors [145] have become available in near real-time (NRT) with an operationally-maintained production chain. However, the direct use of these products within a NWP system is not possible without quality checks given the spatial and temporal discontinuities they may contain. A direct assimilation of LAI and surface albedo products using optimal interpolation analysis with the ECMWF NWP system was explored and evaluated for anomalous years [146]. It was shown that: (i) the assimilation of these products enables detecting/monitoring extreme climate conditions where the LAI anomaly could reach more than 50% and the albedo anomaly could reach 10% (Figure 11), (ii) extreme LAI anomalies have a strong impact on the surface fluxes, while for the albedo, the impact on surface fluxes is small, (iii) neutral to slightly better agreement with in situ surface soil moisture observations and surface energy and CO_2 fluxes from eddy–covariance towers is obtained, and (iv) in forecast mode, the assimilation of LAI reduces the near-surface air temperature and humidity errors both in wet and dry cases, while the albedo has a small impact, mainly in wet cases, when albedo anomalies are more noticeable.

3.2. Land–Atmosphere Fluxes

3.2.1. CO_2 Natural Ecosystem Exchange

More recently, the ECMWF land surface scheme has been extended with a carbon dioxide module based on the A-gs model [147]. The reason for adopting simple vegetation and carbon dioxide schemes, is that these are suitable for the NWP setup where environmental factors are controlled by meteorological forcing and constrained by data assimilation. The model relates photosynthesis to radiation, atmospheric carbon dioxide (CO_2) concentration, soil moisture and temperature. Ecosystem respiration is based on empirical relations dependent on temperature, soil moisture, snow depth and land use [148]. The CO_2 module parameters are optimised by vegetation type considering the Gross Primary Production (GPP) and the Ecosystem Respiration (Reco). Together, they compose the

Net Ecosystem Exchange (NEE) of CO_2 between biosphere and atmosphere. The FLUXNET-based in situ observations from different climate regimes (http://www.fluxdata.org/), combined with a benchmarking systems similar to iLAMB [140], was used for parameter optimization by minimizing flux errors. Subsequently, a different year of the FLUXNET data was used for verification.

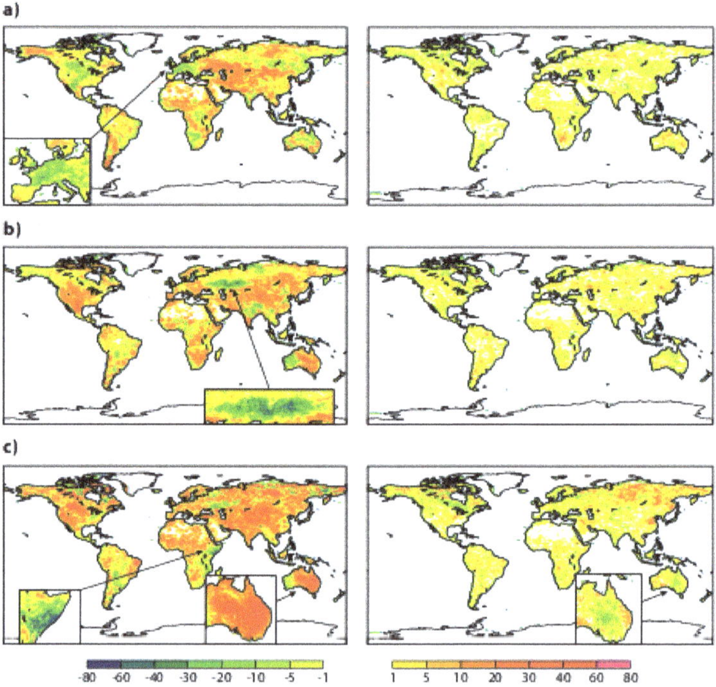

Figure 11. Relative anomaly [%] with respect to mean (1999–2012) climate of Leaf Area Index (LAI) (**left**) and broadband diffuse albedo (**right**) for (**a**) August 2003; (**b**) July 2010 and (**c**) November 2010. Regions of interest are zoomed in.

The seasonal cycle of NEE is illustrated in Figure 12 for six sites with different biomes. Two model configurations are shown: the first uses a stomatal resistance formulation for evaporation that is controlled by the photosynthesis module (C-TESSEL), and the second uses the Jarvis-based stomatal resistance for evapotranspiration (CH-TESSEL). In addition, the CASA climatology (Carnegie–Ames–Stanford Approach, [149]) is shown because it is extensively used in the community and it was previously used as a boundary condition for atmospheric CO_2 in the MACC (Monitoring Atmospheric Composition and Climate) project. Although it is difficult to draw firm conclusions, it is clear that the errors in NEE are large and vary dramatically from site to site, and differences between C-TESSEL and CH-TESSEL are small compared to the errors. The correlation between model NEE and observations averaged over 34 flux tower sites is 0.37 for CASA, 0.68 for C-TESSEL and 0.65 for CH-TESSEL. Both TESSEL versions have a correlation of about 0.80 for sensible and latent heat fluxes [148]. The substantial improvement of C-TESSEL/CH-TESSEL with respect to the CASA climatology is significant because it suggests that the real-time meteorological variability is a key driver of the NEE variability.

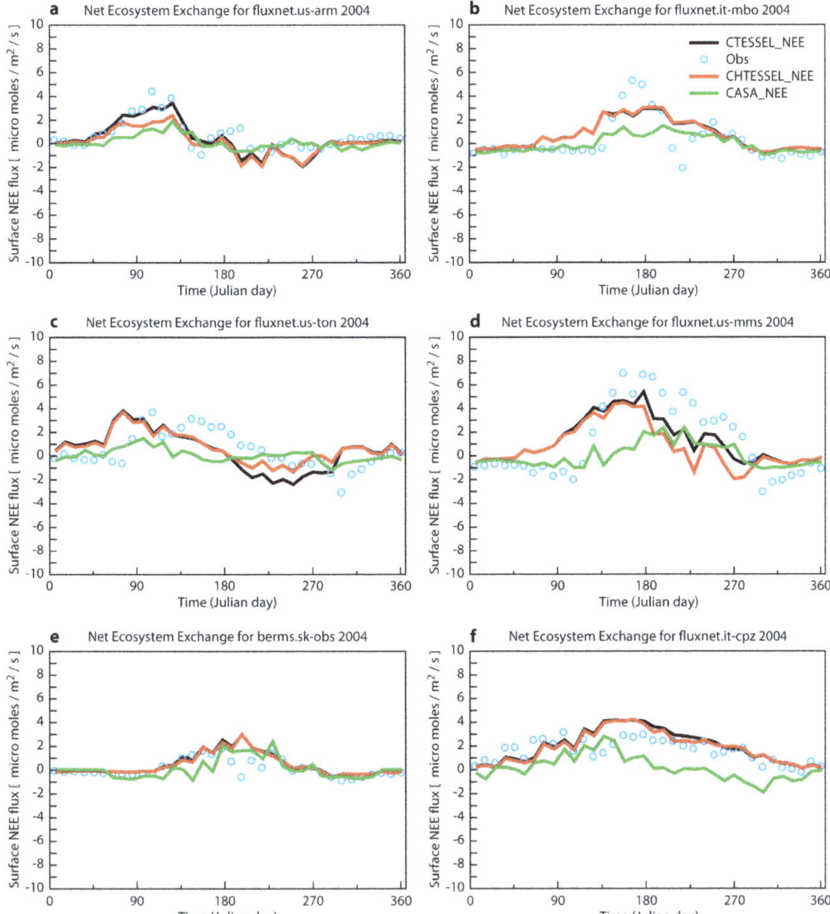

Figure 12. Seasonal cycle (2004) of 10-day averaged offline simulated (lines) and observed (blue dots) Net Ecosystem Exchange μmol m^{-2} s^{-1} for C-TESSEL (with A-gs, black line), CH-TESSEL (with Jarvis-type evaporation, red line) and CASA-GFED3 (green line) at diverse FLUXNET observational sites in the panels (**a**–**f**) representative of different biomes: (**a**) Southern Great Plains site-Lamont, US; (**b**) Monte Bondone, Italy; (**c**) Tonzi Ranch Sierra Nevada, US; (**d**) Morgan Monroe State Forest, US; (**e**) Saskatchewan Western Boreal, Mature Black Spruce forest, Canada; (**f**) Castelporziano, Italy.

Correlations coefficients of the two components of NEE, GPP and Reco, compared with tower observations (0.8 for both fluxes, on average over 34 sites) indicate much higher skill and confirm the robustness of these results. NEE is a small residual of GPP and Reco; therefore, a correlation coefficient above 0.6 is highly relevant. Again, C-TESSEL and CH-TESSEL show similar performance, with site to site variability attributed to representativity of the model grid-box [148]. CH-TESSEL has also been evaluated in coupled integration mode for the 2003 to 2008 period.

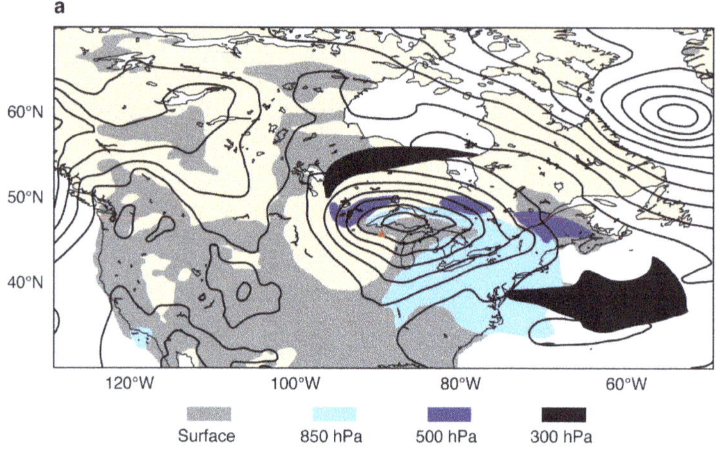

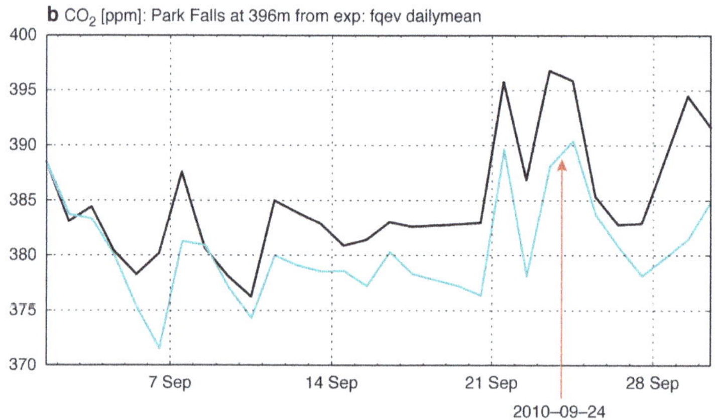

Figure 13. Atmospheric CO_2 anomalies associated with the passage of a low pressure system over N. America: (**a**) CO_2 anomalies above the well-mixed background CO_2 at different levels in the vertical: grey near the surface, cyan at 850 hPa, blue at 500 hPa and dark grey at 300 hPa on 24 September 2010. The anomalies are defined as CO_2 dry molar fraction above the background value of 392 ppm for both near the surface and at the 850 hPa levels, and above the background value of 388 for the 500 and 300 hPa levels. The locations of the observing site of the Park Falls Tower (Wisconsin, USA) is depicted by a red triangle. The black contours of mean sea level pressure show the location of the centre of low pressure systems; (**b**) daily mean dry molar fraction (ppm) of CO_2 from measurements (black) and forecasts (cyan) at Park Falls in September 2010. The observations are courtesy of NOAA Earth System Research Laboratory [150].

The operational CAMS (Copernicus Atmosphere Monitoring Service) atmospheric CO_2 analysis and forecast system has been using the online NEE fluxes from CH-TESSEL ever since its introduction in the IFS, replacing the offline CASA fluxes from the GFED dataset [151]. Extensive testing of this online configuration shows that the global CO_2 atmospheric inter-annual variability is well simulated [152]. The correlation of global CO_2 with observationally based estimates is 0.70. The global CO_2 forecast has high skill in simulating day-to-day synoptic variability, which is crucial in order to be able to assimilate atmospheric CO_2 observations.

Figure 13 illustrates the spatial and temporal CO_2 synoptic anomalies associated with the passage of synoptic weather systems over North America. The CO_2 forecast can represent the peaks of CO_2

observed at Park Falls (Wisconsin, USA) originating mainly from the advection of high CO_2 anomalies generated at the surface within the warm conveyor belt of synoptic low-pressure systems. Modelling day-to-day variability of the CO_2 fluxes from vegetation compared to using equivalent monthly mean fluxes with a diurnal cycle enhances significantly the atmospheric CO_2 variability and skill. Again, this illustrates the advantage of modelling the CO_2 fluxes inside the IFS with real-time meteorology. Despite the synoptic skill provided by the NEE synoptic variability, the model suffers from substantial biases in the biogenic CO_2 fluxes (GPP and Reco). These biases are diagnosed and corrected by comparing the model budget with a reference budget from a flux inversion system [153] based on in situ observations of atmospheric CO_2 at the surface [154]. The biogenic flux adjustment scheme of [153] addresses the important task of reconciling the bottom-up and top-down estimates of CO_2 fluxes. The benefits are a substantial reduction in the atmospheric CO_2 biases as well as a useful diagnostic to guide model developement. New data sets based on FLUXNET data [155] or satellite observations of Solar Induced Chlorophyll Fluorescence [156] can also be used to improve the attribution of the errors associated with GPP and Reco.

3.2.2. CH_4 Natural Methane Fluxes

Recent trends in atmospheric CH_4 are not well understood and attribution using surface networks is underdetermined [157]. The relatively short CH_4 atmospheric lifetime of 9.8 years [158] and the combination of anthropogenic and natural sources make CH_4 a suitable species for spatial and temporal analysis using EO. Near-surface CH_4 sensitivity provided by the GOSAT satellite has been available since 2009 and provides retrievals using the light-path proxy approach from shortwave infrared radiances. To improve attribution and aid model development, retrievals of methane isotopologues, such as 13CH_4, could be used in the future. These retrievals are currently limited to stratospheric retrievals using the Atmospheric Chemistry Experiment (ACE) satellite [159] and remain a challenge for tropospheric instruments.

Wetlands are the most dominant source of CH_4, contributing about 30% to the total flux [160]. Large uncertainties have been identified in both the spatial and temporal distribution of wetland CH_4 fluxes provided by land surface models [161]. Major impacts can be seen when simulating long-term CH_4 emissions taking into account different vegetation classes [162]. Ref. [163] used GOSAT column CH_4 (XCH_4) to evaluate wetland emissions from the Joint UK Land Environment Simulator (JULES, [164]). The detailed spatial and temporal resolution of GOSAT data permitted evaluation of the model performance over specific wetlands, which provided insight into some of the causes of model uncertainty. Bias in the modelled fluxes highlighted a failure to capture large scale, river fed, wetlands in the Pantanal region during high rainfall years. This is most likely caused by a lack of river routing in the model or by a misrepresentation of methanogenesis in high flood waters. In 2009/2010, the Paraná River region experienced a flood event that caused a spike in CH_4 concentration retrieved from GOSAT; this was not reproduced in JULES, highlighting the need to include representation of over-bank inundation in the model.

Figure 14 shows the model failure to transport surface moisture, via river runoff, from the rainfall region in Southern Brazil down the Paraná River towards Paraguay and Argentina. The lack of river routing within the model causes a spatial bias in wetland area in JULES when compared to the GRACE satellite water storage anomaly, resulting in misrepresentation of modelled CH_4 fluxes.

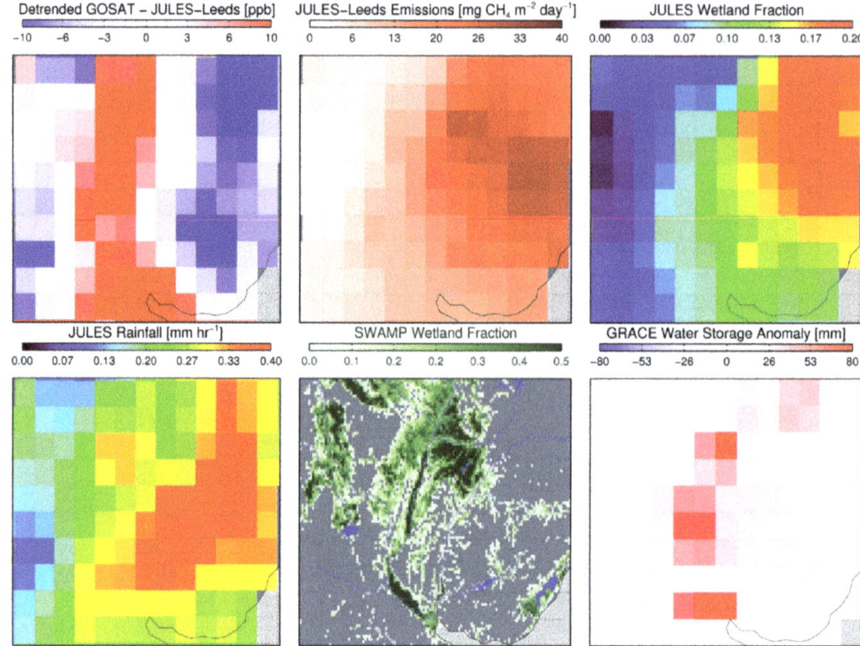

Figure 14. Maps of the Paraná region for November 2009 to February 2010 showing (**top–left**, clockwise): GOSAT-JULES difference, JULES wetland emissions, JULES wetland fraction, GRACE water storage anomaly, Sustainable Wetland Adaptation and Mitigation Program (SWAMP) wetland fraction and JULES rainfall amount [163].

3.2.3. Vegetation Water Fluxes

Transpiration dominates the continental water cycle [165–167]. It is therefore crucial to better represent and constrain vegetation activity and its contribution to the continental energy and water cycles. Historically, continental microwave remote sensing has been focusing on soil moisture retrieval [168,169]. Vegetation was mostly considered during the retrieval of surface soil moisture as a by-product as it affects the signal penetration to the surface [170,171]. The attenuation of the microwave signal when passing through the vegetation layer depends on the so-called Vegetation Optical Depth (VOD). The VOD depends on the frequency of the sensor, on the water content of the plant (trunk, branches, leaves) as well as on the biomass [172–178]). As such, there has been recent interest in using VOD to assess biomass changes [37,39,178–181] or plant water stress strategies [177]. The VOD might prove especially useful in regions where soil moisture retrieval is especially complicated by dense vegetation coverage (e.g., forested landscape). C or X-band do not penetrate much into the canopy and are more sensitive to the top part of the canopy. Figure 15 shows that L band is sensitive up to very dense canopies (300 Mg/hectare) enabling the probing of almost all vegetation canopies [182]. A further advantage of the VOD is that data can be acquired even in the presence of clouds. C-band senses a thicker layer of the canopy and branches so that the signal somewhat differs from the X-band signal especially in humid regions with dense canopy coverage such as the Amazon rainforest (Figure 16). The lower frequency signal, in L-band, senses a thicker vegetation layer corresponding to the above ground biomass such as the ones derived in [183] (Figure 15).

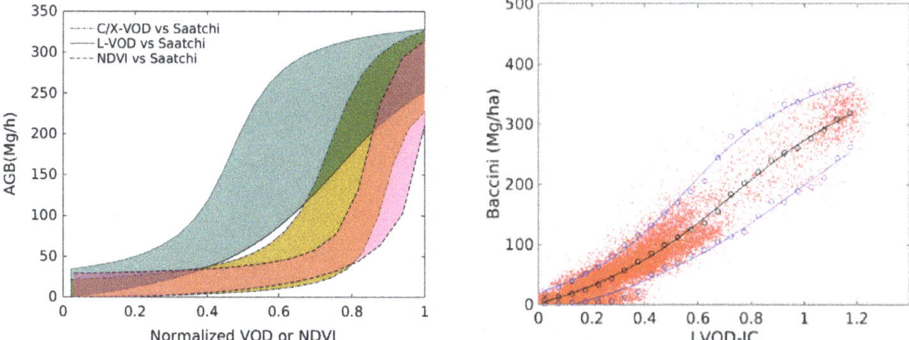

Figure 15. Relationship between Above ground Biomass (AGB) in Mg/hectare and Normalized Difference Vegetation Index (NDVI) data from X-band, C-band and L-band Vegetation Optical Depth (VOD, left panel). The right hand panel shows AGB data versus VOD from SMOS L-band observations (red dots) [183].

Most of our observations of the land surface relate to surface variables (soil moisture, biomass, vegetation coverage among others) but ultimately one of the key objectives is to better constrain the vegetation water, energy and carbon fluxes. Recently, direct observations of a proxy for photosynthesis have been made possible, at the global scale using Solar-Induced Fluorescence (SIF) [184–192]). SIF is a by-product of photosynthesis, which is directly related and even nearly-proportional to photosynthesis in most conditions and is a direct indicator of vegetation stress [156]. One of the advantages of SIF is that it provides information that is not included in the other datasets (e.g., vegetation indices or VOD). Indeed, SIF is able to correctly capture the seasonal cycle across diverse climates dominated by phenology and temperature (Figure 16 top panel the northeastern US), by temperature and water stress (Figure 16 middle panel Mediterranean Spain) or in rainforests, where vegetation indices struggle to correctly define the seasonality which is imposed by leaf aging [193–195]) and radiation structure [196] (Figure 16 lower panel, the Tropical Amazon). SIF thus provides unique information to constrain land-surface fluxes and its response to droughts (e.g., [197]). It is, however, particularly noisy [186] so that averaging in both time and space is needed to extract a meaningful signal. This noise is also a major road block for direct use in either a data assimilation context or to directly assess surface fluxes. As a result, recent approaches have tried to use SIF not directly but as a target of machine learning strategies using MODIS broadband reflectances (Reconstructed SIF, [198]). The advantage of such approach is that it gets rid of most of the large amplitude noise in SIF, while conserving to a large extent the photosynthetic activity dependence across biomes and climates. An alternative to SIF is the Photochemical Reflectance Index (PRI) [199] or combined products that can improve the signal-to-noise ratio.

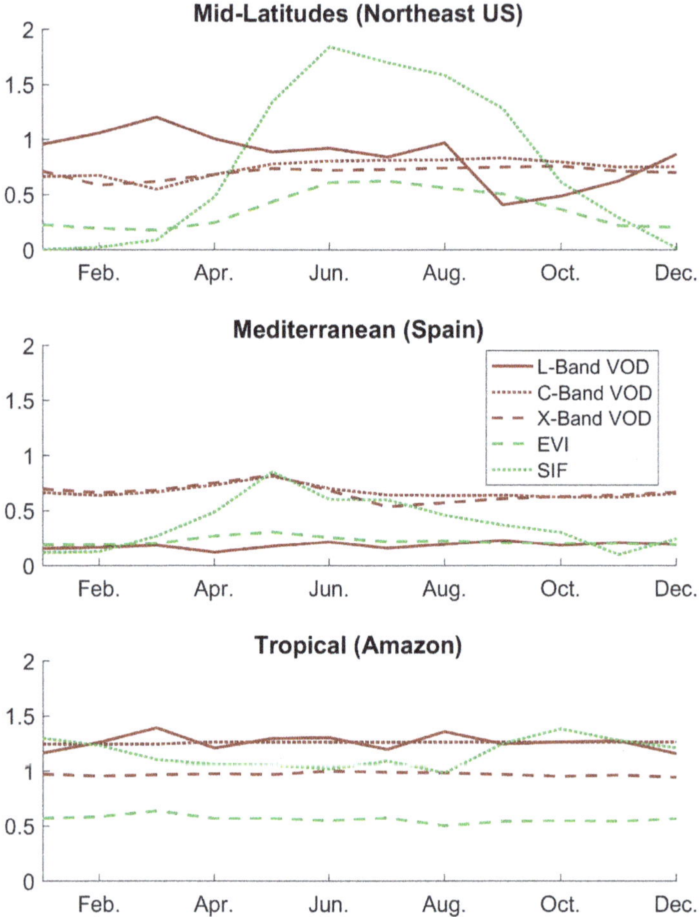

Figure 16. Seasonal cycles over different climates (**top**: energy limited—northeastern US, **middle**: Mediterranean climate in Spain, **bottom**: rainforest in the Amazon) as assessed using different sources of data: Vegetation Optical Depth (VOD) at high (X-band), medium (C-band) or low frequency (L-band), Enhanced Vegetation Index (EVI) and Solar-Induced Fluorescence (SIF) using the GOME-2 satellite data.

3.2.4. CO_2 Anthropogenic Fluxes and Co-Emitters

In the atmospheric source inversion community, the use of satellite data to constrain trace gas fluxes was made possible through the adoption and adaptation of data assimilation techniques already in use at operational NWP centers. As a result, two methods have emerged: Four-Dimensional Variational (4D-Var) and Ensemble Kalman Filter (EnKF) [200] The computational efficiency of 4D-Var to solve high-dimensional inverse problems and its ability to accommodate a large amount of remote-sensing measurements over long periods of time, made 4D-Var a method of choice in the source inversion community. However, the development and maintenance of adjoint models in variational approaches can be cumbersome. On the other hand, EnKF-based methods use a small ensemble of perturbed model simulations to represent and propagate the error statistics explicitly, while similar methodological difficulties characterize DA in NWP and source inversion problems, it is worth noting some specific aspects of the latter. First, while the prior (or background) information in NWP DA systems is associated to prognostic variables of the model that are often directly observed, the source inversion problem involves model parameters that are usually measured only indirectly, making the quality of the optimised sources strongly dependent upon the accuracy of the transport model itself. Additionally, the prior error statistics prescribed for the bottom-up emission sources are often very uncertain owing to the sparsity of the available observational network [201]. These uncertainties in the transport model and the prior error emissions can translate into large errors in the inferred posterior estimates. Although current methodologies used for top-down atmospheric source inversions have reached a high level of sophistication enabling efficient and scalable computation of the posterior estimates along with their information content (i.e., posterior errors, observational constraints), the lack of observations to provide source estimates at accuracies and spatio-temporal resolutions compatible with environmental policy requirements is a key remaining limitation. The advent of satellite missions dedicated to air quality and carbon cycle observations (Sentinel-5P, GeoCarb) will be a milestone toward building such policy-relevant monitoring systems and will provide an unprecedented opportunity to improve the synergy between Earth surface models and numerical simulations of atmospheric dynamics across time scales, from NWP to climate modeling. In turn, such inverse modeling tools can be exploited to conduct instrumental design experiments that can inform mission strategies and requirements [202], and optimize scientific outcomes.

3.3. Land Surface Properties

Land surface processes and parameters strongly depend on geomorphology, land use, water body distribution, vegetation cover and soil type, and therefore the climatological fields describing these characteristics are a key part of any land surface scheme.

3.3.1. Orography

For every grid point, models have associated values for surface elevation (orography), sub-grid orography statistics, land cover (used as land/sea mask), lake cover and depth, glacier cover, low and high vegetation type, low and high vegetation cover, albedo, LAI, and soil texture. Within the ECMWF system, these fields are derived from different external sources (e.g., albedo and LAI from MODIS, vegetation type and cover from GLCC/AVHRR, water bodies from Globcover).

Global datasets come in different resolutions, data formats and projections and need to be interpolated, or merged in the case of non-global coverage. In light of the changing climate signal also affecting surface physiography, multi-year climatological data at 1 km resolution for all surface fields is desirable. One-km resolution represents a challenge for current global NWP and Earth system models, but it is relevant to allow direct comparison with present and future Sentinel missions monitoring the Earth system. Figure 17 shows the orography, ocean and lake bathymetry used operationally at ECMWF. The lake depth is based on [203] and the ocean bathymetry is based on ETOPO1 by [204].

The land-sea mask and orography are based on the following raw data information: ESA's Globcover V2.2 based on Envisat MERIS (300 m resolution) mapping 2005/2006 (ESA, 2010), the Shuttle Radar Topography Mission (SRTM [205] at about 90 m resolution, the Global Land One-kilometer Base Elevation (GLOBE [206], only north of 60N and south of 60S), and specialised DEMs of Greenland [207], Iceland (Icelandic Meteorological Office, 2013) and Antarctica [208], replacing the corresponding data points on the 1 km latitude/longitude grid. The lake mask has been created from the land sea mask and quality controlled by consistency algorithms. All spatial resolutions that are used by the IFS run in fully coupled forecasts [209].

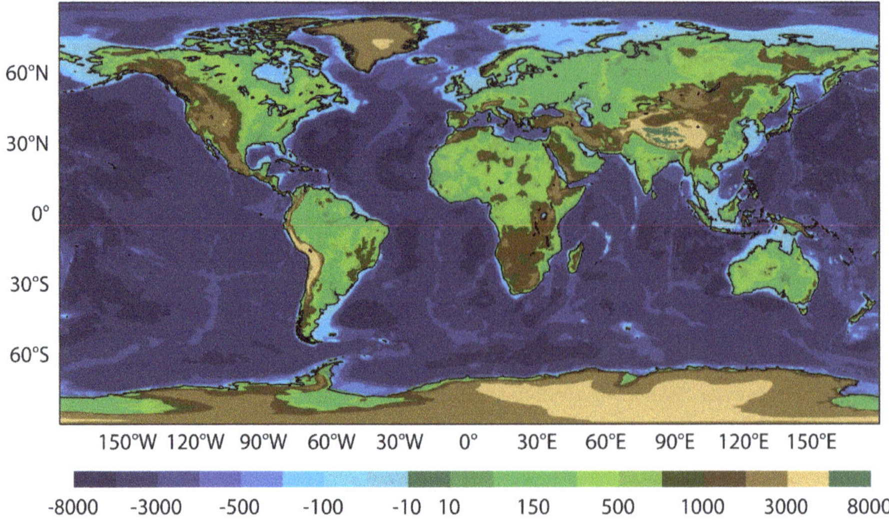

Figure 17. Global map of land orography and ocean+lakes bathymetry in m (above and below sea level or lake shore respectively) at T1279 resolution (9km) as used in the ECMWF—IFS operational system.

3.3.2. Soil Depth

While earth-observing satellite data have been widely used in modeling land processes (e.g., global land cover [210]), in situ measurements as constrained by satellite observations are needed to estimate other variables. One such variable is the Depth To Bedrock (DTB), which is essential for accurate land surface modeling of the energy, water, carbon cycle, and dynamic vegetation. In general, a lower boundary condition is needed to solve the governing equation for vadose zone soil moisture. The presence of bedrock is a player in soil hydrology as can change the water flow and storage [211]. Earlier land models assume a free drainage bottom condition (i.e., without considering groundwater), while more recent models include an unconfined aquifer which implicitly assumes a globally constant bedrock depth. However, no conditions are satisfactory without a DTB estimate [212].

Ref. [213] has recently developed the world's first global 1 km soil and alluvial thickness dataset (Figure 18) by combining geomorphologic theory with the best available data for topography (using satellite elevation measurements), climate, ecosystem [210], and geology as input. This dataset utilised different approaches to separately estimate soil and alluvial thickness for upland hillslopes, upland valley bottoms, and lowlands, as the character of soil depth is fundamentally different for these units: upland hillslopes have relatively shallow soils ($\sim$1 m), while valleys and lowlands have relatively deeper soils/alluvium (10 s of meters or more) (Figure 18). These three units are distinguished at the 90 m pixel scale using a valley network extraction algorithm as well as criteria related to geologic age. On hillslopes, the data set is calibrated and validated using independent data sets of measured soil thicknesses from the U.S. and Europe and on lowlands using DTB observations from groundwater wells in the U.S. This dataset is publicly available at: https://daac.ornl.gov/SOILS/guides/Global_Soil_Regolith_Sediment.html. A similar product based on global spatial ensemble prediction models with various input datasets is also available [214].

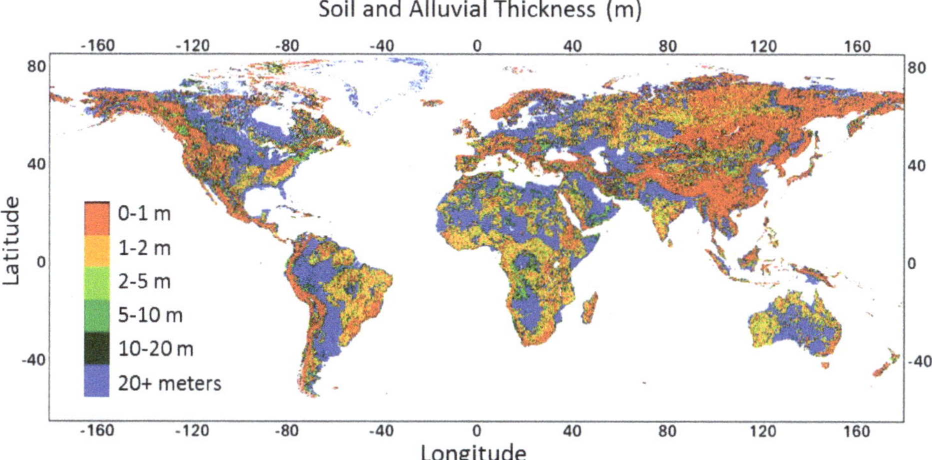

Figure 18. Global map of soil and alluvial thickness [213].

The impact of this DTB dataset on land surface modeling has been demonstrated [215] implementing variable DTB in the Community Land Model (CLM4.5) with 0.9°latitude x 1.25°longitude grid boxes. The greatest changes in the simulation with variable DTB are to baseflow, with the annual minimum generally occurring earlier in the year comparing to fixed soil depth simulations. Smaller changes are seen in latent heat flux and surface runoff primarily as a result of an increase in the annual cycle amplitude. These changes are related to soil moisture changes that are most substantial in locations with shallow bedrock. Total water storage anomalies are not strongly affected over most river basins because they tend to contain deep soils. These anomalies are substantially different for river basins in more mountainous terrain. Additionally, the annual cycle in soil temperature is partially affected by including realistic soil thicknesses resulting from changes in the vertical profile of heat capacity and thermal conductivity. However, the largest changes to soil temperature are introduced by the soil moisture changes in the variable soil thickness simulation.

3.3.3. Soil Texture

Soil properties, particularly the texture of the soil in the near-surface horizon, obviously affect the evolution of soil moisture by controlling the ability of water to move in the soil column. Water retention and conductivity, as governed by soil texture, have secondary effects such as controlling the persistence of soil moisture anomalies and helping determine what type of vegetation can flourish in a location. Less evident is how subsurface properties affect surface soil moisture evolution. This is not well appreciated because it is very difficult to discern the structure of the terrain below the top meter or two of the soil. Weather and climate models typically treat the vadose zone below vegetation roots as terra incognita, assuming no spatial variability except that related to the slope of terrain. If a model simulates the water table, it is often connected directly to the shallow soils without regard to the spatial variability in between. Lumped hydrologic models infer the effect of the vadose zone in their calibration, but this is not possible in ungauged or arid basins. In reality, there can be preferential pathways of drainage that quickly remove water from the reach of direct evaporation or plant transpiration [216]. These correspond to areas of fractured or unconsolidated bedrock, called karst that underlie a significant fraction of the land surface, about a quarter of the continental US [217] and western part of Kazakhstan, where landscape can change quite dramatically, including emergence/disappearance of lakes, in a very short period [218]. Observational and modeling evidence suggests regions overlying shallow karst formations experience very different soil water evolution, affecting surface moisture and heat fluxes, vegetation distribution, and even impacting the evolution of convective precipitation [219]. A direct way to discern the effect of karst is to examine time series of in situ soil moisture measurements from networks using consistent instrumentation spanning both karst formations and less permeable bedrock. In the southern Appalachian region around the Middle Tennessee River Basin, there are extensive shallow carbonate karst formations and a fairly dense network of US Department of Agriculture (USDA) soil moisture stations spanning karst and non-karst substrates. A strong correspondence between weak soil moisture memory (lagged autocorrelation of anomalies) and the presence of karst has been found across a dozen stations [220]. Comparison among closely-located stations helps control for the confounding effects of other climatic, vegetative and hydrologic factors. The methodology developed for in situ data has been applied to remotely sensed soil moisture georeferenced against high-resolution karst information from the US Geological Survey [217] across the southeastern and south-central US. Small-scale (within 2700–3000 sq.km) spatial variability in karst coverage correlates significantly with small-scale variability in soil moisture memory [221].

Figure 19 shows that, where there is strong spatial variation in karst coverage, there is also greater variation in soil moisture memory as derived from NASA/SMAP satellite soil moisture. Correspondingly, where there is little or no variation in bedrock permeability, soil moisture memory is also more uniform.

The fact that the signal of karst appears in satellite soil moisture data yields promise that remote sensing could contribute important hydrologic information in poorly-observed regions of the globe. Radiances in near-infrared and other parts of the electromagnetic spectrum have been shown to correlate to surface soil properties in unvegetated areas [222,223]. Remote sensing of soil moisture is achieving levels of quality and information content that can allow inference of surface and subsurface properties otherwise unobserved or unavailable. Careful assays of remote sensing data could enhance global datasets used for Earth system modeling, whose quality is currently skewed towards developed wealthy nations.

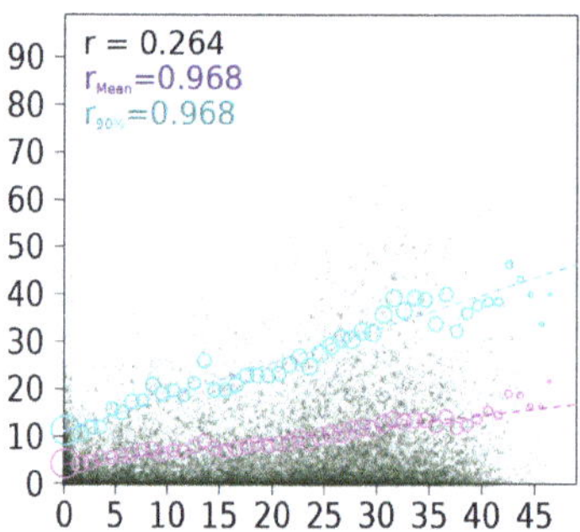

Figure 19. *Cont.*

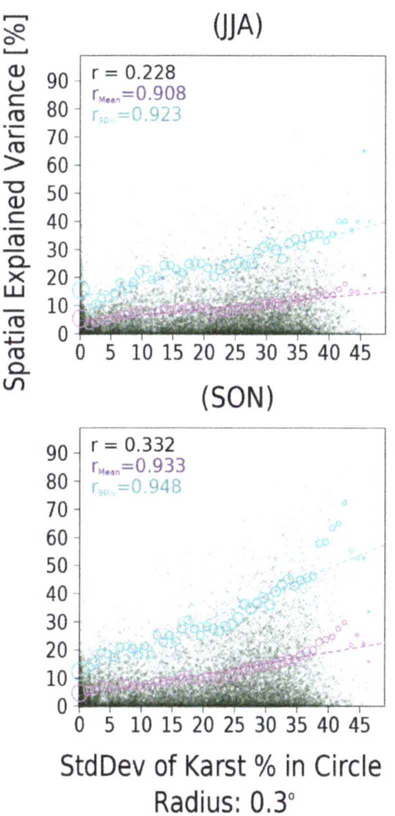

Figure 19. Explained spatial variance in satellite-derived soil moisture memory (e-folding time scale i.e., lag at which the autocorrelation of SMAP soil moisture drops below 1/e) within circles of radius 0.3° as a function of the standard deviation of gridded United States Geological Survey (USGS) shallow carbonate karst coverage in the corresponding circles. The karst data have been interpolated to the 9 km resolution SMAP Level-3 data grid, and the circles for comparison are centered on each land grid cell between 29–40N, 104–83W. Purple circles indicate the means of spatial explained variance in soil moisture memory within bins of width 2.0% centered on each 1.0% step of the abscissa; cyan circles are for the 90th percentile values. Dotted lines are best-fit linear regressions. All correlations are significant at the 99% confidence level.

3.4. Inland-Waters

Globally, there are approximately 117 million lakes (>0.002 sq.km) with a combined surface area of 5×10^6 sq.km, which is approximately 3.7% of the Earth's non-glaciated land surface [224]. Despite their relatively small spatial extent, lakes can have a disproportionately large influence on the climate, in particular within their near-vicinity [225,226]. This occurs as a result of lakes influencing surface energy exchange with the atmosphere, which differs to those from soil or vegetated surfaces [227]. Some lakes also freeze seasonally, resulting in considerable changes in surface albedo and thermal capacity and, as a result, alterations in lake–climate interactions [228].

Although traditionally neglected in land surface models within NWP, primarily because of the computational expense, community efforts in recent years have improved our understanding of the importance of lakes and have highlighted their added-value in simulating accurately regional climatic variations [225,229,230]. Research aimed at introducing inland water bodies (lakes, rivers and coastal waters) into the operational model at ECMWF has started by considering a medium-complexity scheme that satisfies the operational constraint of having low computational cost. FLake [231], a shallow-water scheme originally applied to lakes, was introduced into the IFS in progressive steps. This model is a particularly appropriate choice for NWP as it predicts the vertical temperature structure and mixing conditions in lakes of various depths and on time scales from a few hours to multiple years, while maintaining a relatively low number of prognostics variables (seven in total). The model is intended for use as a lake parameterisation scheme in NWP, climate modelling and other prediction systems for environmental applications. FLake has been implemented in the operational regional weather forecast models of Deutscher Wetterdienst (the German weather service), Finnish Meteorological Institute (FMI), Norwegian Meteorological Institute (Met.no) and Swedish Meteorological and Hydrological Institute (SMHI), and is used for research at several meteorological services across Europe including Météo-France [232] and UK Met Office [233], and is included in several climate models. The most important external variable for lake model FLake is depth (bathymetry or mean depth at least). There were several attempts to derive lake depths from satellite observations. Already in [234], an approach for determining lake depths by using spaceborne Synthetic Aperture Radar (SAR) images was developed. For those lakes that freeze completely to the lake depth some time in the winter, the simulated ice growth curve providing the ice thickness allows to estimate the lake's maximum water depth [234]. In [235], the same idea was used and lake depth of shallow sub-Arctic lakes and ponds was determined by using Landsat Thematic Mapper (TM) and European Remote Sensing (ERS)-1 Synthetic Aperture Radar (SAR) data. Landsat TM image is used to map lake bathymetry and multi-date ERS-1 images acquired during winter are utilised to determine when and which lakes freeze to the bottom during winter. Lake depth estimates computed by this approach correspond well with in situ measurements, especially in the tundra zone; however, the approach needs to be further tested and improved for automatic use [235]. While at the moment still no complete global lake depth dataset exist, local bathymetry dataset can be found at least for large lakes (see Section 6.3.4 of the GCOS report [60]). Beyond a few large lakes, the collection of local datasets is a difficult task; therefore, the most common practise for lake modeling in global NWP is to use in situ measurements and indirect estimates of lake mean depths based for instance on the geological origin of lakes [236].

Prior to its implementation in the IFS, the influence of FLake was evaluated in a series of preparatory studies: first, in an offline experimental framework [227,237], and then extended to fully-coupled lake-atmosphere simulations [230]. More recently, the possibility of treating sub-grid water bodies (lakes and coastal waters) using the land surface tiling methodology has been included.

An important step in the inclusion of lakes in NWP is that the simulated lake surface temperature and lake ice conditions are validated by comparing with observations. The use of remote sensing data for validating lake model outputs in NWP is essential, in particular as in situ observations at a global scale are currently lacking.

With this approach, each grid box is divided into fractions of different land use, each with their own tile. Merits and limitations of the tiling methodology when accommodating lakes and forest within the same model grid-box have been assessed by [238]. They conclude that the tiling method captures well the influence on surface fluxes of the contrast between the "lake" and "forest" surface boundary conditions, on seasonal and diurnal time scales (see Figure 20, upper left and upper right panels, respectively).

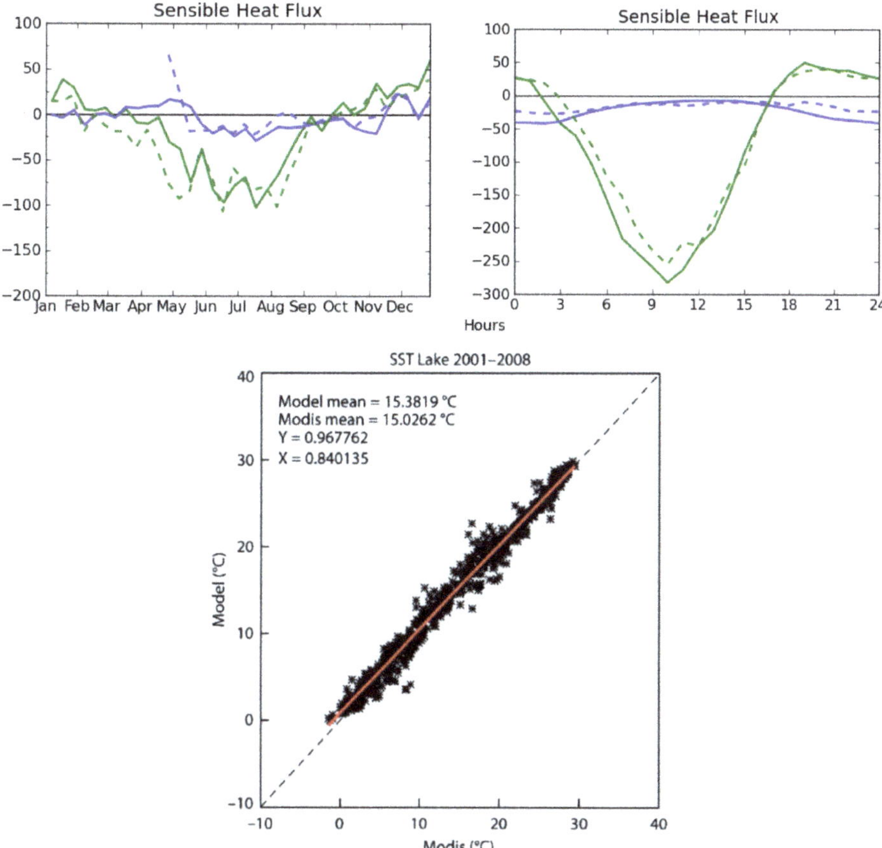

Figure 20. Sensible heat-flux over a lake (blue line) and a near-by forest (green line) in Finland. An annual cycle (upper left) and a mean July diurnal cycle (upper right) are shown for the model (solid line) and the flux-tower observations (dashed line), with a negative sign indicating an upwards heat transfer. Lake surface temperature on global scale simulated by the model FLake and observed with MODIS (bottom).

The behavior of simulated lake surface temperatures (Figure 20, bottom panel) and the period of ice cover in large inland water bodies worldwide was verified using satellite-based products (see [230]). Particularly relevant for validating lake simulations (temperature and ice cover) in NWP is, for example, data provided by the ATSR Reprocessing for Climate, ARC-Lake lake surface water temperature and ice cover dataset [239]. Satellite-derived observations from ARC-Lake have been used in recent years to validate simulations from the Canadian small lake model within the Canadian land surface scheme [240], global lake temperature simulations performed within the CNRM-CM5 climate model [232], and in validating the coupling of FLake to the UK Met Office Unified Model [233], among others.

The impact of introducing interactive inland water bodies in ECMWF's IFS has been examined by a set of dedicated analysis experiments. However, the lake model (FLake, [231]) does not currently consider a mass balance of water (only for the ice components), whereby a static dataset is used to represent the extent and bathymetry of the world's lakes. Monitoring lake depth globally is not currently achievable although inversion methods (e.g., Ref. [237] demonstrates that an effective lake depth can be derived on the basis of the remotely-sensed lake water-surface temperature).

3.5. River Discharge and Hydrological Forecasting

Continental and global scale information on river discharge, and upcoming floods and droughts can be used in many applications including disaster risk reduction initiatives, particularly in preparing for severe events and providing early awareness where local flood models and warning services may not exist [241]. The European and Global Flood Awareness Systems (EFAS and GloFAS) are part of the European Commission's Copernicus Emergency Management Service (CEMS-flood) and provide complementary flood forecast information to relevant stakeholders supporting flood risk management at national, regional and global level at several different timescales ([242–245]; http://www.globalfloods.eu). CEMS-flood is based on an Earth system modelling approach, and the forecasts are derived using in situ and satellite data as well as hydro-meteorological and Earth system models. It is only very recently that improvements in the resolution, precipitation processes, and the improvements in land surface representations discussed in the above subsections have meant that the forecasts of hydrological variables such as precipitation, soil moisture and more recently also runoff, can be considered good enough to make effective forecasts of river flow using Earth System models [241,246]. GloFAS uses the ECMWF Ensemble Prediction System, the operational ensemble forecasting product of ECMWF which consists of a 51 member ensemble of global forecasts. The ECMWF land surface scheme, H-TESSEL, then calculates land surface response to atmospheric forcing, and estimates the surface water and energy fluxes and the evolution of snowpack, soil temperature and soil moisture. Operational ensemble forecasts of surface and sub-surface runoff are resampled to 0.1 degree resolution to be used as input by the river routing model (for more information see [242,245]). River flow climatologies, against which forecasts are compared, have been created using corrected reanalysis products such as ERA-Interim Land [247] which includes precipitation adjustments based on monthly GPCP v2.1 (Global Precipitation Climatology Project). Similar corrected climatologies are planned for the next generation reanalysis product ERA5 (i.e., ERA5-Land). Future needs for hydrological forecasting in terms of data assimilation, model validation/evolution and parameterisation are better information on river flow rates and flood extent (particularly in real time), groundwater contributions, and high resolution river catchment and river channel topography from EO data [248–250].

3.6. Land–Atmosphere Coupling

A promising method of model evaluation is to study the dynamics of the system rather than its states. The response of the land surface to atmospheric drivers repeats after each rainfall event. The repeating wetting–drying cycle can be studied using different observable properties to identify the time response of the surface to the driver which is dependent on evapotranspiration. Refs. [251,252] developed a systematic way of analysing the time-response of evaporation after a rainfall event by analysing the land surface temperature increase. The theory is that a dry soil with low evapotranspiration will warm up more quickly than a wet surface with high-evapotranspiration. Since the dynamical time-response of the system is an inherent property of the land surface, rather than a result of the driving data, the method can be applied to coupled model simulations. The method was applied to evaluate a model in coupled mode by [253]. Two model configurations were tested and Figure 21 shows how the data favours one of the model configurations, at least for the spring months.

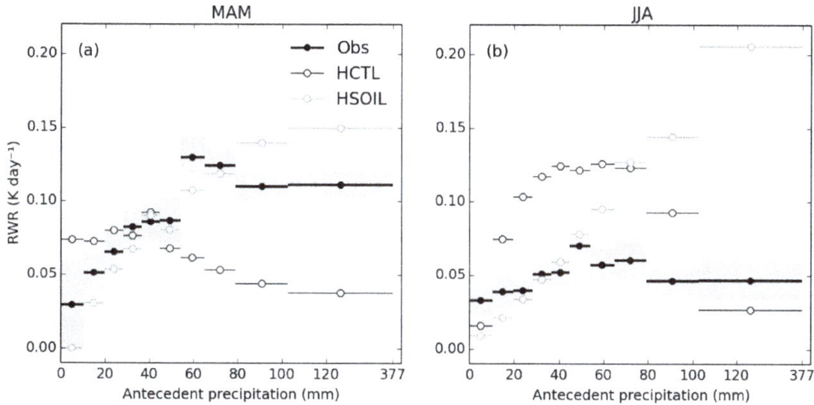

Figure 21. Example of usage of land surface temperature during dry episodes after a precipitation event. The plot shows a composite of the so-called Relative Warming Rate (RWR) as a function of the amount of precipitation during the preceding event for March-April-May (MAM, **a**) and June-July-August (JJA, **b**). RWR quantifies the increase in dry spell land surface temperature relative to air temperature, and is a measure for the evaporation regime of the land surface.

In a similar vein, but at a much slower time-scale, EO observations of Total Water Storage (TWS) have been used [254] at the seasonal time-scale to evaluate evapotranspiration in a suite of land-surface models. Key properties of the models can be obtained. Similarly, a dynamic soil moisture global dataset based on EO data [255] has been obtained by characterizing the dry-down trends. Since EO soil moisture only captures near surface water, it can only be used for two of the three evaporation components (soil surface and interception) and not the larger transpiration. Ref. [256] quantified the power spectra of rainfall, EO soil moisture and modelled soil moisture allowing for inferring the low-pass filter effect of soil hydrology. Neither of these analyses have yet been used to evaluate land-surface model systematically, but the principle shows great promise. These new methods can further be backed up with in situ observations such as drying rates from Flux Tower or soil moisture sensors.

Producing accurate forecasts of near-surface parameters becomes increasingly important as near-surface temperature, humidity and wind are used in a widening range of applications and are themselves highly relevant during extreme events (e.g., wind-storms, droughts and heat-waves). The representation of land-atmosphere interactions depend on a large number of processes. The exchange of energy or moisture between land surface and the atmosphere plays an important role for near-surface temperatures and humidity [70] while the roughness of the surface exerts a major control on near-surface winds. However, the parameterisation of these interactions is hindered by difficulties in estimating the land-atmosphere coupling strength from theory or observations. Thus, the parameterisation relies on a number of parameters, set to poorly constrained values, that depend at most on the vegetation type or sub-grid information (tile fractions). The skin layer conductivity, the roughness length for heat or momentum, the minimum canopy resistance, the dependence of the canopy resistance on the water vapour pressure deficit and the root distribution over the soil layers are just a few examples. In the past years, efforts have been made to (i) evaluate the sensitivity of near-surface variables to the values chosen for such parameters, and (ii) find ways in which to better constrain them [106]. The first obvious candidate was the roughness length for momentum, which has a strong influence on the representation of 10-m wind speeds. For many years, ECMWF forecasts overestimated the near-surface (10 m) wind speed over land. The mean forecast errors, with respect to routine SYNOP observations, range roughly between 0.5 and 1 m/s for various regions/times of the day. A stratification of the 10-m wind speed forecast errors by vegetation type showed that these errors depend on the vegetation type, or more precisely on the momentum roughness length of each vegetation type.

The overestimation of the near-surface wind speed for most of the vegetation types suggests that the values used for the momentum roughness length were too low. This motivated the revision of these values based on theoretical considerations and SYNOP observations of wind speed at 10 m. The basic idea was to search, for each vegetation type, for a new value of the momentum roughness length for which the mean 10-m wind speed forecast error with respect to SYNOP observations drops to zero in near-neutral conditions. This calibration showed that the momentum roughness length values should be increased for nine and decreased for one of the 18 vegetation types characterizing land areas. The newly derived values were introduced in IFS cycle 37r3 (November 2011). As the roughness length for momentum was on average increased, the roughness length for heat was decreased in order to account for terrain heterogeneity. The use of the new momentum roughness length values reduced significantly the 10-m wind speed forecast errors, especially for the predominant vegetation types, but also on average over continental areas. The improvement for Europe was clear both in the pre-operational testing and in the operational verification afterwards. The reduction of the heat roughness length had also a positive impact on the 2-m temperatures, leading to a small warming during night-time and cooling during daytime over the continental regions for both winter and summer, which reduced the forecast biases over Europe [106]. Near surface temperature and humidity biases in weather are handled by assimilating SYNOP temperature and relative humidity and therefore optimising the 2 m forecast performance on given weather conditions [112].

3.7. Ocean–Atmosphere Coupling

Both satellite and in situ observations show evidence of a diurnal cycle in Sea Surface Temperature (SST), with a amplitude in excess of 1K over large regions of the ocean extending from the Tropics to high latitudes (see review by [257]). There is a growing appreciation for the need to resolve coupled physics at time scales shorter than a day. In terms of representing the diurnal cycle in models, a key factor is the difference between the temperature at a depth of 10m (foundation SST), the temperature at shallower depths which is generally warmer (bulk SST) and the temperature in the ocean's thin skin layer (<1 mm thick) which can be cooler at night and under certain wind conditions.

Observations of the diurnal cycle have motivated the development of ocean models to represent the diurnal cycle. This can be achieved in one of two ways: the first approach is to provide sufficient vertical resolution (O(1m)) in the near surface to explicitly resolve the diurnal cycle [258,259] which models the bulk SST. The second approach is to use coarser vertical resolution but provide a sub-model within the mixed layer scheme which parameterises the bulk and skin temperature [260]. Both approaches agree well with the observed diurnal amplitude of the bulk SST (equivalent to 1 m). There is a subtle difference in the approaches which is taken into account when calculating surface fluxes from bulk formulae; [260] were able to parameterise the skin temperature of the ocean while [259] resolve the bulk SST with a 1 m vertical resolution and three hourly coupling. Ref. [260] show differences in the air–sea fluxes of up to 10 W/m^2 in the Tropics when the diurnal cycle of SST is included.

The diurnal cycle of solar radiation can lead to stabilization of the surface ocean during daytime and to nighttime convective mixing. As a consequence, when winds are weak, daytime solar warming can create near-surface temperature stratification, referred to as a "warm layer", with skin temperatures up to 2C (or more) warmer than the foundation SST below the warm layer. This stratification and the peak skin temperature, however, cannot be simulated in an ocean model forced by daily-averaged air–sea fluxes [261]. Critically, the very large daytime surface heat fluxes associated with solar radiation, acting on the thin daytime surface mixed layer, result in a rectified SST that is warmer than the SST produced by the averaged fluxes. Without the diurnal cycle coupling, the skin temperature would be equivalent to the bulk foundation SST. The rectified SST can then feedback to affect the air–sea fluxes. Ref. [262] show that, when foundation SST is used instead of the diurnally varying skin temperature, the computed net surface heat flux can differ by more than 5–10 W/m^2 in parts of the tropics when averaged over the 10-year period. The afternoon near-surface stratification can also cause trapping of wind-momentum resulting in a thin diurnal jet [263] and enhanced surface freshening associated with daytime rainfall under light winds [264]. The resulting rectified SST, enhanced near-surface stratification, shear, and advection can lead to a cascade of errors if not properly represented.

The benefits of coupling a dynamical ocean to the ECMWF medium-range forecasts have been evaluated in a set of dedicated forecast sensitivity experiments and from June 2018 all forecasting systems running operationally now include the ocean model (NEMO v3.4) in the ORCA025Z75 configuration with 75 levels in the vertical and a quarter of degree horizontal resolution to represent the 3D ocean and includes a dynamical sea-ice model (LIM2) and a wave-model (EC-WAM). These coupled forecasts, in which the ocean interactively responds to the meteorology, are better performing particularly for the prediction of tropical cyclones and their intensity [265]. The use of all-weather SST from AMSR-2 microwave sensor is shown to validate the cold wake 5-day forecast of the Pacific's tropical cyclone Neoguri in July 2014 with the ECMWF high resolution model coupled and uncoupled with ocean (Figure 22).

Moreover, the use of in situ SSTs observations has allowed to assess the diurnal variability in the ECMWF coupled forecasts and the advantages of ocean-atmosphere coupling [266].

Indeed, GCMs that resolve the air–sea flux diurnal variability tend to perform significantly better than ones which do not [259,267–269]. In the tropics, the mean SST cool bias [270] is reduced, with warmer SST leading to increased precipitation that compares better to climatology [259]. Further motivations for the inclusion of the diurnal cycle in SST is the potential improvement it might provide in predictions of the Madden–Julian Oscillation (MJO, [271]). The results of [259] actually show a reduction in the performance of the MJO index although there is indication that the diurnal coupling produces a stronger and more coherent MJO. Increasing the coupling frequency to resolve the diurnal cycle also enhanced the intensity of MJO convection [272] and changed the magnitude of the El Niño–Southern Oscillation (ENSO) amplitude [268,273]. In the Southern Ocean, resolving diurnal and intra-diurnal variations leads to a colder mean SST, and stronger westerlies [269], probably due to the resolution of fast-moving storms. It is therefore likely that resolving the diurnal cycle in SST and coupling on these timescales may provide at least part of the picture in improved extended-range predictions.

Air–sea exchanges of properties are highly sensitive to wind speed. However, if the wind is gusty or associated with a fast-moving storm, so that the wind direction is variable over the course of the day, then the magnitude of the mean wind, computed from the average zonal and meridional wind components, will be less than the average "scalar" wind speed computed from the variable wind speed magnitude. In the state-of-the-art Coupled Ocean Atmosphere Response Experiment (COARE) bulk flux algorithm [274], averaged hourly wind speeds are increased by a gustiness parameter to account for this additional wind speed variance at hourly time scales [275]. Using high resolution Tropical Atmosphere and Ocean (TAO) mooring data [276] showed that the diurnal 'mesoscale' gustiness was more than twice as large as the parameterised 'convective' gustiness and that neglecting this diurnal wind variance could cause a mean bias of up to 8 W/m^2 in the latent heat flux component alone. For scaling purposes, a 20 m layer subject to 10 W/m^2 net surface heat flux error over a 1-year period would produce a temperature error in that layer of 4C. In an operational context [277], the wave forecasting and the coupling with atmosphere and ocean models is progressing with detectable benefits both on near surface winds and sea-surface temperatures. The impact of waves on the coupled climate system is also highlighted in [278] for the continuous exchange of heat, energy, water vapor, and carbon dioxide that are relevant from diurnal cycles to extended prediction ranges.

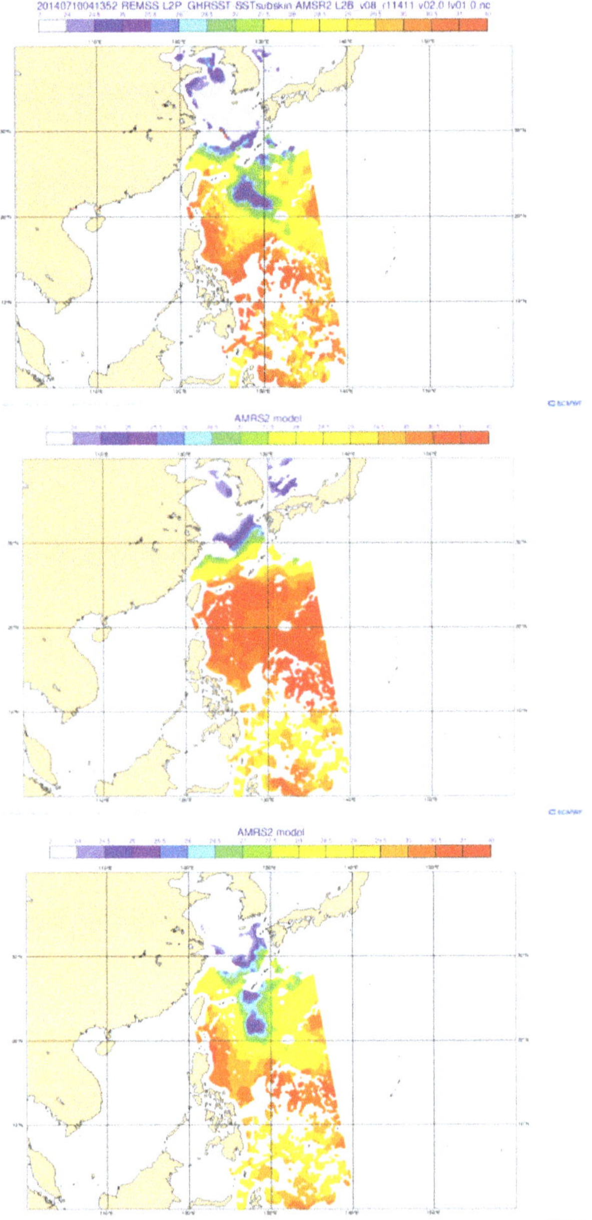

Figure 22. AMSR2 Observations of SST on 10 July 2014 after the passage of Tropical Cyclone Neoguri (**upper**), ECMWF uncoupled day-5 forecast (**middle**), ECMWF coupled ocean forecast (**lower**) valid at observation time.

3.8. Ocean Cryosphere: Snow and Ice

The presence of sea ice on the ocean drastically alters air–sea fluxes of heat and momentum. Hence, sea ice is an important component of Earth system models at high latitudes. As operational centres incorporate the ocean cryosphere within seamless prediction systems (i.e., in forecast systems spanning days to decades), there is an increasing demand for high-latitude coverage of observations to compare models for assessment of forecast performance and support ongoing development of the model, its coupling and data assimilation systems. As the polar regions have relatively sparse in situ observations, model development relies heavily on the earth observation data. It is important to note that this also requires the EO products to have a good estimate of the uncertainty as there are limited independent data sets to validate against. Recent advances in EO products are allowing model developments to be compared across the entire Arctic domain rather than just using limited in situ data at a single point. For the Arctic cryosphere, this is also important as the region is changing so rapidly. Field campaigns that have provided large datasets for model development such as SHEBA [279] may well be less representative of the ice conditions in recent years as much of the multi-year ice has been lost.

There are several products that are available operationally to describe the sea ice field. They are derived from brightness temperatures from various instruments and scatterometer data. These describe: the cover in terms of concentration, type, and edge; physical properties such as emissivity, skin temperature and thickness; and the dynamic properties in the form of motion (drift).

Of these EO measurements, the most established for use in NWP is sea ice concentration. It is one of the few fields routinely assimilated by operational centres (e.g., [280–282]) and also the main type of observation that the models are verified against (e.g., [283–287]). An example case study is shown in Figure 23 of an unusual event occurring to the north of Greenland that was captured in the model and also observed. Here, the performance of the coupled ECMWF HRES forecast is compared to the sea ice concentration as shown in the SAR image.

In recent years, it has become increasingly clear that thickness, i.e., the vertical distance from the air–ice interface at the top to the ice–water interface at the bottom, is essential for sea-ice modelling and prediction [288–290]. This is because thin ice allows much stronger conductive heat fluxes than thick ice, and it evolves much more quickly than thick ice because it is more susceptible to dispersion or compression by winds, and it can melt or grow more rapidly.

One way to observe the thickness of sea ice from space is to measure the surface emissivity in the L-band, which is sensitive to sea-ice thickness if the surface is dry, the ice temperature is well below freezing, and the ice is thinner than about 1 m [291]. This allows detection of areas which are covered by thin ice even though the sea-ice area fraction might be close to 100%. Typically, this occurs close to the ice edge, where new ice forms during seasonal ice advance. However, thin ice is also present in polynyas and fracture zones within the ice pack. These are frequent in winter where the old ice breaks up under the force of winds, and new ice forms rapidly on the exposed water surface (see Figure 24). This thin ice is difficult to model because it requires a good representation of the complex mechanical properties of sea ice combined with a good representation of its sub-grid-scale thickness distribution. Commonly used sea-ice models, like the LIM2 sea ice model currently operational at ECMWF, feature a simple visco-plastic rheology and a single sea-ice thickness category. This can be improved upon e.g., by introducing multiple sea-ice categories that allow treatment of sub-grid-scale variability in the ice thickness, but large-scale observations of sea-ice thickness are urgently needed to validate these new model developments.

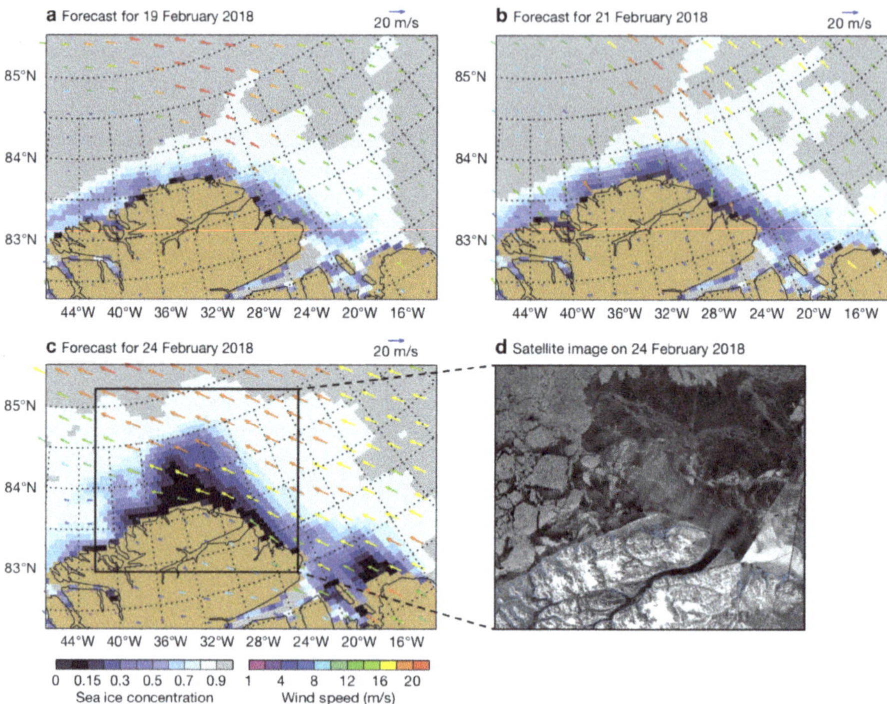

Figure 23. Evolution of sea ice concentration and the preceding 24-h mean 10-m wind field for forecasts initialised on 18 February 2018 and valid on (**a**) 19 February; (**b**) 21 February and (**c**) 24 February; (**d**) shows the Synthetic Aperture Radar (SAR) image from the Sentinel-1B satellite for 24 February around Cape Morris Jesup. (Satellite image: European Space Agency via Danish Meteorological Institute).

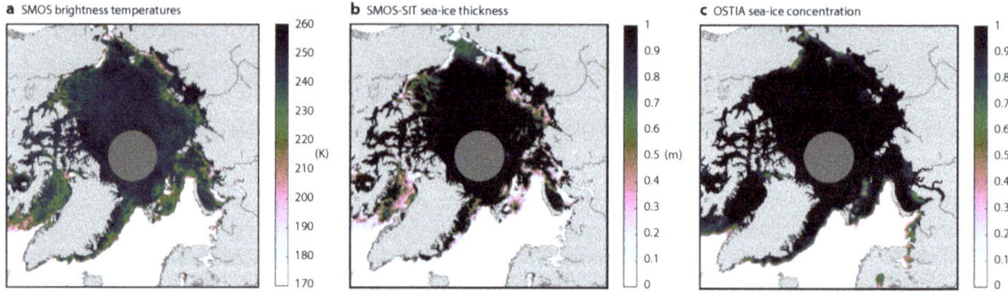

Figure 24. Daily-mean fields for 16 April 2017 showing (**a**) brightness temperature in L-band from SMOS (K); (**b**) sea-ice thickness from the University of Hamburg SMOS sea-ice thickness product (m); and (**c**) observed sea-ice concentration from the EUMETSAT Ocean and Sea–Ice Satellite Application Facility (OSI-SAF). Brightness temperatures over land are masked out. The greyed-out area around the North Pole does not have observations due to the characteristics of orbit and instrumentation of the SMOS satellite.

At present, there are large discrepancies in the representation of thickness of thin sea ice between observational products derived from L-band and ocean analyses with a prognostic ice model, assimilating sea-ice concentration but not sea-ice thickness [292]. This is not entirely due to limitations in the sea-ice model or data assimilation methods, but systematic biases in the observational data set also play a role. Further improvements in the retrieval methods are clearly needed [293], but the current sea-ice thickness data sets from L-band radiance have already proven highly valuable for exposing weaknesses in currently used sea-ice models. Hence, sea-ice thickness observations from L-band are essential to further improve prognostic sea-ice models to the benefit of advancing characterization and prediction of polar regions.

Another sea ice model development has been the parametrization of melt ponds, a feature essential for correct representation of the surface albedo of the ice and therefore capturing the correct evolution of the sea ice melt during the summer. In the past melt pond fractions have been estimated from field campaigns (e.g., [294]), but coverage of the whole Arctic basin are now possible, such as the data set produced by [295], which enables assessment of model parametrization schemes (e.g., [296])

For processes and parameters that are particularly important for Earth system model development, such as near surface temperature over polar oceans, there is a reliance on reanalysis data which can only be assessed in terms of performance with relatively limited in situ observational data within the ice [297] or using land based in situ data [298]. Sea ice and snow surface temperature data sets derived from different satellite products are starting to be used in data assimilation and are allowing the ongoing development of the sea ice model themselves [299].

Another example from the coupled system development is that data from CALIPSO has helped to identify model weakness in representing marine stratus cloud. It also helped to identify areas of model physics that would benefit from further development [300].

4. Towards Enhanced Global-Scale Local-Relevant Monitoring and Forecasting

Global monitoring and forecasting for environmental, weather and climate applications needs an accurate specification of the Earth surface as boundary condition to the atmosphere. This need can be satisfied at various levels of precision and complexity. However, some properties do not receive enough attention or they are simply neglected and therefore introduce errors in the surface representation. Others are approximated, due lack of information, by drastically simplified biogeophysical processes, typically when satellite datasets are not used. To enhance global-scale forecasting that maintains high relevance at local scale, it is necessary to pay more attention to the mapping of surface properties and parameters, in association with the parameterisation schemes used. In this section, we highlight some of the important mapping efforts and their relevance for modelling capabilities.

4.1. Anthropogenic Surface Modifications

NWP and climate models need lower boundary conditions to calculate the evolution of dynamic processes in the atmosphere and to produce a usable weather forecast, such as skin temperature, surface fluxes of heat, moisture and momentum. To compute them accurately, detailed plant functional maps are required. Previously, ecosystem maps were produced by collecting regional geodesic information at the national level and then merging the data in global maps. The main issue was that those maps were rapidly outdated because of changes in land use. Nowadays, human activity influences the Earth surface very rapidly by changing the landscape from forest to crop fields, breeding cattle in more profitable locations, providing water for arid places, or building homes for an expanding population.

4.1.1. Urban Areas

Although urban areas occupy 2.7% of the world's land excluding Antarctica (according to the Columbia University Socioeconomic Data and Applications Center Gridded Population of the World and the Global Rural-Urban Mapping Project (GRUMP)), they hosted 54% of the world's population in 2014 (according to United Nations report World Urbanization Prospects, 2014 revision) and forecasts for these areas are particularly crucial. The surface energy balance in urban areas is quite different from the one over vegetated surfaces, due to the complex three-dimensional building structure and construction materials, and therefore a special treatment for heating processes related to urban canyons is needed [301]. An urban model can be implemented as an additional tile or as a geo-located high-resolution subsurface. Inter-comparison studies involving several urban parameterisations [302,303] with simplified approaches to urban modelling were shown to be well suited for NWP and permit to represent the heat-island effect that characterize the modified diurnal cycle in large cities [304]. This is another good example where continuous satellite observations are extremely useful to detect anthropogenic influence in an urban or human settlement ecosystem type. In places of their habitat, humans develop built-up areas. The main issue with built-up areas is that they do not release their heat as fast as the surrounding country side, which leads to increased daytime temperatures, reduced night-time cooling and higher air pollution levels—Urban Heat Island effect (UHI) [304]. For instance, over Berlin the mean annual difference in UHI between simulations with without the city is 0.5 °K, and in winter daily differences are as large as several degrees [305]. Significant advances in monitoring urban areas temperature with remote sensing have recently allowed to characterise the different fluxes including the urban anthropogenic heat flux [306].

4.1.2. Changing Water Bodies and Irrigated Areas

Inland waters have been dramatically modified by humans in the last decades. Recent satellite observations demonstrate that the world's surface water bodies are far from static [307,308]. In particular, by analysing more than 3 million satellite images between 1984 and 2015 by the USGS/NASA Landsat satellite programme, new global maps of surface water occurrence and change have been produced with 30-m resolution [307]. These global maps demonstrate that, over the past 30 years, 90,000 square kilometres of what was thought to be permanent surface water has disappeared from the Earth's surface, linked with climatic variability and human influence (e.g., river diversions or damming and unregulated withdrawal). Elsewhere, 184,000 square kilometres of new permanent bodies of surface water have appeared, some man-made (e.g., reservoirs) and others as a result of climate change (e.g., melting of glaciers). Satellite observations provide a means for defining spatially varying lakes in NWP, but this concept has not yet been realised fully. Future studies should investigate how NWP can benefit from the real-time monitoring of inland surface water bodies. A more dynamic representation of lakes should be a future ambition of forecast centres. One of the recent possibilities to observe changes in ecosystems and take them into account in NWP is to use the 30 m (1″) horizontal resolution Global Water Surface Dataset (GWSD) created by Joint Research Centre (JRC). JRC used Landsat 5, 7 and 8 individual images over the past 32 years to map the spatial and temporal variability of global surface water and its long-term changes.

Figure 25 shows different places on the globe and their evolution in recent years. Upper plots show Aral Sea area. After massive diversion of water for cotton and rice cultivation in the 1960s shrinking of water surface area accelerated and in 1998 became 28,990 sq.km (less than a half of its initial size) [309]. Due to major recovery program launched in 2001 by Kazakhstan President Nursultan Nazarbayev and supported by the World Bank, Aral Sea water surface area started to stabilise and in 2014 became 7660 sq.km (almost 10 times less of its initial size) [309].

Lower plots show emergence of the new, largest in Western Europe, man-made Alqueva reservoir (Figure 25). It was built in 2002, and reached its present water surface area of 210 sq.km in 2006. Newly built reservoirs can be found all over the globe. Influence of inland water bodies on local climate is over 1 °K in temperature [225,230]. Local weather influence can be more pronounced—correct lake surface state (ice/no ice) in winter conditions can lead up to 10 °K [310] difference in 2 m temperature. At the moment, there are several satellite products that have lake ice cover variables.

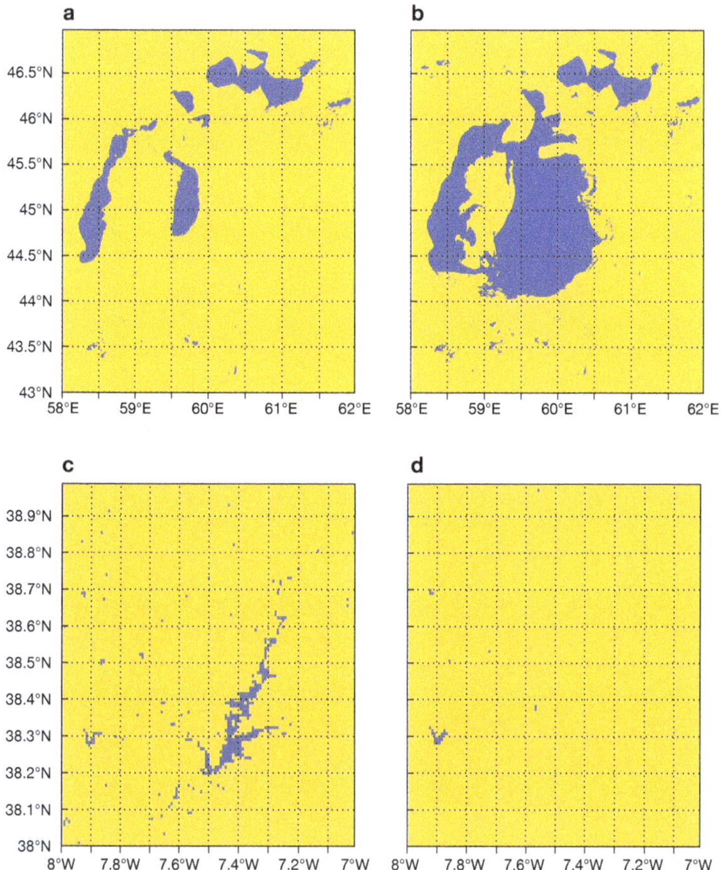

Figure 25. Aral sea (Kazakhstan/Uzbekistan, **top panels, a and b**), and Alqueva reservoir (Portugal, **lower panels, c and d**): water distribution in 2015 (**left, a and c**) and 1998 (**right, b and d**).

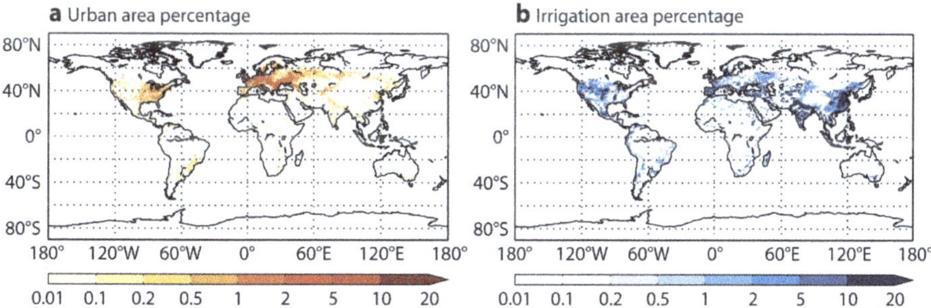

Figure 26. Urban areas and irrigated areas.

Irrigated areas have also been considered in preliminary tests, following the approach by [311,312]. Ref. [313] have demonstrated the relevance of including irrigation in a 20th century reanalysis. Figure 26 illustrates the fractional cover of irrigation and urban areas as dominant surface anthropogenic land use modifications, that are yet neglected in most global NWP models.

4.2. Relevance for Atmospheric Composition

Satellite retrievals of gases and aerosols in the global atmosphere carry the imprint of the surface fluxes of these compounds. The latter can therefore be inferred from the former by inversion of the atmospheric transport and of the atmospheric chemistry within a Bayesian approach. In recent years, Earth observations from space-based instruments have played a key role in improving our understanding of surface processes driving air composition, like for air quality [314].

The need to complement or evaluate the diverse bottom-up estimations of these fluxes by a top-down approach that would be inclusive of all source and sink processes has dramatically increased the interest in inverse modelling over the last decade (see, e.g., item 59 in SBSTA 2017, https://unfccc. int/sites/default/files/resource/docs/2017/sbsta/eng/07.pdf). In some cases, like for greenhouse gases, surface flux estimation has become a primary motivation for some satellite programmes, like NASA's Orbiting Carbon Observatory. A step further consists in optimizing some key parameters of Earth surface models based on inferred surface fluxes. For carbon monoxide, such an optimization could target fire emission factors to resolve the long-lasting mismatch in the peak fire month in Africa between models and satellite observations [315]. For ammonia, it could improve the representation of fertilisers spreading practices in inventories [316]. For methane, it could allow wetland models to better represent the response to extreme events [317]. For carbon dioxide, it could facilitate the sophistication of carbon processes representation outside the peak growing season, such as leaf flushing [318]. For aerosols, it could improve binding energy or other parameters in mineral dust emission models [319].

However, in practice, detecting enhanced or depleted plumes from space does not imply that the associated surface sources and sinks can be quantified to sufficiently useful accuracy and localization. The modesty of many plume signals when averaged vertically (particularly in the case of long-lived tracers), their fuzzy boundary within their background and with other plumes, joined to systematic errors (for given atmospheric or surface conditions) in satellite retrievals, the coarse resolution of EO systems, and in atmospheric transport and chemistry models, have slowed down progress and keep inverse modelling for Earth observation as a major scientific challenge. Some of the solutions found so far are trivial: an empirical bias correction of the retrievals before or within the inversion systems; a rise in computer resources allocated to chemistry-transport models in inversion systems in order to increase horizontal and vertical numerical resolutions. More robust solutions require dedicated spectroscopic or transport-process studies. For chemistry, measurements of new tracers that could serve as proxies may be necessary. In the next decade, satellite imagers for greenhouse gases and pollutants are expected and will bring more emphasis on flux estimation at the local scale for usage outside the scientific community. They will likely require optimizing local meteorological variables together with surface fluxes. Similarly, there is a growing need to merge information from various sensors for a given species, or for species emitted by common processes, like fire or fossil fuel burning. Properly quantifying the uncertainty of each data-stream, of the correlations between these, and of underlying models is prerequisite for these developments.

4.3. Improved Diurnal Cycle for Assimilation Purposes

The availability of new satellite data informative about superficial water over continental surfaces, particularly at low microwave frequencies (L-band and C-band), requires improvements in the description of land surfaces in atmospheric models for the simulation of satellite products, as a preliminary step towards land data assimilation. This is illustrated in the following by few examples.

Satellite pixels in the microwave frequency range have footprints of several kilometres typically encompass many surface types. It requires horizontal heterogeneities to be accounted for in the forward modelling since most NWP models have horizontal resolutions below 10 km. Indeed, at low microwave frequencies, the emission of a vegetated surface is very different from that of an open water surface (e.g., lake, river) or of a snow covered area. Radiometers dedicated to probe water in the soil are also sensitive to liquid water present in other media (vegetation, lakes, oceans, snow).

An improved description of inland water bodies has been proposed by [230] for the ECMWF model by using the prognostic lake scheme «FLake». Another important constraint on land surface schemes with this type of observation comes from the fact that remote-sensing instruments probe at most the first 5 cm of soil. However, the water contained in the top soil layer is not the main driver of surface evapotranspiration that is of primary interest in NWP models. What dominates the strength of water losses in the atmosphere or in rivers is the water content over a deeper layer including the root-zone (tens of centimetres to few metres). The usefulness of these satellite data relies on the capacity of land surface schemes to transfer background departures located in the superficial soil layer into relevant increments in the root-zone layer. The simplicity of force-restore equations from the ISBA scheme [320] allows the estimation of analytical Jacobians that relate sensitivity of the superficial soil moisture at a given time T to changes in deep soil moisture at the beginning of an assimilation window [321].

However, a contradiction appears between the "analytical expression soil moisture Jacobians" and the physical understanding of how information can be transferred from the superficial to the deep layer. This unexpected behavior is explained by the use of a single energy balance for bare soil and vegetation, together with the nonlinear behavior of vegetation transpiration near the wilting point. Indeed, an increase in root-zone soil moisture, when above wilting point, enhances vegetation transpiration that cools the surface. A decrease in surface temperature reduces bare soil evaporation and in turn reduces water losses of the superficial layer. Therefore, in a data assimilation system, satellite observations sensitive to the surface soil moisture could appear more informative than they actually are. Studies described in [322] and in [323] reveal that two-layer schemes artificially enhance deep layer corrections because the actual physical transfers (through vertical gradients of water potential) are not explicitly described.

In conclusion, multi-layer surface schemes with separate surface energy balances are expected to lead to a more realistic extraction of information from remotely sensed superficial soil moisture in land surface models. This work is particularly relevant to illustrate that approaching data assimilation with simplified models may impair the transfer of information from and to the deeper soil layers that control predictive skills in medium and extended range forecasting.

5. Towards Enhanced Use of High Resolution EO Data in Earth System Modelling

In this section, an analysis of data availability with the most common challenges in utilising the data is presented with the aim of identifying virtuous pathways to data inclusion in global modelling for forecasting and monitoring applications. The volume of Earth Observations of the surface creates an issue which calls for data reduction and data streaming with a clear set of applications in mind. A too passive role of EO data users (awaiting new observations without listing modelling needs), or assuming that a web portal will multiply by itself the use of satellite products, has the risk of limited user-uptake of EO data. There is the potential of creating a gap between availability and exploitation of data. Closer connection between EO surface data and Earth surface modelling, as part of weather and environmental prediction applications, can multiply users of the data by orders of magnitude.

Earth Observations ultimately drive model development, via data assimilation and via exposing systematic model errors. However, the balance between model resolution, the degree of sophistication of represented processes and the size of ensemble configurations (Figure 27) is a trade-off considering that computing resources are limited. The Earth surface cannot be easily observed nor modelled if coarse resolution sensors or models are utilised because the subgrid variability becomes too large. The use of ensembles with a large number of members enables representing the uncertainty of the forecasts [16] attributed to the modelling steps as well as to the analyses used as initial conditions that encompass EO observability and representativity issued.

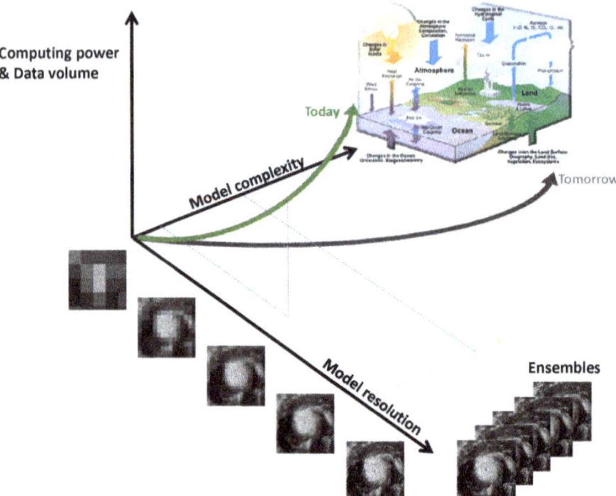

Figure 27. Model complexity, model resolution and ensemble, and High Performance Computing and data volume trade-offs. Courtesy of Peter Bauer, ECMWF.

5.1. Towards More Comprehensive Model Improvement through Joint Use of Multivariate EO Data

As illustrated throughout various examples in this paper, satellite-based EO data products are essential to improving and informing model development. However, the potential of the growing suite of EO datasets could be better exploited through the joint use of multiple EO datasets. So far, this is only done in classical data assimilation even though the approach holds promising potential also for improving model structures, and for deriving more accurate, more unique, parameter value sets.

For example, Ref. [13] showed that robust improvements in the parameter calibration of the H-TESSEL model are only possible with multiple observation-based reference datasets. For this purpose, they used several EO datasets in conjunction with in situ measurements. This allows overcoming their individual shortcomings; e.g., the satellite-based soil moisture offers unprecedented spatial coverage but is available only during relatively short time periods and is constrained to the surface layer, whereas in situ soil moisture measurements cover longer time periods and multiple soil depths. Such a comprehensive approach helps to advance model development as it can capture and adjust various couplings within a model. In contrast, calibration against individual datasets such as land surface temperature results in degraded model performance with respect to other variables such as soil moisture. This is due to an imperfect representation of the energy-water coupling in the model. In this way, the use of different combinations of reference datasets for training and validation enables a targeted identification of flawed couplings within a model.

While the suite of EO datasets is growing, the complexity of state-of-the-art (land surface and hydrological) models is also increasing. The related increasing number of parameters renders it difficult to sufficiently constrain such models. The potential of a comprehensive model calibration through an adequate ratio between the numbers of model parameters and employed reference datasets has been illustrated in [8]. They used a conceptual water balance model with few parameters and calibrated it thoroughly against a range of EO and in situ datasets. As a result, this simplistic model reached similar (and partly better) performance in simulating key hydrological variables such as soil moisture and runoff.

Despite the success of a simplistic model in this particular case, the increased complexity of state-of-the-art models is required in the context of coupled weather forecasting systems, or Earth system models [324]. This indicates that the use of emerging new EO datasets is absolutely required to better constrain those complex operational models. As a result, the improved calibration of individual modules, such as H-TESSEL, within a weather forecasting system, can contribute to advance temperature and precipitation forecasting skills, as exemplified in [325].

Following this avenue, it is critical to better link model development efforts—for example to enable to explicit representation of (to be) observed EO variables—with the planning and design of upcoming EO missions. This way, the growing suite of EO data products can contribute to an improved physical understanding of land–atmosphere interactions, and furthermore to more accurate weather predictions as well as climate projections.

5.2. Delivering EO-Driven Research to Services

Buizza and colleagues at ECMWF [326,327] have presented a typical Research to Operation (R2O) cycle for new model and assimilation versions. It includes internal updates, but it does not address yet the inter-agency planning. The latter is achieved via a variety of collaborations, externally funded projects, and internal projects. We present a sketch of the R2O in a given organization specialised in global atmospheric data assimilation and discuss ways in which cooperative work can extend to multiple organizations particularly to tackle the evolution towards a kilometer-scale representation of the Earth surface.

Earth Observations have a dual role in Earth-system monitoring and prediction applications. Firstly, they are required to define the initial state of the Earth-system components, i.e., the state from where forecasts are generated by integrating numerically the Earth-system equations. Secondly, they provide an estimate of the 'truth' (the true state of the Earth), that is required to evaluate model accuracy and forecast performance. Good quality observations are essential in the testing and evaluation phases. Diagnostics that allow objective evaluation of improvements in the development and testing phases, and routine monitoring of forecast performance are essential. The latter can trigger investigations that feed into subsequent model upgrades.

The establishment of focused multi-institute projects and working-groups e.g., around international field-campaigns, is a common practice to link remote-sensing and Earth system modelling.

5.3. Satellite-Focused Field Campaign: The Concordiasi Example

Concordiasi was a field campaign, part of the Thorpex experiment [328] dedicated to the study of the atmosphere above Antarctica as well as the land surface of the Antarctic continent and the surrounding sea-ice. In 2010, an innovative constellation of balloons, including 13 driftsondes, provided a unique set of measurements covering both a large volume and time. The balloons drifted for several months on isopycnic surfaces in the lowermost stratosphere around 18 km, circling over Antarctica in the winter vortex, performing, on command, hundreds of soundings of the troposphere. Many dropsondes were released to coincide with overpasses of the Meteorological Operation (MetOp) satellite, allowing comparison with data from the Infrared Atmospheric Sounding Interferometer (IASI). IASI is an advanced infrared sounder that has a large impact on NWP systems in general. However, there are some difficulties in its use over polar areas because the extremely cold polar environment makes it more difficult to extract temperature information from infrared spectra and makes it difficult to detect cloud properties.

A number of results have been described in the Concordiasi workshop report [329,330]. Although the performance of NWP analyses and forecasts have dramatically improved over the last decade, large systematic differences remain in analyses from various models for temperature over Antarctica and for winds on the surrounding oceans. A comparison between short-range forecasts and the Concordiasi data was investigated for various NWP centres. Results show that models suffer from deficiencies in representing near-surface temperature over the Antarctic high terrain. The very strong thermal inversion observed in the data is a challenge in numerical modeling because models need both a very good representation of turbulent exchanges in the atmosphere and of snow processes to be able to simulate this extreme atmospheric behavior. It has been shown that satellite retrievals also have problems representing the sharp inversions over the area.

An example of comparison of such profiles is provided in the Figure 28 (courtesy of T August, EUMETSAT). At the surface, particular attention has been paid to observing and modeling the interaction between snow and the atmosphere. This research has led to an improvement of snow representation over Antarctica in the Integrated Forecasting System (IFS) model at ECMWF [331]. Coupled snow–atmosphere simulations performed at Météo-France with the Crocus and Applications of Research to Operations at Mesoscale (AROME) models have been shown to realistically reproduce the internal and surface temperatures of the snow as well as boundary layer characteristics [332]. In this example of the Concordiasi campaign, datasets obtained from the field experiment over snow and sea ice have proven invaluable to document the quality and deficiencies of both satellite retrievals and NWP model fields. This has helped to provide directions in which to explore the improvement of models, which will be beneficial for weather prediction and climate monitoring. The validation and diagnostics of numerical weather modelling over snow and sea-ice with Earth observations pose challenges because of the lack of regular in situ data and because satellite observations can be difficult to interpret in those areas. Field experiments can help to provide precious information to assist scientists in this validation work.

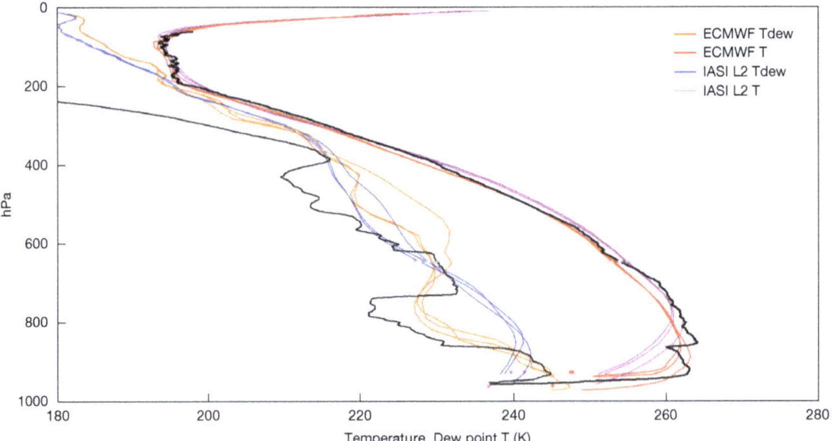

Figure 28. This Figure shows an example of a dropsonde vertical profile of temperatures (in black) on 23/10/2010 together with nearest ECMWF model profiles (in red and orange) and EUMETSAT IASI retrievals (in magenta and blue). Courtesy of Thomas August, EUMETSAT.

5.4. Links with the World Weather / Climate Research WWRP/WCRP Programmes

In the context of a more seamless prediction from day to decades, the World Weather Research Programme (WWRP) and the World Climate Research Programme (WCRP) have both recognised the importance of EO for the evaluation of climate models and development of climate research. For instance, the Observations for Climate Model Intercomparisons (Obs4MIP) project under the governance of WCRP (https://esgf-node.llnl.gov/projects/obs4mips/) is aiming at developing a database of existing observational products that can be directly used for model validation. GEWEX, the core project of WCRP focusing on land–climate interactions (http://www.gewex.org), has a particularly strong emphasis in this area, and oversees the GEWEX Data and Assessment Panel (GDAP), which provides regular assessments and evaluation of existing EO products. The use of EO products to improve the understanding and simulations of global water and energy processes is substantially highlighted in the GEWEX Science Questions [333].

The GEWEX Strategy 2019–2028 reviews the scope of research in the context of increasing greenhouse gases and changes in climate extremes. Examples are heat waves, heavy precipitation, droughts, or floods, which all are of particular concern [87].

Several recent studies have shown that land-surface processes play an essential role for climate extremes, and thus EO of land surface variables are likely to help constrain the representation of these extremes in current climate models. This is particularly the case for heatwaves and changes in temperature variability, which have been shown to be strongly affected by soil moisture availability (e.g., [334–340]).

In particular, recent results show that much of the changes in temperature extremes in mid-latitude regions can be understood as the combination of the global climate sensitivity—i.e., the global temperature response to greenhouse gas forcing—and regional effects due to soil moisture [341]. Climate simulations also suggest that typical changes in land albedo (on the order of 0.1) can substantially affect temperature extremes on local to regional scales [342,343].

Soil moisture–precipitation feedback have been shown to be relevant as well [76], although climate models possibly present biases in the representation of these relationships [344]. There are only a few studies of the possible impact of soil moisture on precipitation extremes, but there is evidence that land evapotranspiration was a strong contributor to the 2010 Pakistan floods [345] and climate model simulations also suggest that soil moisture–precipitation feedback impact projected changes in precipitation extremes [339,346]. Besides the effects of land–climate interactions for temperature and precipitation extremes, soil moisture states also play an essential role in the development of droughts (e.g., [347]) and floods, and combined with high resolution imagers can reach km-scale resolution (e.g., SMOS/MODIS combination [348]). Soil moisture products at km-scale resolution will in turn benefit from dedicated field campaigns, such as for instance the AirMOSS (https://airmoss.ornl.gov [349]) as well as in situ ground based network for calibration/validation activities.

5.5. Links with EO Satellite Data Providers

Many important land surface geophysical parameters can be inferred from satellites including: soil moisture, snow cover, surface temperature and LAI. Due to their indirect nature and given the currently used assimilation systems, remotely sensed observations are still significantly under exploited. Used products are generated by complex external processing levels (L2 to L4) which can potentially introduce inconsistencies with other atmospheric observations and processes. Coupled data assimilation systems would address these issues by allowing a consistent handling of these observations across systems, provided systematic differences (e.g., due to calibration) are removed by bias correction schemes, that can be integral part of the data assimilation system [350]. Moreover, a stronger link between observations and data assimilation, will by design enable a quality control system with reduced latency.

5.6. An International Surface Working Group

The International Surface Working Group (ISWG) has been active since 2016 to gather requirements specific to surface observations, and to enhance both our understanding and ability to monitor the components of the Earth system. Land, vegetation, snow, ice, coastal waters and open waters are included. An international science working group focused on land surface would integrate the other active scientific groups, gathering requirements that are specific to land and surface-water monitoring. A coordinated effort to bring concerns and advances from the scientific community could bring a clear focus by reviewing what technology has reached a maturity level to be directly used in land surface monitoring and modeling applications and feed into weather and climate. Usage of readily available observing capacity that has not been exploited (such as several geostationary platforms) could be fostered by dedicated workshops and a gap analysis for observing needs from this community could provide a link to future missions. The recognition of ISWG with its focus on existing international scientific coordination can be critical to support operational applications in forecasting and monitoring of weather and climate. In particular, the satellite based estimation of soil moisture, snow, land surface temperature and surface water body extents are particularly poorly considered for continuity in current satellite missions (with few exceptions). Unfortunately, when satellite data is available, users' uptake is often confined to local applications not encompassing global weather and environmental forecasting or climate monitoring, which would naturally scale the number of users to a broader community. The primary task of ISWG should be to bring forward a continuity plan using existing systems and ones which would fill application gaps in future. ISWG would provide the community a point of focus for applications of EO data that would be particularly relevant to bridge gaps between monitoring and mapping of terrestrial surfaces that have been historically developed in different communities.

6. Conclusions

This review presents some examples of how Earth Observations can successfully drive model development for Earth monitoring and prediction systems, such as those used for weather and climate. Often the link between observing a process, reaching a physical understanding and designing parameterisations that can be used in models is not explicitly made but is nonetheless fundamental.

Recent land surface modelling developments, aimed at increased hydrological consistency, have been substantially guided by water-dedicated missions such as SMOS and SMAP. Surface temperatures have begun to be used in models despite the resolution barriers and large systematic model errors, which had previously limited the observation uptake.

This list is non-exhaustive and aims to exemplify the great opportunities offered by remote sensing of the Earth surface combined with in situ observations. Emphasizing the link between observations and model development is important as global Earth system modelling moves to the kilometer scale. Existing models need to evolve to reduce model-data discrepancies at the surface, and data assimilation needs to evolve to coupled surface-atmosphere data assimilation.

Scientists working in the field can benefit from more than a decade of research in regional weather and climate models. The data volume required for a high resolution representation of the Earth surface (close to 1 billion points if 1/120 degree resolution is considered) requires smart data reductions and prioritisation of benchmarks. Surface temperature-sensitive and water-sensitive remote sensing products are key data to drive model improvements.

Products that combine unique remote sensing capabilities at coarser resolution (e.g., SMOS/SMAP for soil moisture, GOME-2 for vegetation) with high resolution imagers are promising for the support and further development of the next generation of weather and climate models which will be approaching kilometer-scale resolution at the surface. To ensure continuity of the existing observation capabilities to preserve and extend complementarity and diverse and covariant information, i.e., for low frequency passive microwave measurements a follow-on mission for SMOS or SMAP is needed.

While the examples given here clearly demonstrate the benefit of investing further in EO-driven model development, the gaps are substantial. In addition, the lack of coordinated efforts to attain such a goal using current and future observations can be a major obstacle. Forming application-focused model development working groups such as the ISWG will help to channel efforts and gather modelling requirements for a more effective data uptake. Three recommendations include:

- The improvement of parameterisation schemes over land and water surfaces to reduce systematic model errors that are clearly associated to missing processes (e.g., lack or spatial resolution in horizontal and vertical dimension, inadequate representation of local physiography) that impair the realism of surface fluxes partitioning.
- The development of observation operators that map the observed satellite radiance into physical quantities that are represented in models, particularly for thermal infrared imagers (e.g., for LST) and low-frequency microwave radiometer/radars (e.g., for L-band brightness temperature).
- The improvement of assimilation schemes capable of handling the models and observations uncertainties related to surface heterogeneities and processes and associated to the diverse remote

sensing resolutions (e.g., LST and L-band). Data assimilation shall make use of observations for constraining both surface state-variables and fluxes.

To ensure progress in those areas, the synergetic use of in situ observations and satellite data is of paramount importance and should be developed in parallel so that ground-based observations can provide a solid anchor for the physical interpretation of satellite radiances in diverse biomes, ecosystems and climates.

Advantages for an increase accuracy in monitoring and prediction at the surface can be anticipated to be sizeable. An Earth system model which is capable of reproducing the observed range of surface temperatures in accordance with the amplitude of the diurnal and the seasonal cycle, and which reproduces synoptic disturbances or persistent weather, will have greater skill in predicting temperature extremes and therefore increase the relevance for climate studies. Similarly, the water cycle can benefit from more realistic precipitation and evaporation rates when approaching scales that are relevant to convection.

An accurate and up-to-date land-use description is necessary for obtaining modelled soil moisture that is more directly comparable to kilometer scale remote-sensing products. Attention to the energy and water cycles realism is a crucial ingredient to support simulations of the natural and anthropogenic carbon exchanges at the surface.

Representing the anthropogenic influence on the Earth surface and its effects within Earth system models is a challenge for the next generation of models. In these models, there will be an increased role for Earth observations to map and monitor the natural and human-driven surface changes in the Earth system. Their aim will be to integrate weather, climate and environmental applications.

Author Contributions: This review paper is the result of a collective effort from all authors. G.B. looked after the structure and coordination of the review and the introductions to each section. A.B. (Anton Beljaars) contributed to the sections' harmonisation. The authors contributions clustered by topic are indicatively as follows: On EO insitu and satellite datasets M.D. and M.F.C. for surface networks; C.L., I.F.T. and R.O. on LST satellites; J.M.-S. and Y.K. on L-band satellites. On ESM aspects: P.A.D. and X.Z. on soil; A.A.-P., P.G., J.M. and S.B. on carbon/vegetation ; G.A. and E.N.D. on snow and land cryosphere; I.S. and N.W. on drag; H.H., J.B., K.M., S.P.E.K. and S.T. on ocean and marine cryosphere; C.A., F.P., H.C. and R.U. on soil hydrology; M.C. and R.I.W. on lakes. On assimilation aspects B.R., F.C., J.-F.M., N.B., P.D.R., R.R. and S.K. On EO ESM research transfer A.B. (Andy Brown), C.B., F.R., R.B., M.E., M.D., S.I.S. and S.M. All authors have substantially contributed to the critical thinking, text redaction and/or figures preparation.

Funding: Funding from the European Union CO2 Human Emissions H2020 project grant agreement No 776186 (Nicolas Bousserez, Margarita Choulga, Joe McNorton) and United States NOAA-PMEL contribution No 4808 (Meghan F. Cronin) are acknowledged.

Acknowledgments: The authors wish to thank the supporting institutes of the respective affiliation. Adrian Simmons and Alan Betts are thanked for their precious advice and review of early versions of this manuscript. We thank three anonymous reviewers and the assigned Remote Sensing editors for their valuable comments. At ECMWF, Georg Lentze, Anabel Bowen and Simon Witter editorial help and assistance have been greatly appreciated.

Conflicts of Interest: The authors declare no conflict of interest.

Abbreviations

The following abbreviations are used in this manuscript:

AIRS	Atmospheric InfraRed Sounder
AMSR-E	Advanced Microwave Scanning Radiometer for EOS (Earth Observing System)
AMSR-2	Advanced Microwave Scanning Radiometer 2
AMSU-A	Advanced Microwave Sounding Unit-A
AROME	Action de Recherche Petite Echelle Grande Echelle
ASTER	Advanced Spaceborne Thermal Emission and Reflection Radiometer
ATSR	Along Track Scanning Radiometers
AVHRR	Advanced Very High Resolution Radiometer

BEVAP	Bare-ground Evaporation
BSWB	Basin Scale Water Balance
C3S	Copernicus Climate Change Service
CALIPSO	Cloud-Aerosol Lidar and Infrared Pathfinder Satellite Observation
CAMS	Copernicus Atmosphere Monitoring Service
CASA	Carnegie–Ames–Stanford Approach
CEMS	Copernicus Emergency Monitoring Service
CERES	Clouds and the Earth's Radiant Energy System
CGLS	Copernicus Global Land Service
CH-TESSEL	Carbon and Hydrology—Tiled ECMWF Scheme for Surface Exchanges over Land
CLM	Community Land Model
CMA	China Meteorological Agency
CNRM	Centre National de Recherches Météorologique
COARE	Center for Oceanic Awareness, Research, and Education
DEM	Digital Elevation Model
ECMWF	European Centre for Medium-Range Weather Forecasts
EC-WAM	ECMWF Wave Model
ENVISAT	ENVIronment SATellite
ENS	Ensemble System
ENSO	El Niño–Southern Oscillation
EO	Earth Observations
ERS	European Remote-sensing Satellites
ESA	European Space Agency
ESM	Earth System Modelling
EUMETSAT	European Organization for the Exploitation of Meteorological Satellites
EVI	Enhanced Vegetation Index
FMI	Finnish Meteorological Institute
GEWEX	Global Energy and Water Exchanges
GCOS	Global Climate Observing System
GDAP	GEWEX Data and Analysis Panel
GFED	Global Fire Emissions Database
GLCC	Global Land Cover Characterization
GLEAM	Global Land Evaporation Amsterdam Model
GLOBE	Global Land One-kilometer Base Elevation
GloFAS	Global Flood Awareness System
GOES	Geostationary Operational Environmental Satellite
GOOS	Global Ocean Observing System
GOME	Global Ozone Monitoring Experiment
GOSAT	Greenhouse Gases Observing Satellite
GPCP	Global Precipitation Climatology Project
GPP	Gross Primary Production
GRACE	Gravity Recovery and Climate Experiment
GRUMP	Global Rural-Urban Mapping Project
GSWP	Global Soil Wetness Project
GWSD	Global Water Surface Dataset
HRES	High Resolution System
HSB	Humidity Sounder for Brazil
IASI	Infrared Atmospheric Sounding Interferometer
IFS	Integrated Forecasting System
ISRO	Indian Space Research Organisation
ISWG	International Surface Working Group
JMA	Japan Meteorological Agency
JULES	Joint UK Land Environment Simulator
LAI	Leaf Area Index
LANDSAT	Land Remote-Sensing Satellite (System)
LIM	Louvain-la-Neuve Sea Ice Model

LST	Land Surface Temperature
MACC	Monitoring Atmospheric Composition and Climate
MERIS	MEdium Resolution Imaging Spectrometer
METAR	METeorological Aerodrome Report
MetOp	Meteorological Operational Satellite
MetOp-SG	Meteorological Operational Satellite - Second Generation
MISR	Multi-angle Imaging SpectroRadiometer
MJO	Madden–Julian Oscillation
MODIS	Moderate Resolution Imaging Spectroradiometer
MOPITT	Measurements of Pollution in the Troposphere
MTSAT	Multifunctional Transport Satellites
NASA	National Aeronautics and Space Administration
NEE	Net Ecosystem Exchange
NEMO	Nucleus for European Modelling of the Ocean
NDVI	Normalized Difference Vegetation Index
NOAA	National Oceanic and Atmospheric Administration
NWP	Numerical Weather Prediction
Obs4MIP	Observations for Model Inter-Comparison Project
OLI	Operational Land Imager
OSI-SAF	Ocean and Sea Ice Satellite Application Facility
QuickSCAT	Quick Scatterometer
R2O	Research to Operation
Reco	Ecosystem respiration
SAR	Synthetic Aperture Radar
SSP	SubSatellite Point
SBSTA	Subsidiary Body for Scientific and Technological Advice
SEASAT	Sea Satellite
SEBS	Surface Energy Balance System
SHEBA	Surface Heat Budget of the Arctic Ocean
SIF	Solar-Induced Fluorescence
SMAP	Soil Moisture Active Passive
SMHI	Swedish Meteorological and Hydrological Institute
SMOS	Soil Moisture Ocean Salinity
SST	Sea Surface Temperature
SRTM	Shuttle Radar Topography Mission
SYNOP	Synoptic Operations
TAO	Tropical Atmosphere-Ocean
TISR	Thermal Infrared Sensor
TOPEX/POSEIDON	Topography Experiment—Positioning, Ocean, Solid Earth, Ice Dynamics, Orbital NavigatorAIRS
TWS	Terrestrial Water Storage
UHI	Urban Heat Island
VOD	Vegetation Optical Depth
WECANN	Water, Energy, and Carbon with Artificial Neural Networks
WCRP	World Climate Research Programme
WWRP	World Weather Research Programme

References

1. Bierkens, M.F.P.; Bell, V.A.; Burek, P.; Chaney, N.; Condon, L.E.; David, C.H.; Roo, A.; Döll, P.; Drost, N.; Famiglietti, J.S.; et al. Hyper-resolution global hydrological modelling: What is next? *Hydrol. Process.* **2015**, *29*, 310–320. [CrossRef]
2. Singh, R.S.; Reager, J.T.; Miller, N.L.; Famiglietti, J.S. Toward hyper-resolution land-surface modeling: The effects of fine-scale topography and soil texture on CLM4.0 simulations over the Southwestern U.S. *Water Resour. Res.* **2015**, *51*, 2648–2667. [CrossRef]

3. Wood, E.F.; Roundy, J.K.; Troy, T.J.; van Beek, L.P.H.; Bierkens, M.F.P.; Blyth, E.; de Roo, A.; Döll, P.; Ek, M.; Famiglietti, J.; et al. Hyperresolution global land surface modeling: Meeting a grand challenge for monitoring Earth's terrestrial water. *Water Resour. Res.* **2011**, *47*, W05301. [CrossRef]

4. Beven, K.; Cloke, H.; Pappenberger, F.; Lamb, R.; Hunter, N. Hyperresolution information and hyperresolution ignorance in modelling the hydrology of the land surface. *Sci. China Earth Sci.* **2015**, *58*, 25–35. [CrossRef]

5. Melsen, L.A.; Teuling, A.J.; Torfs, P.J.J.F.; Uijlenhoet, R.; Mizukami, N.; Clark, M.P. HESS Opinions: The need for process-based evaluation of large-domain hyper-resolution models. *Hydrol. Earth Syst. Sci.* **2016**, *20*, 1069–1079. [CrossRef]

6. Orth, R.; Staudinger, M.; Seneviratne, S.I.; Seibert, J.; Zappa, M. Does model performance improve with complexity? A case study with three hydrological models. *J. Hydrol.* **2015**, *523*, 147–159. [CrossRef]

7. Gudmundsson, L.; Seneviratne, S.I. Towards observation-based gridded runoff estimates for Europe. *Hydrol. Earth Syst. Sci.* **2015**, *19*, 2859–2879. [CrossRef]

8. Orth, R.; Seneviratne, S.I. Introduction of a simple-model-based land surface dataset for Europe. *Environ. Res. Lett.* **2015**, *10*, 044012. [CrossRef]

9. Mizielinski, M.S.; Roberts, M.J.; Vidale, P.L.; Schiemann, R.; Demory, M.E.; Strachan, J.; Edwards, T.; Stephens, A.; Lawrence, B.N.; Pritchard, M.; et al. High-resolution global climate modelling: The UPSCALE project, a large-simulation campaign. *Geosci. Model Dev.* **2014**, *7*, 1629–1640. [CrossRef]

10. Palmer, T.N. A personal perspective on modelling the climate system. *Proc. R. Soc. Lond. A Math. Phys. Eng. Sci.* **2016**, *472*, [CrossRef]

11. Bauer, P.; Thorpe, A.; Brunet, G. The quiet revolution of numerical weather prediction. *Nature* **2015**, *525*, 47. [CrossRef]

12. Garnaud, C.; Bélair, S.; Berg, A.; Rowlandson, T. Hyperresolution Land Surface Modeling in the Context of SMAP Cal–Val. *J. Hydrometeorol.* **2016**, *17*, 345–352. [CrossRef]

13. Orth, R.; Dutra, E.; Trigo, I.F.; Balsamo, G. Advancing land surface model development with satellite-based Earth observations. *Hydrol. Earth Syst. Sci.* **2017**, *21*, 2483–2495. [CrossRef]

14. National Academies of Sciences Engineering and Medicine. *Next Generation Earth System Prediction: Strategies for Subseasonal to Seasonal Forecasts*; The National Academies Press: Washington, DC, USA, 2016.

15. Clark, M.P.; Bierkens, M.F.P.; Samaniego, L.; Woods, R.A.; Uijlenhoet, R.; Bennett, K.E.; Pauwels, V.R.N.; Cai, X.; Wood, A.W.; Peters-Lidard, C.D. The evolution of process-based hydrologic models: Historical challenges and the collective quest for physical realism. *Hydrol. Earth Syst. Sci.* **2017**, *21*, 3427–3440. [CrossRef]

16. Martin, L.; Sarah-Jane, L.; Pirkka, O.; Lang, S.T.; Gianpaolo, B.; Peter, B.; Massimo, B.; Hannah, M.C.; Michail, D.; Emanuel, D.; et al. Stochastic representations of model uncertainties at ECMWF: State of the art and future vision. *Q. J. R. Meteorol. Soc.* **2017**, *143*, 2315–2339. [CrossRef]

17. Giuliano, D.B.; Alberto, V.; Gemma, C.; Linda, K.; Kun, Y.; Luigia, B.; Günter, B. Debates—Perspectives on socio-hydrology: Capturing feedback between physical and social processes. *Water Resour. Res.* **2014**, *51*, 4770–4781. [CrossRef]

18. Hoegh-Guldberg O. Jacob, D.; Taylor, M.; Bindi, M.; Brown, S.; Camilloni, I.; Diedhiou, A.; Djalante, R.; Ebi, K.; Engelbrecht F.; Guiot J.; et al. Impacts of 1.5 °C Global Warming on Natural and Human Systems. Available online: https://www.ipcc.ch/site/assets/uploads/sites/2/2018/12/SR15_Chapter3_Low_Res.pdf (accessed on 1 January 2018).

19. Diffenbaugh, N.S.; Singh, D.; Mankin, J.S. Unprecedented climate events: Historical changes, aspirational targets, and national commitments. *Sci. Adv.* **2018**, *4*, eaao3354. [CrossRef]

20. Ding, Q.; Schweiger, A.; L'Heureux, M.; Battisti, D.; Po-Chedley, S.; Johnson, N.; Blanchard-Wrigglesworth, E.; Harnos, K.; Zhang, Q.; Eastman, R.; et al. Influence of high-latitude atmospheric circulation changes on summertime Arctic seaice. *Nat. Clim. Chang.* **2017**, *7*, 289. [CrossRef]

21. Comiso, J.C.; Meier, W.N.; Gersten, R. Variability and trends in the Arctic Sea ice cover: Results from different techniques. *J. Geophys. Res. Oceans* **2017**, *122*, 6883–6900. [CrossRef]

22. Mudryk, L.R.; Derksen, C.; Howell, S.; Laliberté, F.; Thackeray, C.; Sospedra-Alfonso, R.; Vionnet, V.; Kushner, P.J.; Brown, R. Canadian snow and sea ice: Historical trends and projections. *Cryosphere* **2018**, *12*, 1157–1176. [CrossRef]

23. Calvet, J.C.; de Rosnay, P.; Barbu, A.L.; Boussetta, S. 12—Satellite Data Assimilation: Application to the Water and Carbon Cycles. In *Land Surface Remote Sensing in Continental Hydrology*; Baghdadi, N., Zribi, M., Eds.; Elsevier: Amsterdam, The Netherlands, 2016; pp. 401–428.

24. Drusch, M.; Viterbo, P. Assimilation of Screen-Level Variables in ECMWF's Integrated Forecast System: A Study on the Impact on the Forecast Quality and Analyzed Soil Moisture. *Mon. Weather Rev.* **2007**, *135*, 300–314. [CrossRef]

25. Bokhorst, S.; Pedersen, S.H.; Brucker, L.; Anisimov, O.; Bjerke, J.W.; Brown, R.D.; Ehrich, D.; Essery, R.L.H.; Heilig, A.; Ingvander, S.; et al. Changing Arctic snow cover: A review of recent developments and assessment of future needs for observations, modelling, and impacts. *Ambio* **2016**, *45*, 516–537. [CrossRef] [PubMed]

26. National Academies of Sciences Engineering and Medicine. *Thriving on Our Changing Planet: A Decadal Strategy for Earth Observation from Space*; The National Academies Press: Washington, DC, USA, 2018.

27. GCOS Global Climate Observing System. *The Global Observing System for Climate: Implementation Needs*; World Meteorological Organisation: Geneva, Switzerland, 2016; Volume GCOS-200, (GOOS-214), p. 325.

28. McCabe, M.F.; Rodell, M.; Alsdorf, D.E.; Miralles, D.G.; Uijlenhoet, R.; Wagner, W.; Lucieer, A.; Houborg, R.; Verhoest, N.E.C.; Franz, T.E.; et al. The future of Earth observation in hydrology. *Hydrol. Earth Syst. Sci.* **2017**, *21*, 3879–3914. [CrossRef] [PubMed]

29. Kerr, Y.H.; Al-Yaari, A.; Rodriguez-Fernandez, N.; Parrens, M.; Molero, B.; Leroux, D.; Bircher, S.; Mahmoodi, A.; Mialon, A.; Richaume, P.; et al. Overview of SMOS performance in terms of global soil moisture monitoring after six years in operation. *Remote Sens. Environ.* **2016**, *180*, 40–63. [CrossRef]

30. Kerr, Y.H.; Waldteufel, P.; Wigneron, J.P.; Delwart, S.; Cabot, F.; Boutin, J.; Escorihuela, M.J.; Font, J.; Reul, N.; Gruhier, C.; et al. The SMOS Mission: New Tool for Monitoring Key Elements ofthe Global Water Cycle. *Proc. IEEE* **2010**, *98*, 666–687. [CrossRef]

31. Kerr, Y.H.; Waldteufel, P.; Wigneron, J.P.; Martinuzzi, J.; Font, J.; Berger, M. Soil moisture retrieval from space: The Soil Moisture and Ocean Salinity (SMOS) mission. *IEEE Trans. Geosci. Remote Sens.* **2001**, *39*, 1729–1735. [CrossRef]

32. Rodriguez-Fernandez, N.J.; Aires, F.; Richaume, P.; Kerr, Y.H.; Prigent, C.; Kolassa, J.; Cabot, F.; Jimenez, C.; Mahmoodi, A.; Drusch, M. Soil Moisture Retrieval Using Neural Networks: Application to SMOS. *IEEE Trans. Geosci. Remote Sens.* **2015**, *53*, 5991–6007. [CrossRef]

33. Rodriguez-Fernandez, N.J.; Sabater, J.M.; Richaume, P.; de Rosnay, P.; Kerr, Y.H.; Albergel, C.; Drusch, M.; Mecklenburg, S. SMOS near-real-time soil moisture product: Processor overview and first validation results. *Hydrol. Earth Syst. Sci.* **2017**, *21*, 5201–5216. [CrossRef]

34. Al Bitar, A.; Mialon, A.; Kerr, Y.H.; Cabot, F.; Richaume, P.; Jacquette, E.; Quesney, A.; Mahmoodi, A.; Tarot, S.; Parrens, M.; et al. The global SMOS Level 3 daily soil moisture and brightness temperature maps. *Earth Syst. Sci. Data* **2017**, *9*, 293–315. [CrossRef]

35. Molero, B.; Merlin, O.; Malbeteau, Y.; Al Bitar, A.; Cabot, F.; Stefan, V.; Kerr, Y.; Bacon, S.; Cosh, M.H.; Bindlish, R.; et al. SMOS disaggregated soil moisture product at 1 km resolution: Processor overview and first validation results. *Remote Sens. Environ.* **2016**, *180*, 361–376. [CrossRef]

36. Tomer, S.K.; Al Bitar, A.; Sekhar, M.; Zribi, M.; Bandyopadhyay, S.; Kerr, Y. MAPSM: A Spatio-Temporal Algorithm for Merging Soil Moisture from Active and Passive Microwave Remote Sensing. *Remote Sens.* **2016**, *8*, 990. [CrossRef]

37. Brandt, M.; Wigneron, J.P.; Chave, J.; Tagesson, T.; Penuelas, J.; Ciais, P.; Rasmussen, K.; Tian, F.; Mbow, C.; Al-Yaari, A.; et al. Satellite passive microwaves reveal recent climate-induced carbon losses in African drylands. *Nat. Ecol. Evol.* **2018**, *2*, 827–835. [CrossRef] [PubMed]

38. Vittucci, C.; Ferrazzoli, P.; Richaume, P.; Kerr, Y. Effective Scattering Albedo of Forests Retrieved by SMOS and a Three-Parameter Algorithm. *IEEE Geosci. Remote Sens. Lett.* **2017**, *14*, 2260–2264. [CrossRef]

39. Fan, L.; Wigneron, J.P.; Xiao, Q.; Al-Yaari, A.; Wen, J.; Martin-StPaul, N.; Dupuy, J.L.; Pimont, F.; Al Bitar, A.; Fernandez-Moran, R.; et al. Evaluation of microwave remote sensing for monitoring live fuel moisture content in the Mediterranean region. *Remote Sens. Environ.* **2018**, *205*, 210–223. [CrossRef]

40. Kaleschke, L.; Tian-Kunze, X.; Maass, N.; Beitsch, A.; Wernecke, A.; Miernecki, M.; Mueller, G.; Fock, B.H.; Gierisch, A.M.U.; Schluenzen, K.H.; et al. SMOS sea ice product: Operational application and validation in the Barents Sea marginal ice zone. *Remote Sens. Environ.* **2016**, *180*, 264–273. [CrossRef]

41. Roman-Cascon, C.; Pellarin, T.; Gibon, F.; Brocca, L.; Cosme, E.; Crow, W.; Fernandez-Prieto, D.; Kerr, Y.H.; Massari, C. Correcting satellite-based precipitation products through SMOS soil moisture data assimilation in two land-surface models of different complexity: API and SURFEX. *Remote Sens. Environ.* **2017**, *200*, 295–310. [CrossRef]

42. Entekhabi, D.; Njoku, E.G.; O'Neill, P.E.; Kellogg, K.H.; Crow, W.T.; Edelstein, W.N.; Entin, J.K.; Goodman, S.D.; Jackson, T.J.; Johnson, J.; et al. The Soil Moisture Active Passive (SMAP) Mission. *Proc. IEEE* **2010**, *98*, 704–716. [CrossRef]

43. Fore, A.G.; Yueh, S.H.; Tang, W.; Stiles, B.W.; Hayashi, A.K. Combined active/passive retrievals of ocean vector wind and sea surface salinity with SMAP. *IEEE Trans. Geosci. Remote Sens.* **2016**, *54*, 7396–7404. [CrossRef]

44. Zhou, X.; Chong, J.; Yang, X.; Li, W.; Guo, X. Ocean Surface Wind Retrieval using SMAP L-Band SAR. *IEEE J. Sel. Top. Appl. Earth Obs. Remote Sens.* **2017**, *10*, 65–74. [CrossRef]

45. Reichle, R.H.; Lannoy, G.J.M.D.; Liu, Q.; Ardizzone, J.V.; Colliander, A.; Conaty, A.; Crow, W.; Jackson, T.J.; Jones, L.A.; Kimball, J.S.; et al. Assessment of the SMAP Level-4 Surface and Root-Zone Soil Moisture Product Using In Situ Measurements. *J. Hydrometeorol.* **2017**, *18*, 2621–2645. [CrossRef]

46. Donlon, C.; Berruti, B.; Buongiorno, A.; Ferreira, M.H.; Féménias, P.; Frerick, J.; Goryl, P.; Klein, U.; Laur, H.; Mavrocordatos, C.; et al. The Global Monitoring for Environment and Security (GMES) Sentinel-3 mission. *Remote Sens. Environ.* **2012**, *120*, 37–57. [CrossRef]

47. Freitas, S.C.; Trigo, I.F.; Macedo, J.; Barroso, C.; Silva, R.; Perdigão, R. Land surface temperature from multiple geostationary satellites. *Int. J. Remote Sens.* **2013**, *34*, 3051–3068. [CrossRef]

48. Trigo, I.F.; Boussetta, S.; Viterbo, P.; Balsamo, G.; Beljaars, A.; Sandu, I. Comparison of model land skin temperature with remotely sensed estimates and assessment of surface-atmosphere coupling. *J. Geophys. Res. Atmos.* **2015**, *120*, 12096–12111. [CrossRef]

49. Gentine, P.; Entekhabi, D.; Polcher, J. The Diurnal Behavior of Evaporative Fraction in the Soil–Vegetation–Atmospheric Boundary Layer Continuum. *J. Hydrometeorol.* **2011**, *12*, 1530–1546. [CrossRef]

50. Molero, B.; Leroux, D.J.; Richaume, P.; Kerr, Y.H.; Merlin, O.; Cosh, M.H.; Bindlish, R. Multi-Timescale Analysis of the Spatial Representativeness of In Situ Soil Moisture Data within Satellite Footprints. *J. Geophys. Res. Atmos* **2018**, *123*, 3–21. [CrossRef]

51. Dorigo, W.A.; Wagner, W.; Hohensinn, R.; Hahn, S.; Paulik, C.; Xaver, A.; Gruber, A.; Drusch, M.; Mecklenburg, S.; van Oevelen, P.; et al. The International Soil Moisture Network: A data hosting facility for global in situ soil moisture measurements. *Hydrol. Earth Syst. Sci.* **2011**, *15*, 1675–1698. [CrossRef]

52. Albergel, C.; de Rosnay, P.; Balsamo, G.; Isaksen, L.; Muñoz-Sabater, J. Soil Moisture Analyses at ECMWF: Evaluation Using Global Ground-Based In Situ Observations. *J. Hydrometeorol.* **2012**, *13*, 1442–1460. [CrossRef]

53. Albergel, C.; de Rosnay, P.; Gruhier, C.; Muñoz-Sabater, J.; Hasenauer, S.; Isaksen, L.; Kerr, Y.; Wagner, W. Evaluation of remotely sensed and modelled soil moisture products using global ground-based in situ observations. *Remote Sens. Environ.* **2012**, *118*, 215–226. [CrossRef]

54. Albergel, C.; Balsamo, G.; de Rosnay, P.; Muñoz Sabater, J.; Boussetta, S. A bare ground evaporation revision in the ECMWF land-surface scheme: evaluation of its impact using ground soil moisture and satellite microwave data. *Hydrol. Earth Syst. Sci.* **2012**, *16*, 3607–3620. [CrossRef]

55. Albergel, C.; Dutra, E.; noz-Sabater, J.M.; Haiden, T.; Balsamo, G.; Beljaars, A.; Isaksen, L.; Rosnay, P.D.; Sandu, I.; Wedi, N. Soil temperature at ECMWF: An assessment using ground-based observations. *J. Geophys. Res. Atmos.* **2014**, *120*, 1361–1373. [CrossRef]

56. Beck, H.E.; Vergopolan, N.; Pan, M.; Levizzani, V.; van Dijk, A.I.J.M.; Weedon, G.P.; Brocca, L.; Pappenberger, F.; Huffman, G.J.; Wood, E.F. Global-scale evaluation of 22 precipitation datasets using gauge observations and hydrological modeling. *Hydrol. Earth Syst. Sci.* **2017**, *21*, 6201–6217. [CrossRef]

57. Levizzani, V.; Kidd, C.; Aonashi, K.; Bennartz, R.; Ferraro, R.R.; Huffman, G.J.; Roca, R.; Turk, F.J.; Wang, N.Y. The activities of the International Precipitation Working Group. *Q. J. R. Meteorol. Soc.* **2017**. [CrossRef]

58. Menne, M.J.; Durre, I.; Vose, R.S.; Gleason, B.E.; Houston, T.G. An Overview of the Global Historical Climatology Network-Daily Database. *J. Atmos. Ocean. Technol.* **2012**, *29*, 897–910. [CrossRef]

59. Hannah, D.M.; Demuth, S.; van Lanen, H.A.; Looser, U.; Prudhomme, C.; Rees, G.; Stahl, K.; Tallaksen, L.M. Large-scale river flow archives: Importance, current status and future needs. *Hydrol. Process.* **2011**, *25*, 1191–1200. [CrossRef]

60. GCOS Global Climate Observing System. *Status of the Global Observing System for Climate*; World Meteorological Organisation: Geneva, Switzerland, 2015; Volume GCOS-195, p. 353.

61. Lindstrom, E.; Gunn, J.; Fischer, A.; McCurdy, A.; Glover, L. *A Framework for Ocean Observing. By the Task Team for an Integrated Framework for Sustained Ocean Observing (revised in 2017)*; IOC/INF-1284 rev. 2; UNESCO: Paris, France, 2012.

62. Argo. *Argo Float Data and Metadata from Global Data Assembly Centre (Argo GDAC)*; SEANOE: 2000. Available online: https://www.seanoe.org/data/00311/42182/ (accessed on 1 January 2018). [CrossRef]

63. Roemmich, D.; Johnson, G.C.; Riser, S.; Davis, R.; Gilson, J.; Owens, W.B.; Garzoli, S.L.; Schmid, C.; Ignaszewski, M. The Argo Program: Observing the global ocean with profiling floats. *Oceanography* **2009**, *22*, 34–43. [CrossRef]

64. Freeland, H.J.; Roemmich, D.; Garzoli, S.L.; Le Traon, P.Y.; Ravichandran, M.; Riser, S.; Thierry, V.; Wijffels, S.; Belbéoch, M.; Gould, J.; et al. Argo—A decade of progress. In Proceedings of the OceanObs' 09: Sustained Ocean Observations and Information for Society, Venice, Italy, 21–25 September 2009; Volume 2.

65. Lumpkin, R.; Pazos, M. Measuring surface currents with Surface Velocity Program drifters: The instrument, its data, and some recent results. In *Lagrangian Analysis and Prediction of Coastal and Ocean Dynamics*; Cambridge University Press: Cambridge, UK, 2007; pp. 39–67.

66. Manabe, S. Climate and the Ocean Circulation. *Mon. Weather Rev.* **1969**, *97*, 739–774.<0739:CATOC>2.3.CO;2. [CrossRef]

67. Shukla, J.; Mintz, Y. Influence of Land-Surface Evapotranspiration on the Earth's Climate. *Science* **1982**, *215*, 1498–1501. [CrossRef] [PubMed]

68. Delworth, T.; Manaba, S. Climate variability and land-surface processes. *Adv. Water Resour.* **1993**, *16*, 3–20. [CrossRef]

69. Dirmeyer, P.A. The Role of the Land Surface Background State in Climate Predictability. *J. Hydrometeorol.* **2003**, *4*, 599–610.<0599:TROTLS>2.0.CO;2. [CrossRef]

70. Pierre, G.; Alix, G.; Seung-Bu, P.; Ji, N.; Giuseppe, T.; Zhiming, K. Role of surface heat fluxes underneath cold pools. *Geophys. Res. Lett.* **2015**, *43*, 874–883. [CrossRef]

71. Léo, L.; Pierre, G.; Marc, S.; Philippe, D.; Simone, F. Modification of land-atmosphere interactions by CO_2 effects: Implications for summer dryness and heat wave amplitude. *Geophys. Res. Lett.* **2016**, *43*, 10240–10248. [CrossRef]

72. Lemordant, L.; Gentine, P.; Swann, A.S.; Cook, B.I.; Scheff, J. Critical impact of vegetation physiology on the continental hydrologic cycle in response to increasing CO_2. *Proc. Natl. Acad. Sci. USA* **2018**, 4093–4098. [CrossRef] [PubMed]

73. Beljaars, A.C.; Viterbo, P.; Miller, M.J.; Betts, A.K. The anomalous rainfall over the United States during July 1993: Sensitivity to land surface parameterization and soil moisture anomalies. *Mon. Weather Rev.* **1996**, *124*, 362–383. [CrossRef]

74. Koster, R.D.; Suarez, M.J. Modeling the land surface boundary in climate models as a composite of independent vegetation stands. *J. Geophys. Res. Atmos.* **1992**, *97*, 2697–2715. [CrossRef]

75. Wang, W.; Kumar, A. A GCM assessment of atmospheric seasonal predictability associated with soil moisture anomalies over North America. *J. Geophys. Res. Atmos.* **1998**, *103*, 28637–28646. [CrossRef]

76. Koster, R.D.; Dirmeyer, P.A.; Guo, Z.; Bonan, G.; Chan, E.; Cox, P.; Gordon, C.T.; Kanae, S.; Kowalczyk, E.; Lawrence, D.; et al. Regions of Strong Coupling Between Soil Moisture and Precipitation. *Science* **2004**, *305*, 1138–1140. [CrossRef]

77. Green, J.K.; Konings, A.G.; Alemohammad, S.H.; Berry, J.; Entekhabi, D.; Kolassa, J.; Lee, J.E.; Gentine, P. Regionally strong feedback between the atmosphere and terrestrial biosphere. *Nat. Geosci.* **2017**, *10*, 410. [CrossRef]

78. Betts, A.; Fisch, G.; Von Randow, C.; Silva Dias, M.; Cohen, J.; Da Silva, R.; Fitzjarrald, D. The Amazonian boundary layer and mesoscale circulations. In *Amazonia and Global Change*; Michael, K., Mercedes Bustamante, J.G., Dias, P.S., Eds.; Geophysical Monograph Series 186; AGU: Washington, DC, USA, 2009; pp. 163–181.

79. Betts, A.K.; Desjardins, R.; Worth, D.; Cerkowniak, D. Impact of land use change on the diurnal cycle climate of the Canadian Prairies. *J. Geophys. Res. Atmos.* **2013**, *118*, 11–996. [CrossRef]

80. Betts, A.K.; Desjardins, R.; Worth, D.; Beckage, B. Climate coupling between temperature, humidity, precipitation, and cloud cover over the Canadian Prairies. *J. Geophys. Res. Atmos.* **2014**, *119*, 13–305. [CrossRef]

81. Gentine, P.; D'Odorico, P.; Lintner, B.R.; Sivandran, G.; Salvucci, G. Interdependence of climate, soil, and vegetation as constrained by the Budyko curve. *Geophys. Res. Lett.* **2012**, *39*. [CrossRef]

82. Gentine, P.; Betts, A.K.; Lintner, B.R.; Findell, K.L.; Van Heerwaarden, C.C.; D'andrea, F. A probabilistic bulk model of coupled mixed layer and convection. Part II: Shallow convection case. *J. Atmos. Sci.* **2013**, *70*, 1557–1576. [CrossRef]

83. Taylor, C.M.; Parker, D.J.; Harris, P.P. Interdependence of Climate, Soil, and Vegetation as Constrained by the Budyko Curve. Available online: https://agupubs.onlinelibrary.wiley.com/doi/10.1029/2012GL053492 (accessed on 9 October 2012).

84. Hohenegger, C.; Brockhaus, P.; Bretherton, C.S.; Schär, C. The soil moisture–precipitation feedback in simulations with explicit and parameterized convection. *J. Clim.* **2009**, *22*, 5003–5020. [CrossRef]

85. Guillod, B.P.; Orlowsky, B.; Miralles, D.G.; Teuling, A.J.; Seneviratne, S.I. Reconciling spatial and temporal soil moisture effects on afternoon rainfall. *Nat. Commun.* **2015**, *6*, 6443. [CrossRef] [PubMed]

86. Hohenegger, C.; Stevens, B. The role of the permanent wilting point in controlling the spatial distribution of precipitation. *Proc. Natl. Acad. Sci. USA* **2018**, 5692–5697. [CrossRef]

87. Seneviratne, S.I.; Easterling, D.; Goodess, C.M.; Kanae, S.; Kossin, J.; Luo, Y.; Marengo, J.; McInnes, K.; Rahimi, M.; Reichstein, M.; et al. Changes in climate extremes and their impacts on the natural physical environment. In *Managing the Risks of Extreme Events and Disasters To Advance Climate Change Adaptation: Special Report of the Intergovernmental Panel on Climate Change*; Cambridge University Press: Cambridge, UK, 2012; pp. 109–230.

88. Seneviratne, S.I.; Donat, M.G.; Mueller, B.; Alexander, L.V. No pause in the increase of hot temperature extremes. *Nat. Clim. Chang.* **2014**, *4*, 161. [CrossRef]

89. Balsamo, G.; Beljaars, A.; Scipal, K.; Viterbo, P.; van den Hurk, B.; Hirschi, M.; Betts, A.K. A Revised Hydrology for the ECMWF Model: Verification from Field Site to Terrestrial Water Storage and Impact in the Integrated Forecast System. *J. Hydrometeorol.* **2009**, *10*, 623–643. [CrossRef]

90. Koster, R.D.; Suarez, M.J.; Ducharne, A.; Stieglitz, M.; Kumar, P. Basin scale water-balance estimates of terrestrial water storage variations from ECMWF operational forecast analysis. *Geophys. Res. Lett.* **2006**, *33*. [CrossRef]

91. Koster, R.D.; Suarez, M.J.; Ducharne, A.; Stieglitz, M.; Kumar, P. A catchment-based approach to modeling land surface processes in a general circulation model: 1. Model structure. *J. Geophys. Res. Atmos.* **2000**, *105*, 24809–24822. [CrossRef]

92. Gelaro, R.; McCarty, W.; Suárez, M.J.; Todling, R.; Molod, A.; Takacs, L.; Randles, C.A.; Darmenov, A.; Bosilovich, M.G.; Reichle, R.; et al. The modern-era retrospective analysis for research and applications, version 2 (MERRA-2). *J. Clim.* **2017**, *30*, 5419–5454. [CrossRef]

93. De Lannoy, G.J.; Koster, R.D.; Reichle, R.H.; Mahanama, S.P.; Liu, Q. An updated treatment of soil texture and associated hydraulic properties in a global land modeling system. *J. Adv. Model. Earth Syst.* **2014**, *6*, 957–979. [CrossRef]

94. De Lannoy, G.J.; Reichle, R.H.; Pauwels, V.R. Global calibration of the GEOS-5 L-band microwave radiative transfer model over nonfrozen land using SMOS observations. *J. Hydrometeorol.* **2013**, *14*, 765–785. [CrossRef]

95. Reichle, R.H.; De Lannoy, G.J.M.; Liu, Q.; Koster, R.D.; Kimball, J.S.; Crow, W.T.; Ardizzone, J.V.; Chakraborty, P.; Collins, D.W.; Conaty, A.L.; et al. Global Assessment of the SMAP Level-4 Surface and Root-Zone Soil Moisture Product Using Assimilation Diagnostics. *J. Hydrometeorol.* **2017**, *18*, 3217–3237. [CrossRef] [PubMed]

96. Mahanama, S.P.; Koster, R.D.; Walker, G.K.; Takacs, L.L.; Reichle, R.H.; De Lannoy, G.; Liu, Q.; Zhao, B.; Suarez, M.J. *Land Boundary Conditions for the Goddard Earth Observing System Model Version 5 (GEOS-5) Climate Modeling System: Recent Updates and Data File Descriptions*; NASA Technical Report Series on Global Modeling and Data Assimilation, NASA/TM-2015-104606; Technical Report 2; National Aeronautics and Space Administration, Goddard Space Flight Center: Greenbelt, MD, USA, 2015; Volume 39, p. 55.

97. Reichle, R.H.; Liu, Q.; Koster, R.D.; Ardizzone, J.V.; Colliander, A.; Crow, W.T.; Lannoy, G.J.M.D.; Kimball, J.S. *Soil Moisture Active Passive (SMAP) Project Assessment Report for Version 4 of the L4-SM Data Product*; NASA Technical Report Series on Global Modeling and Data Assimilation, NASA/TM-2018-104606; NASA Goddard Space Flight Center: Greenbelt, MD, USA, 2018; Volume 52, pp. 1–67.
98. O'Neill, P.; Chan, S.; Njoku, E.; Jackson, T.; Bindlish, R. *SMAP L2 Radiometer Half-Orbit 36 km EASE-Grid Soil Moisture*, version 3; National Snow and Ice Data Center Distributed Active Archive Center: Boulder, CO, USA, October 2016.
99. Koster, R.D.; Liu, Q.; Mahanama, S.P.; Reichle, R.H. Improved Hydrological Simulation Using SMAP Data: Relative Impacts of Model Calibration and Data Assimilation. *J. Hydrometeorol.* **2018**, *19*, 727–741. [CrossRef] [PubMed]
100. Betts, A.K.; Desjardins, R.; Worth, D.; Wang, S.; Li, J. Coupling of winter climate transitions to snow and clouds over the Prairies. *J. Geophys. Res. Atmos.* **2013**, *119*, 1118–1139. [CrossRef]
101. Islam, S.U.; Dery, S.J.; Werner, A.T. Future Climate Change Impacts on Snow and Water Resources of the Fraser River Basin, British Columbia. *J. Hydrometeorol.* **2017**, *18*, 473–496. [CrossRef]
102. Groisman, P.Y.; Karl, T.R.; Knight, R.W. Observed impact of snow cover on the heat balance and the rise of continental spring temperatures. *Science* **1994**, *263*, 198–200. [CrossRef] [PubMed]
103. Viterbo, P.; Betts, A.K. Impact on ECMWF forecasts of changes to the albedo of the boreal forests in the presence of snow. *J. Geophys. Res. Atmos.* **1999**, *104*, 27803–27810. [CrossRef]
104. Cook, B.I.; Bonan, G.B.; Levis, S.; Epstein, H.E. The thermoinsulation effect of snow cover within a climate model. *Clim. Dyn.* **2008**, *31*, 107–124. [CrossRef]
105. Viterbo, P.; Beljaars, A.; Mahfouf, J.F.; Teixeira, J. The representation of soil moisture freezing and its impact on the stable boundary layer. *Q. J. R. Meteorol. Soc.* **1999**, *125*, 2401–2426. [CrossRef]
106. Sandu, I.; Beljaars, A.; Bechtold, P.; Mauritsen, T.; Balsamo, G. Why is it so difficult to represent stably stratified conditions in numerical weather prediction (NWP) models? *J. Adv. Model. Earth Syst.* **2013**, *5*, 117–133. [CrossRef]
107. Gentine, P.; Steeneveld, G.J.; Heusinkveld, B.G.; Holtslag, A.A. Coupling Between Radiative Flux Divergence and Turbulence Near the Surface. Available online: https://rmets.onlinelibrary.wiley.com/doi/10.1002/qj.3333 (accessed on 26 June 2018).
108. Groisman, P.Y.; Knight, R.W.; Karl, T.R.; Easterling, D.R.; Sun, B.; Lawrimore, J.H. Contemporary changes of the hydrological cycle over the contiguous United States: Trends derived from in situ observations. *J. Hydrometeorol.* **2004**, *5*, 64–85. [CrossRef]
109. Douville, H.; Royer, J.F.; Mahfouf, J.F. A new snow parameterization for the Meteo-France climate model. *Clim. Dyn.* **1995**, *12*, 21–35. [CrossRef]
110. Rutter, N.; Essery, R.; Pomeroy, J.; Altımır, N.; Andreadis, K.; Baker, I.; Barr, A.; Bartlett, P.; Boone, A.; Deng, H.; et al. Evaluation of Forest Snow Processes Models (SnowMIP2). *J. Geophys. Res. Atmos.* **2009**, *114*. Available online: https://agupubs.onlinelibrary.wiley.com/doi/full/10.1029/2008JD011063 (accessed on 25 March 2009). [CrossRef]
111. Dutra, E.; Balsamo, G.; Viterbo, P.; Miranda, P.M.; Beljaars, A.; Schär, C.; Elder, K. An improved snow scheme for the ECMWF land surface model: Description and offline validation. *J. Hydrometeorol.* **2010**, *11*, 899–916. [CrossRef]
112. De Rosnay, P.; Balsamo, G.; Albergel, C.; Muñoz-Sabater, J.; Isaksen, L. Initialisation of Land Surface Variables for Numerical Weather Prediction. *Surv. Geophys.* **2014**, *35*, 607–621. [CrossRef]
113. Clifford, D. Global estimates of snow water equivalent from passive microwave instruments: History, challenges and future developments. *Int. J. Remote Sens.* **2010**, *31*, 3707–3726. [CrossRef]
114. Luojus, K.; Pulliainen, J.; Cohen, J.; Ikonen, J.; Derksen, C.; Mudryk, L.; Nagler, T.; Bojkov, B. Assessment of Northern Hemisphere Snow Water Equivalent Datasets in ESA SnowPEx project. Paper pretented at the EGU General Assembly Conference Abstracts, Vienna Austria, 17–22 April 2016; Volume 18, p. 2941.
115. Lemmetyinen, J.; Derksen, C.; Rott, H.; Macelloni, G.; King, J.; Schneebeli, M.; Wiesmann, A.; Leppänen, L.; Kontu, A.; Pulliainen, J. Retrieval of Effective Correlation Length and Snow Water Equivalent from Radar and Passive Microwave Measurements. *Remote Sens.* **2018**, *10*, 170. [CrossRef]
116. Andreadis, K.M.; Lettenmaier, D.P. Assimilating remotely sensed snow observations into a macroscale hydrology model. *Adv. Water Resour.* **2006**, *29*, 872–886. [CrossRef]

117. Foster, J.L.; Sun, C.; Walker, J.P.; Kelly, R.; Chang, A.; Dong, J.; Powell, H. Quantifying the uncertainty in passive microwave snow water equivalent observations. *Remote Sens. Environ.* **2005**, *94*, 187–203. [CrossRef]

118. Picard, G.; Sandells, M.; Löwe, H. SMRT: An active–passive microwave radiative transfer model for snow with multiple microstructure and scattering formulations (v1. 0). *Geosci. Model Dev.* **2018**, *11*, 2763–2788. [CrossRef]

119. Lemmetyinen, J.; Schwank, M.; Rautiainen, K.; Kontu, A.; Parkkinen, T.; Matzler, C.; Wiesmann, A.; Wegmuller, U.; Derksen, C.; Toose, P.; et al. Snow density and ground permittivity retrieved from L-band radiometry: Application to experimental data. *Remote Sens. Environ.* **2016**, *180*, 377–391. [CrossRef]

120. Schwank, M.; Naderpour, R. Snow Density and Ground Permittivity Retrieved from L-Band Radiometry: Melting Effects. *Remote Sens.* **2018**, *10*, 354. [CrossRef]

121. Dutra, E.; Kotlarski, S.; Viterbo, P.; Balsamo, G.; Miranda, P.M.; Schär, C.; Bissolli, P.; Jonas, T. Snow cover sensitivity to horizontal resolution, parameterizations, and atmospheric forcing in a land surface model. *J. Geophys. Res. Atmos.* **2011**, *116*. [CrossRef]

122. Malik, M.J.; van der Velde, R.; Vekerdy, Z.; Su, Z. Assimilation of satellite-observed snow albedo in a land surface model. *J. Hydrometeorol.* **2012**, *13*, 1119–1130. [CrossRef]

123. Macelloni, G.; Leduc-Leballeur, M.; Brogioni, M.; Ritz, C.; Picard, G. Analyzing and modeling the SMOS spatial variations in the East Antarctic Plateau. *Remote Sens. Environ.* **2016**, *180*, 193–204. [CrossRef]

124. Leduc-Leballeur, M.; Picard, G.; Mialon, A.; Arnaud, L.; Lefebvre, E.; Possenti, P.; Kerr, Y. Modeling L-Band Brightness Temperature at Dome C in Antarctica and Comparison With SMOS Observations. *IEEE Trans. Geosci. Remote Sens.* **2015**, *53*, 4022–4032. [CrossRef]

125. Hall, D.K.; Box, J.E.; Casey, K.A.; Hook, S.J.; Shuman, C.A.; Steffen, K. Comparison of satellite-derived and in situ observations of ice and snow surface temperatures over Greenland. *Remote Sens. Environ.* **2008**, *112*, 3739–3749. [CrossRef]

126. Fréville, H.; Brun, E.; Picard, G.; Tatarinova, N.; Arnaud, L.; Lanconelli, C.; Reijmer, C.; Van den Broeke, M. Using MODIS land surface temperatures and the Crocus snow model to understand the warm bias of ERA-Interim reanalyses at the surface in Antarctica. *Cryosphere* **2014**, *8*, 1361–1373. [CrossRef]

127. Dutra, E.; Sandu, I.; Balsamo, G.; Beljaars, A.; Freville, H.; Vignon, E.; Brun, E. Understanding the ECMWF Winter Surface Temperature Biases over Antarctica *ECMWF Tech. Memo.* **2015**, *762*, 1 18.

128. Mu, Q.; Heinsch, F.A.; Zhao, M.; Running, S.W. Development of a global evapotranspiration algorithm based on MODIS and global meteorology data. *Remote Sens. Environ.* **2007**, *111*, 519–536. [CrossRef]

129. Miralles, D.G.; Holmes, T.R.H.; De Jeu, R.A.M.; Gash, J.H.; Meesters, A.G.C.A.; Dolman, A.J. Global land-surface evaporation estimated from satellite-based observations. *Hydrol. Earth Syst. Sci.* **2011**, *15*, 453–469. [CrossRef]

130. Miralles, D.G.; De Jeu, R.A.M.; Gash, J.H.; Holmes, T.R.H.; Dolman, A.J. Magnitude and variability of land evaporation and its components at the global scale. *Hydrol. Earth Syst. Sci.* **2011**, *15*, 967–981. [CrossRef]

131. Su, Z. The Surface Energy Balance System (SEBS) for estimation of turbulent heat fluxes. *Hydrol. Earth Syst. Sci.* **2002**, *6*, 85–100. [CrossRef]

132. Fisher, J.; Tu, K.P.; Baldocchi, D.D. Global estimates of the land–atmosphere water flux based on monthly AVHRR and ISLSCP-II data, validated at 16 FLUXNET sites. *Remote Sens. Environ.* **2008**, *112*, 901–919. [CrossRef]

133. Jung, M.; Reichstein, M.; Schwalm, C.R.; Huntingford, C.; Sitch, S.; Ahlström, A.; Arneth, A.; Camps-Valls, G.; Ciais, P.; Friedlingstein, P.; et al. Compensatory water effects link yearly global land CO_2 sink changes to temperature. *Nature* **2017**, *541*, 516. [CrossRef] [PubMed]

134. Alemohammad, S.H.; Fang, B.; Konings, A.G.; Aires, F.; Green, J.K.; Kolassa, J.; Gonzalez Miralles, D.; Prigent, C.; Gentine, P. Water, Energy, and Carbon with Artificial Neural Networks (WECANN): A statistically based estimate of global surface turbulent fluxes and gross primary productivity using solar-induced fluorescence. *Biogeosciences* **2017**, *14*, 4101–4124. [CrossRef] [PubMed]

135. Jiménez, C.; Prigent, C.; Mueller, B.; Seneviratne, S.I.; McCabe, M.F.; Wood, E.F.; Rossow, W.B.; Balsamo, G.; Betts, A.K.; Dirmeyer, P.A.; et al. Global intercomparison of 12 land surface heat flux estimates. *J. Geophys. Res. Atmos.* **2011**, *116*. [CrossRef]

136. Mueller, B.; Seneviratne, S.I.; Jimenez, C.; Corti, T.; Hirschi, M.; Balsamo, G.; Ciais, P.; Dirmeyer, P.; Fisher, J.; Guo, Z.; et al. Evaluation of global observations-based evapotranspiration datasets and IPCC AR4 simulations. *Geophys. Res. Lett.* **2011**, *38*. [CrossRef]

137. Miralles, D.; Jiménez, C.; Jung, M.; Michel, D.; Ershadi, A.; McCabe, M.; Hirschi, M.; Martens, B.; Dolman, A.; Fisher, J.; et al. The WACMOS-ET project-Part 2: Evaluation of global terrestrial evaporation data sets. *Hydrol. Earth Syst. Sci.* **2016**, *20*, 823–842. [CrossRef]

138. Michel, D.; Jiménez, C.; Miralles, D.G.; Jung, M.; Hirschi, M.; Ershadi, A.; Martens, B.; McCabe, M.; Fisher, J.B.; Mu, Q.; et al. The WACMOS-ET project–Part 1: Tower-scale evaluation of four remote sensing-based evapotranspiration algorithms. *Hydrol. Earth Syst. Sci.* **2015**, *20*, 803–822. [CrossRef]

139. Kumar, S.; Holmes, T.; Mocko, D.M.; Wang, S.; Peters-Lidard, C. Attribution of Flux Partitioning Variations between Land Surface Models over the Continental U.S. *Remote Sens.* **2018**, *10*, 751. [CrossRef]

140. Luo, Y.Q.; Randerson, J.T.; Abramowitz, G.; Bacour, C.; Blyth, E.; Carvalhais, N.; Ciais, P.; Dalmonech, D.; Fisher, J.B.; Fisher, R.; et al. A framework for benchmarking land models. *Biogeosciences* **2012**, *9*, 3857–3874. [CrossRef]

141. Schellekens, J.; Dutra, E.; Martínez-de la Torre, A.; Balsamo, G.; van Dijk, A.; Weiland, F.S.; Minvielle, M.; Calvet, J.C.; Decharme, B.; Eisner, S.; et al. A global water resources ensemble of hydrological models: The eartH2Observe Tier-1 dataset. *Earth Syst. Sci. Data* **2017**, *9*, 389–413. [CrossRef]

142. Van den Hurk, B.J.J.M.; Viterbo, P.; Beljaars, A.C.M.; Betts, A.K. Offline validation of the ERA40 surface scheme. *ECMWF Tech. Memo.* **2000**, *295*, 1–42.

143. Boussetta, S.; Balsamo, G.; Beljaars, A.; Kral, T.; Jarlan, L. Impact of a satellite-derived leaf area index monthly climatology in a global numerical weather prediction model. *Int. J. Remote Rens.* **2013**, *34*, 3520–3542. [CrossRef]

144. Myneni, R.B.; Hoffman, S.; Knyazikhin, Y.; Privette, J.; Glassy, J.; Tian, Y.; Wang, Y.; Song, X.; Zhang, Y.; Smith, G.; et al. Global products of vegetation leaf area and fraction absorbed PAR from year one of MODIS data. *Remote Sens. Environ.* **2002**, *83*, 214–231. [CrossRef]

145. Baret, F.; Weiss, M.; Lacaze, R.; Camacho, F.; Makhmara, H.; Pacholcyzk, P.; Smets, B. GEOV1: LAI and FAPAR essential climate variables and FCOVER global time series capitalizing over existing products. Part1: Principles of development and production. *Remote Sens. Environ.* **2013**, *137*, 299–309. [CrossRef]

146. Boussetta, S.; Balsamo, G.; Dutra, E.; Beljaars, A.; Albergel, C. Assimilation of surface albedo and vegetation states from satellite observations and their impact on numerical weather prediction. *Remote Sens. Environ.* **2015**, *163*, 111–126. [CrossRef]

147. Calvet, J.C.; Noilhan, J.; Roujean, J.L.; Bessemoulin, P.; Cabelguenne, M.; Olioso, A.; Wigneron, J.P. An interactive vegetation SVAT model tested against data from six contrasting sites. *Agric. For. Meteorol.* **1998**, *92*, 73–95. [CrossRef]

148. Boussetta, S.; Balsamo, G.; Beljaars, A.; Panareda, A.A.; Calvet, J.C.; Jacobs, C.; Van den Hurk, B.; Viterbo, P.; Lafont, S.; Dutra, E.; et al. Natural land carbon dioxide exchanges in the ECMWF Integrated Forecasting System: Implementation and offline validation. *J. Geophys. Res. Atmos.* **2013**, *118*, 5923–5946. [CrossRef]

149. Potter, C.S.; Randerson, J.T.; Field, C.B.; Matson, P.A.; Vitousek, P.M.; Mooney, H.A.; Klooster, S.A. Terrestrial ecosystem production: A process model based on global satellite and surface data. *Glob. Biogeochem. Cycles* **1993**, *7*, 811–841. [CrossRef]

150. Andrews, A.; Kofler, J.; Trudeau, M.; Williams, J.; Neff, D.; Masarie, K.; Chao, D.; Kitzis, D.; Novelli, P.; Zhao, C.; et al. CO_2, CO, and CH_4 measurements from tall towers in the NOAA Earth System Research Laboratory's Global Greenhouse Gas Reference Network: Instrumentation, uncertainty analysis, and recommendations for future high-accuracy greenhouse gas monitoring efforts. *Atmos. Meas. Tech.* **2014**, *7*, 647–687. [CrossRef]

151. Van der Werf, G.R.; Randerson, J.T.; Giglio, L.; van Leeuwen, T.T.; Chen, Y.; Rogers, B.M.; Mu, M.; van Marle, M.J.E.; Morton, D.C.; Collatz, G.J.; Yokelson, R.J.; Kasibhatla, P.S. Global fire emissions estimates during 1997–2016. *Earth Syst. Sci. Data* **2017**, *9*, 697–720. [CrossRef]

152. Agusti-Panareda, A.; Massart, S.; Chevallier, F.; Boussetta, S.; Balsamo, G.; Beljaars, A.; Ciais, P.; Deutscher, N.; Engelen, R.; Jones, L.; et al. Forecasting global atmospheric CO_2. *Atmos. Chem. Phys.* **2014**, *14*, 11959–11983. [CrossRef]

153. Agustí-Panareda, A.; Massart, S.; Chevallier, F.; Balsamo, G.; Boussetta, S.; Dutra, E.; Beljaars, A. A biogenic CO_2 flux adjustment scheme for the mitigation of large-scale biases in global atmospheric CO_2 analyses and forecasts. *Atmos. Chem. Phys.* **2016**, *2016*, 10399–10418. [CrossRef]

154. Chevallier, F.; Ciais, P.; Conway, T.; Aalto, T.; Anderson, B.; Bousquet, P.; Brunke, E.; Ciattaglia, L.; Esaki, Y.; Fröhlich, M.; et al. CO_2 surface fluxes at grid point scale estimated from a global 21 year reanalysis of atmospheric measurements. *J. Geophys. Res. Atmos.* **2010**, *115*, D21. [CrossRef]

155. Jung, M.; Reichstein, M.; Margolis, H.A.; Cescatti, A.; Richardson, A.D.; Arain, M.A.; Arneth, A.; Bernhofer, C.; Bonal, D.; Chen, J.; et al. Global patterns of land-atmosphere fluxes of carbon dioxide, latent heat, and sensible heat derived from eddy covariance, satellite, and meteorological observations. *J. Geophys. Res. Biogeosci.* **2011**, *116*, G3. [CrossRef]

156. Guanter, L.; Zhang, Y.; Jung, M.; Joiner, J.; Voigt, M.; Berry, J.A.; Frankenberg, C.; Huete, A.R.; Zarco-Tejada, P.; Lee, J.E.; et al. Global and time-resolved monitoring of crop photosynthesis with chlorophyll fluorescence. *Proc. Natl. Acad. Sci. USA* **2014**, 201320008. Available online: https://www.pnas.org/content/111/14/E1327/tab-article-info (accessed on 8 April 2014). [CrossRef]

157. Turner, A.J.; Frankenberg, C.; Wennberg, P.O.; Jacob, D.J. Ambiguity in the causes for decadal trends in atmospheric methane and hydroxyl. *Proc. Natl. Acad. Sci. USA* **2017**, *114*, 5367–5372. [CrossRef]

158. Voulgarakis, A.; Naik, V.; Lamarque, J.F.; Shindell, D.T.; Young, P.J.; Prather, M.J.; Wild, O.; Field, R.D.; Bergmann, D.; Cameron-Smith, P.; et al. Analysis of present day and future OH and methane lifetime in the ACCMIP simulations. *Atmos. Chem. Phys.* **2013**, *13*, 2563–2587. [CrossRef]

159. Buzan, E.M.; Beale, C.A.; Boone, C.D.; Bernath, P.F. Global stratospheric measurements of the isotopologues of methane from the Atmospheric Chemistry Experiment Fourier transform spectrometer. *Atmos. Meas. Tech.* **2016**, *9*, 1095–1111. [CrossRef]

160. Saunois, M.; Bousquet, P.; Poulter, B.; Peregon, A.; Ciais, P.; Canadell, J.G.; Dlugokencky, E.J.; Etiope, G.; Bastviken, D.; Houweling, S.; et al. The global methane budget 2000–2012. *Earth Syst. Sci. Data* **2016**, *8*, 697–751. [CrossRef]

161. Melton, J.R.; Wania, R.; Hodson, E.L.; Poulter, B.; Ringeval, B.; Spahni, R.; Bohn, T.; Avis, C.A.; Beerling, D.J.; Chen, G.; et al. Present state of global wetland extent and wetland methane modelling: conclusions from a model inter-comparison project (WETCHIMP). *Biogeosciences* **2013**, *10*, 753–788. [CrossRef]

162. Li, T.; Raivonen, M.; Alekseychik, P.; Aurela, M.; Lohila, A.; Zheng, X.; Zhang, Q.; Wang, G.; Mammarella, I.; Rinne, J.; et al. Importance of vegetation classes in modeling CH_4 emissions from boreal and subarctic wetlands in Finland. *Sci. Total Environ.* **2016**, *572*, 1111–1122. [CrossRef] [PubMed]

163. Parker, R.J.; Boesch, H.; McNorton, J.; Comyn-Platt, E.; Gloor, M.; Wilson, C.; Chipperfield, M.P.; Hayman, G.D.; Bloom, A.A. Evaluating year-to-year anomalies in tropical wetland methane emissions using satellite CH_4 observations. *Remote Sens. Environ.* **2018**, *211*, 261–275. [CrossRef]

164. McNorton, J.; Gloor, E.; Wilson, C.; Hayman, G.D.; Gedney, N.; Comyn-Platt, E.; Marthews, T.; Parker, R.J.; Boesch, H.; Chipperfield, M.P. Role of regional wetland emissions in atmospheric methane variability. *Geophys. Res. Lett.* **2016**, *43*, 11433–11444. [CrossRef]

165. Evaristo, J.; Jasechko, S.; McDonnell, J.J. Global separation of plant transpiration from groundwater and streamflow. *Nature* **2015**, *525*, 91. [CrossRef]

166. Jasechko, S.; Sharp, Z.D.; Gibson, J.J.; Birks, S.J.; Yi, Y.; Fawcett, P.J. Terrestrial water fluxes dominated by transpiration. *Nature* **2013**, *496*, 347. [CrossRef]

167. Schlesinger, W.H.; Jasechko, S. Transpiration in the global water cycle. *Agric. For. Meteorol.* **2014**, *189*, 115–117. [CrossRef]

168. Entekhabi, D.; Nakamura, H.; Njoku, E.G. Solving the inverse problem for soil moisture and temperature profiles by sequential assimilation of multifrequency remotely sensed observations. *IEEE Trans. Geosci. Remote Sens.* **1994**, *32*, 438–448. [CrossRef]

169. Reichle, R.H.; Entekhabi, D.; McLaughlin, D.B. Downscaling of radio brightness measurements for soil moisture estimation: A four-dimensional variational data assimilation approach. *Water Resour. Res.* **2001**, *37*, 2353–2364. [CrossRef]

170. Kurum, M.; Lang, R.H.; O'Neill, P.E.; Joseph, A.T.; Jackson, T.J.; Cosh, M.H. A first-order radiative transfer model for microwave radiometry of forest canopies at L-band. *IEEE Trans. Geosci. Remote Sens.* **2011**, *49*, 3167–3179. [CrossRef]

171. Kurum, M.; O'Neill, P.E.; Lang, R.H.; Joseph, A.T.; Cosh, M.H.; Jackson, T.J. Effective tree scattering and opacity at L-band. *Remote Sens. Environ.* **2012**, *118*, 1–9. [CrossRef]

172. Meesters, A.G.; De Jeu, R.A.; Owe, M. Analytical derivation of the vegetation optical depth from the microwave polarization difference index. *IEEE Geosci. Remote Sens. Lett.* **2005**, *2*, 121–123. [CrossRef]

173. Jones, M.O.; Jones, L.A.; Kimball, J.S.; McDonald, K.C. Satellite passive microwave remote sensing for monitoring global land surface phenology. *Remote Sens. Environ.* **2011**, *115*, 1102–1114. [CrossRef]

174. Liu, Y.Y.; de Jeu, R.A.; McCabe, M.F.; Evans, J.P.; van Dijk, A.I. Global long-term passive microwave satellite-based retrievals of vegetation optical depth. *Geophys. Res. Lett.* **2011**, *38*. [CrossRef]

175. Grant, J.; Wigneron, J.P.; De Jeu, R.; Lawrence, H.; Mialon, A.; Richaume, P.; Al Bitar, A.; Drusch, M.; Van Marle, M.; Kerr, Y. Comparison of SMOS and AMSR-E vegetation optical depth to four MODIS-based vegetation indices. *Remote Sens. Environ.* **2016**, *172*, 87–100. [CrossRef]

176. Konings, A.G.; Piles, M.; Rötzer, K.; McColl, K.A.; Chan, S.K.; Entekhabi, D. Vegetation optical depth and scattering albedo retrieval using time series of dual-polarized L-band radiometer observations. *Remote Sens. Environ.* **2016**, *172*, 178–189. [CrossRef]

177. Konings, A.G.; Gentine, P. Global variations in ecosystem-scale isohydricity. *Glob. Chang. Biol.* **2017**, *23*, 891–905. [CrossRef]

178. Tian, F.; Brandt, M.; Liu, Y.Y.; Verger, A.; Tagesson, T.; Diouf, A.A.; Rasmussen, K.; Mbow, C.; Wang, Y.; Fensholt, R. Remote sensing of vegetation dynamics in drylands: Evaluating vegetation optical depth (VOD) using AVHRR NDVI and in situ green biomass data over West African Sahel. *Remote Sens. Environ.* **2016**, *177*, 265–276. [CrossRef]

179. Fernandez-Moran, R.; Al-Yaari, A.; Mialon, A.; Mahmoodi, A.; Al Bitar, A.; De lannoy, G.; Rodriguez-Fernandez, N.; Lopez-Baeza, E.; Kerr, Y.; Wigneron, J.P. SMOS-IC: An Alternative SMOS Soil Moisture and Vegetation Optical Depth Product. *Remote Sens.* **2017**, *9*, 457. [CrossRef]

180. Zhou, L.; Tian, Y.; Myneni, R.B.; Ciais, P.; Saatchi, S.; Liu, Y.Y.; Piao, S.; Chen, H.; Vermote, E.F.; Song, C.; et al. Widespread decline of Congo rainforest greenness in the past decade. *Nature* **2014**, *509*, 86. [CrossRef] [PubMed]

181. Smith, W.K.; Reed, S.C.; Cleveland, C.C.; Ballantyne, A.P.; Anderegg, W.R.; Wieder, W.R.; Liu, Y.Y.; Running, S.W. Large divergence of satellite and Earth system model estimates of global terrestrial CO_2 fertilization. *Nat. Clim. Chang.* **2016**, *6*, 306. [CrossRef]

182. Rodríguez-Fernández, N.J.; Mialon, A.; Mermoz, S.; Bouvet, A.; Richaume, P.; Al Bitar, A.; Al-Yaari, A.; Brandt, M.; Kaminski, T.; Le Toan, T.; et al. An evaluation of SMOS L-band vegetation optical depth (L-VOD) data sets: High sensitivity of L-VOD to above-ground biomass in Africa. *Biogeosciences* **2018**, *15*, 4627–4645. [CrossRef]

183. Baccini, A.; Goetz, S.; Walker, W.; Laporte, N.; Sun, M.; Sulla-Menashe, D.; Hackler, J.; Beck, P.; Dubayah, R.; Friedl, M.; et al. Estimated carbon dioxide emissions from tropical deforestation improved by carbon-density maps. *Nat. Clim. Chang.* **2012**, *2*, 182. [CrossRef]

184. Guanter, L.; Alonso, L.; Gómez-Chova, L.; Meroni, M.; Preusker, R.; Fischer, J.; Moreno, J. Developments for vegetation fluorescence retrieval from spaceborne high-resolution spectrometry in the O2-A and O2-B absorption bands. *J. Geophys. Res. Atmos.* **2010**, *115*. Available online: https://agupubs.onlinelibrary.wiley.com/doi/full/10.1029/2009JD013716 (accessed on 6 October 2010). [CrossRef]

185. Frankenberg, C.; Fisher, J.B.; Worden, J.; Badgley, G.; Saatchi, S.S.; Lee, J.E.; Toon, G.C.; Butz, A.; Jung, M.; Kuze, A.; et al. New global observations of the terrestrial carbon cycle from GOSAT: Patterns of plant fluorescence with gross primary productivity. *Geophys. Res. Lett.* **2011**, *38*. Available online: https://agupubs.onlinelibrary.wiley.com/doi/10.1029/2011GL048738 (accessed on 14 September 2011). [CrossRef]

186. Joiner, J.; Yoshida, Y.; Vasilkov, A.; Middleton, E. First observations of global and seasonal terrestrial chlorophyll fluorescence from space. *Biogeosciences* **2011**, *8*, 637–651. [CrossRef]

187. Frankenberg, C.; O'Dell, C.; Guanter, L.; McDuffie, J. Remote sensing of near-infrared chlorophyll fluorescence from space in scattering atmospheres: Implications for its retrieval and interferences with atmospheric CO_2 retrievals. *Atmos. Meas. Tech.* **2012**, *5*, 2081–2094. [CrossRef]

188. Guanter, L.; Frankenberg, C.; Dudhia, A.; Lewis, P.E.; Gómez-Dans, J.; Kuze, A.; Suto, H.; Grainger, R.G. Retrieval and global assessment of terrestrial chlorophyll fluorescence from GOSAT space measurements. *Remote Sens. Environ.* **2012**, *121*, 236–251. [CrossRef]

189. Joiner, J.; Guanter, L.; Lindstrot, R.; Voigt, M.; Vasilkov, A.; Middleton, E.; Huemmrich, K.; Yoshida, Y.; Frankenberg, C. Global monitoring of terrestrial chlorophyll fluorescence from moderate-spectral-resolution near-infrared satellite measurements: methodology, simulations, and application to GOME-2. *Atmos. Meas. Tech.* **2013**, *6*, 2803–2823. [CrossRef]

190. Frankenberg, C.; O'Dell, C.; Berry, J.; Guanter, L.; Joiner, J.; Köhler, P.; Pollock, R.; Taylor, T.E. Prospects for chlorophyll fluorescence remote sensing from the Orbiting Carbon Observatory-2. *Remote Sens. Environ.* **2014**, *147*, 1–12. [CrossRef]

191. Schimel, D.; Pavlick, R.; Fisher, J.B.; Asner, G.P.; Saatchi, S.; Townsend, P.; Miller, C.; Frankenberg, C.; Hibbard, K.; Cox, P. Observing terrestrial ecosystems and the carbon cycle from space. *Glob. Chang. Biol.* **2015**, *21*, 1762–1776. [CrossRef] [PubMed]

192. Sun, Y.; Fu, R.; Dickinson, R.; Joiner, J.; Frankenberg, C.; Gu, L.; Xia, Y.; Fernando, N. Drought onset mechanisms revealed by satellite solar-induced chlorophyll fluorescence: Insights from two contrasting extreme events. *J. Geophys. Res. Biogeosci.* **2015**, *120*, 2427–2440. [CrossRef]

193. Bi, J.; Knyazikhin, Y.; Choi, S.; Park, T.; Barichivich, J.; Ciais, P.; Fu, R.; Ganguly, S.; Hall, F.; Hilker, T.; et al. Sunlight mediated seasonality in canopy structure and photosynthetic activity of Amazonian rainforests. *Environ. Res. Lett.* **2015**, *10*, 064014. [CrossRef]

194. Lopes, A.P.; Nelson, B.W.; Wu, J.; de Alencastro Graça, P.M.L.; Tavares, J.V.; Prohaska, N.; Martins, G.A.; Saleska, S.R. Leaf flush drives dry season green-up of the Central Amazon. *Remote Sens. Environ.* **2016**, *182*, 90–98. [CrossRef]

195. Saleska, S.R.; Wu, J.; Guan, K.; Araujo, A.C.; Huete, A.; Nobre, A.D.; Restrepo-Coupe, N. Dry-season greening of Amazon forests. *Nature* **2016**, *531*, E4. [CrossRef] [PubMed]

196. Morton, D.C.; Nagol, J.; Carabajal, C.C.; Rosette, J.; Palace, M.; Cook, B.D.; Vermote, E.F.; Harding, D.J.; North, P.R. Amazon forests maintain consistent canopy structure and greenness during the dry season. *Nature* **2014**, *506*, 221. [CrossRef] [PubMed]

197. Giardina, F.; Konings, A.G.; Kennedy, D.; Alemohammad, S.H.; Oliveira, R.S.; Uriarte, M.; Gentine, P. Tall Amazonian forests are less sensitive to precipitation variability. *Nat. Geosci.* **2018**, *11*, 405–409. [CrossRef]

198. Gentine, P.; Alemohammad, S. Reconstructed Solar-Induced Fluorescence: A Machine Learning Vegetation Product Based on MODIS Surface Reflectance to Reproduce GOME-2 Solar-Induced Fluorescence. *Geophys. Res. Lett.* **2018**, *45*, 3136–3146. [CrossRef] [PubMed]

199. Sukhova, E.; Sukhov, V. Connection of the Photochemical Reflectance Index (PRI) with the Photosystem II Quantum Yield and Nonphotochemical Quenching Can Be Dependent on Variations of Photosynthetic Parameters among Investigated Plants: A Meta-Analysis. *Remote Sens.* **2018**, *10*, 771. [CrossRef]

200. Asch, M.; Bocquet, M.; Nodet, M. *Data Assimilation: Methods, Algorithms, and Applications*; SIAM, 2016; Volume 11. Available online: https://hal.inria.fr/hal-01402885 (accessed on 25 November 2016). [CrossRef]

201. Chevallier, F.; Viovy, N.; Reichstein, M.; Ciais, P. On the assignment of prior errors in Bayesian inversions of CO_2 surface fluxes. *Geophys. Res. Lett.* **2006**, *33*. [CrossRef]

202. Bousserez, N.; Henze, D.K.; Rooney, B.; Perkins, A.; Wecht, K.J.; Turner, A.J.; Natraj, V.; Worden, J.R. Constraints on methane emissions in North America from future geostationary remote-sensing measurements. *Atmos. Chem. Phys.* **2016**, *16*, 6175–6190. [CrossRef]

203. Kourzeneva, E. External data for lake parameterization in Numerical Weather Prediction and climate modeling. *Boreal Environ. Res.* **2010**, *15*, 165–177.

204. Amante, C.; Eakins, B. *ETOPO1 Global Relief Model Converted to PanMap Layer Format*; NOAA-National Geophysical Data Center: Boulder, CO, USA, 2009; Volume 10.

205. Farr, T.G.; Rosen, P.A.; Caro, E.; Crippen, R.; Duren, R.; Hensley, S.; Kobrick, M.; Paller, M.; Rodriguez, E.; Roth, L.; et al. The Shuttle Radar Topography Mission. *Rev. Geophys.* **2007**, *45*. [CrossRef]

206. Hastings, D.A.; Dunbar, P.K.; Elphingstone, G.M.; Bootz, M.; Murakami, H.; Maruyama, H.; Masaharu, H.; Holland, P.; Payne, J.; Bryant, N.A.; et al. *The Global Land One-Kilometer Base Elevation (GLOBE) Digital Elevation Model*, version 1.0; National Oceanic and Atmospheric Administration, National Geophysical Data Center: Boulder, CO, USA, 1999; Volume 325, pp. 80305–83328.

207. Scambos, T.A.; Haran, T. An image-enhanced DEM of the Greenland ice sheet. *Ann. Glaciol.* **2002**, *34*, 291–298. [CrossRef]

208. Liu, H.; Jezek, K.; Li, B.; Zhao, Z. *Radarsat Antarctic Mapping Project Digital Elevation Model*, version 2; Digital Media; National Snow and Ice Data Center: Boulder, CO, USA, 2001.

209. Wedi, N.P. Increasing horizontal resolution in numerical weather prediction and climate simulations: Illusion or panacea? *Philos. Trans. R. Soc. A* **2014**, *372*, 20130289. [CrossRef]

210. Broxton, P.D.; Zeng, X.; Sulla-Menashe, D.; Troch, P.A. A global land cover climatology using MODIS data. *J. Appl. Meteorol. Climatol.* **2014**, *53*, 1593–1605. [CrossRef]

211. Rempe, D.M.; Dietrich, W.E. A bottom-up control on fresh-bedrock topography under landscapes. *Proc. Natl. Acad. Sci. USA* **2014**, *111*, 6576–6581. [CrossRef]

212. Zeng, X.; Decker, M. Improving the numerical solution of soil moisture–based Richards equation for land models with a deep or shallow water table. *J. Hydrometeorol.* **2009**, *10*, 308–319. [CrossRef]

213. Pelletier, J.D.; Broxton, P.D.; Hazenberg, P.; Zeng, X.; Troch, P.A.; Niu, G.Y.; Williams, Z.; Brunke, M.A.; Gochis, D. A gridded global data set of soil, intact regolith, and sedimentary deposit thicknesses for regional and global land surface modeling. *J. Adv. Model. Earth Syst.* **2016**, *8*, 41–65. [CrossRef]

214. Shangguan, W.; Hengl, T.; de Jesus, J.M.; Yuan, H.; Dai, Y. Mapping the global depth to bedrock for land surface modeling. *J. Adv. Model. Earth Syst.* **2017**, *9*, 65–88. [CrossRef]

215. Brunke, M.A.; Broxton, P.; Pelletier, J.; Gochis, D.; Hazenberg, P.; Lawrence, D.M.; Leung, L.R.; Niu, G.Y.; Troch, P.A.; Zeng, X. Implementing and evaluating variable soil thickness in the community land model, version 4.5 (CLM4. 5). *J. Clim.* **2016**, *29*, 3441–3461. [CrossRef]

216. Good, S.P.; Noone, D.; Bowen, G. Hydrologic connectivity constrains partitioning of global terrestrial water fluxes. *Science* **2015**, *349*, 175–177. [CrossRef]

217. Weary, D.J.; Doctor, D.H. *Karst in the United States: A Digital Map Compilation And Database*; US Department of the Interior, US Geological Survey, 2014. Available online: http://earth.eoas.fsu.edu/~mye/2017KarstSymposium/Doctor3.pdf (accessed on 1 January 2014). [CrossRef]

218. Akhmedenov, K.; Iskaliev, D.; Petrishev, V. Karst and Pseudokarst of the West Kazakhstan (Republic of Kazakhstan). *Int. J. Geosci.* **2014**, *5*, 131–136. [CrossRef]

219. Johnson, C.M.; Fan, X.; Mahmood, R.; Groves, C.; Polk, J.S.; Yan, J. Evaluating Weather Research and Forecasting Model Sensitivity to Land and Soil Conditions Representative of Karst Landscapes. *Bound.-Layer Meteorol.* **2018**, *166*, 503–530. [CrossRef]

220. Sobocinski-Norton, H.E.; Dirmeyer, P. Soil moisture memory in karst and non-karst terrains. *Geophys. Res. Lett.* **2018**, in review.

221. Dirmeyer, P.A.; Norton, H.E. Indications of Surface and Sub-Surface Hydrologic Properties from SMAP Soil Moisture Retrievals. *Hydrology* **2018**, *5*, 36. [CrossRef]

222. Barnes, E.M.; Sudduth, K.A.; Hummel, J.W.; Lesch, S.M.; Corwin, D.L.; Yang, C.; Daughtry, C.S.; Bausch, W.C. Remote-and ground-based sensor techniques to map soil properties. *Photogramm. Eng. Remote Sens.* **2003**, *69*, 619–630. [CrossRef]

223. Steinberg, A.; Chabrillat, S.; Stevens, A.; Segl, K.; Foerster, S. Prediction of common surface soil properties based on Vis-NIR airborne and simulated EnMAP imaging spectroscopy data: Prediction accuracy and influence of spatial resolution. *Remote Sens.* **2016**, *8*, 613. [CrossRef]

224. Verpoorter, C.; Kutser, T.; Seekell, D.A.; Tranvik, L.J. A global inventory of lakes based on high-resolution satellite imagery. *Geophys. Res. Lett.* **2014**, *41*, 6396–6402. [CrossRef]

225. Samuelsson, P.; Kourzeneva, E.; Mironov, D. The impact of lakes on the European climate as simulated by a regional climate model. *Boreal Environ. Res.* **2010**, *15*, 113–129.

226. Thiery, W.; Davin, E.L.; Panitz, H.J.; Demuzere, M.; Lhermitte, S.; Van Lipzig, N. The impact of the African Great Lakes on the regional climate. *J. Clim.* **2015**, *28*, 4061–4085. [CrossRef]

227. Dutra, E.; Stepanenko, V.M.; Balsamo, G.; Viterbo, P.; Miranda, P.; Mironov, D.; Schär, C. An offline study of the impact of lakes on the performance of the ECMWF surface scheme. *Boreal Environ. Res.* **2010**, *15*, 100–112.

228. Brown, L.C.; Duguay, C.R. The response and role of ice cover in lake-climate interactions. *Prog. Phys. Geogr.* **2010**, *34*, 671–704. [CrossRef]

229. Bonan, G.B. Sensitivity of a GCM simulation to inclusion of inland water surfaces. *J. Clim.* **1995**, *8*, 2691–2704. [CrossRef]

230. Balsamo, G.; Salgado, R.; Dutra, E.; Boussetta, S.; Stockdale, T.; Potes, M. On the contribution of lakes in predicting near-surface temperature in a global weather forecasting model. *Tellus A Dyn. Meteorol. Oceanogr.* **2012**, *64*, 15829. [CrossRef]

231. Mironov, D.; Heise, E.; Kourzeneva, E.; Ritter, B.; Schneider, N.; Terzhevik, A. Implementation of the lake parameterisation scheme FLake into the numerical weather prediction model COSMO. *Boreal Environ. Res.* **2010**, *15*, 218–230.

232. Le Moigne, P.; Colin, J.; Decharme, B. Impact of lake surface temperatures simulated by the FLake scheme in the CNRM-CM5 climate model. *Tellus A Dyn. Meteorol. Oceanogr.* **2016**, *68*, 31274. [CrossRef]

233. Rooney, G.G.; Bornemann, F.J. The performance of FLake in the Met Office Unified Model. *Tellus A Dyn. Meteorol. Oceanogr.* **2013**, *65*, 21363. [CrossRef]

234. Jeffries, M.; Morris, K.; Liston, G.E. A Method to Determine Lake Depth and Water Availability on the North Slope of Alaska With Spaceborne Imaging Radar and Numerical Ice Growth Modeling. *Arctic* **1996**, *49*, 367–374. [CrossRef]

235. Duguay, C.R.; Lafleur, P.M. Determining depth and ice thickness of shallow sub-Arctic lakes using space-borne optical and SAR data. *Int. J. Remote Sens.* **2003**, *24*, 475–489. [CrossRef]

236. Choulga, M.; Kourzeneva, E.; Zakharova, E.; Doganovsky, A. Estimation of the mean depth of boreal lakes for use in numerical weather prediction and climate modelling. *Tellus A Dyn. Meteorol. Oceanogr.* **2014**, *66*, 21295. [CrossRef]

237. Balsamo, G.; Dutra, E.; Stepanenko, V.; Viterbo, P.; Miranda, P.; Mironov, D. Deriving an Effective Lake Depth from Satellite Lake Surface Temperature: A Feasibility Study with MODIS Data. *Boreal Environ. Res.* **2010**, *15*, 178–190.

238. Manrique-Suñén, A.; Nordbo, A.; Balsamo, G.; Beljaars, A.; Mammarella, I. Representing land surface heterogeneity: Offline analysis of the tiling method. *J. Hydrometeorol.* **2013**, *14*, 850–867. [CrossRef]

239. MacCallum, S.N.; Merchant, C.J. Surface water temperature observations of large lakes by optimal estimation. *Can. J. Remote Sens.* **2012**, *38*, 25–45. [CrossRef]

240. Verseghy, D.L.; MacKay, M.D. Offline Implementation and Evaluation of the Canadian Small Lake Model with the Canadian Land Surface Scheme over Western Canada. *J. Hydrometeorol.* **2017**, *18*, 1563–1582. [CrossRef]

241. Emerton, R.E.; Stephens, E.M.; Pappenberger, F.; Pagano, T.C.; Weerts, A.H.; Wood, A.W.; Salamon, P.; Brown, J.D.; Hjerdt, N.; Donnelly, C.; et al. Continental and global scale flood forecasting systems. *Wiley Interdiscip. Rev. Water* **2016**, *3*, 391–418. [CrossRef]

242. Alfieri, L.; Burek, P.; Dutra, E.; Krzeminski, B.; Muraro, D.; Thielen, J.; Pappenberger, F. GloFAS-global ensemble streamflow forecasting and flood early warning. *Hydrol. Earth Syst. Sci.* **2013**, *17*, 1161. [CrossRef]

243. Smith, P.; Pappenberger, F.; Wetterhall, F.; del Pozo, J.T.; Krzeminski, B.; Salamon, P.; Muraro, D.; Kalas, M.; Baugh, C. On the operational implementation of the European Flood Awareness System (EFAS). In *Flood Forecasting*; Elsevier: Amsterdam, The Netherlands, 2016; pp. 313–348.

244. Arnal, L.; Cloke, H.L.; Stephens, E.; Wetterhall, F.; Prudhomme, C.; Neumann, J.; Krzeminski, B.; Pappenberger, F. Skilful seasonal forecasts of streamflow over Europe? *Hydrol. Earth Syst. Sci.* **2018**, *22*, 2057–2072. [CrossRef]

245. Emerton, R.; Zsoter, E.; Arnal, L.; Cloke, H.L.; Muraro, D.; Prudhomme, C.; Stephens, E.M.; Salamon, P.; Pappenberger, F. Developing a global operational seasonal hydro-meteorological forecasting system: GloFAS v2. 2 Seasonal v1. 0. *Geosci. Model Dev.* **2018**, *11*, 3327–3346. [CrossRef]

246. Cloke, H.L.; Pappenberger, F.; Smith, P.J.; Wetterhall, F. How do I know if I've improved my continental scale flood early warning system? *Environ. Res. Lett.* **2017**, *12*, 044006. [CrossRef]

247. Balsamo, G.; Albergel, C.; Beljaars, A.; Boussetta, S.; Brun, E.; Cloke, H.; Dee, D.; Dutra, E.; Muñoz-Sabater, J.; Pappenberger, F.; et al. ERA-Interim/Land: A global land surface reanalysis data set. *Hydrol. Earth Syst. Sci.* **2015**, *19*, 389–407. [CrossRef]

248. Schumann, G.J.P.; Bates, P.D.; Neal, J.C.; Andreadis, K.M. Technology: Fight floods on a global scale. *Nature* **2014**, *507*, 169. [CrossRef] [PubMed]

249. Grimaldi, S.; Li, Y.; Pauwels, V.R.; Walker, J.P. Remote sensing-derived water extent and level to constrain hydraulic flood forecasting models: Opportunities and challenges. *Surv. Geophys.* **2016**, *37*, 977–1034. [CrossRef]

250. Sichangi, A.W.; Wang, L.; Yang, K.; Chen, D.; Wang, Z.; Li, X.; Zhou, J.; Liu, W.; Kuria, D. Estimating continental river basin discharges using multiple remote sensing data sets. *Remote Sens. Environ.* **2016**, *179*, 36–53. [CrossRef]

251. Gallego-Elvira, B.; Taylor, C.M.; Harris, P.P.; Ghent, D.; Veal, K.L.; Folwell, S.S. Global observational diagnosis of soil moisture control on the land surface energy balance. *Geophys. Res. Lett.* **2016**, *43*, 2623–2631. [CrossRef]

252. Folwell, S.S.; Harris, P.P.; Taylor, C.M. Large-scale surface responses during European dry spells diagnosed from land surface temperature. *J. Hydrometeorol.* **2016**, *17*, 975–993. [CrossRef]

253. Harris, P.P.; Folwell, S.S.; Gallego-Elvira, B.; Rodríguez, J.; Milton, S.; Taylor, C.M. An evaluation of modeled evaporation regimes in Europe using observed dry spell land surface temperature. *J. Hydrometeorol.* **2017**, *18*, 1453–1470. [CrossRef]
254. Levine, P.A.; Randerson, J.T.; Swenson, S.C.; Lawrence, D.M. Evaluating the strength of the land–atmosphere moisture feedback in Earth system models using satellite observations. *Hydrol. Earth Syst. Sci. (Online)* **2016**, *20*, 4837–4856. [CrossRef]
255. McColl, K.A.; Wang, W.; Peng, B.; Akbar, R.; Short Gianotti, D.J.; Lu, H.; Pan, M.; Entekhabi, D. Global characterization of surface soil moisture drydowns. *Geophys. Res. Lett.* **2017**, *44*, 3682–3690. [CrossRef]
256. Polcher, J.; Piles, M.; Gelati, E.; Barella-Ortiz, A.; Tello, M. Comparing surface-soil moisture from the SMOS mission and the ORCHIDEE land-surface model over the Iberian Peninsula. *Remote Sens. Environ.* **2016**, *174*, 69–81. [CrossRef]
257. Kawai, Y.; Wada, A. Diurnal sea surface temperature variation and its impact on the atmosphere and ocean: A review. *J. Oceanogr.* **2007**, *63*, 721–744. [CrossRef]
258. Bernie, D.J.; Woolnough, S.J.; Slingo, J.M.; Guilyardi, E. Modeling Diurnal and Intraseasonal Variability of the Ocean Mixed Layer. *J. Clim.* **2005**, *18*, 1190–1202. [CrossRef]
259. Bernie, D.J.; Guilyardi, E.; Madec, G.; Slingo, J.M.; Woolnough, S.J.; Cole, J. Impact of resolving the diurnal cycle in an ocean–atmosphere GCM. Part 2: A diurnally coupled CGCM. *Clim. Dyn.* **2008**, *31*, 909–925. [CrossRef]
260. Large, W.; Caron, J. Diurnal cycling of sea surface temperature, salinity, and current in the CESM coupled climate model. *J. Geophys. Res. Oceans* **2015**, *120*, 3711–3729. [CrossRef]
261. Bernie, D.; Guilyardi, E.; Madec, G.; Slingo, J.; Woolnough, S. Impact of resolving the diurnal cycle in an ocean–atmosphere GCM. Part 1: A diurnally forced OGCM. *Clim. Dyn.* **2007**, *29*, 575–590. [CrossRef]
262. Clayson, C.A.; Bogdanoff, A.S. The effect of diurnal sea surface temperature warming on climatological air–sea fluxes. *J. Clim.* **2013**, *26*, 2546–2556. [CrossRef]
263. Cronin, M.F.; Kessler, W.S. Near-surface shear flow in the tropical Pacific cold tongue front. *J. Phys. Oceanogr.* **2009**, *39*, 1200–1215. [CrossRef]
264. Drushka, K.; Sprintall, J.; Gille, S.T. Subseasonal variations in salinity and barrier-layer thickness in the eastern equatorial Indian Ocean. *J. Geophys. Res. Oceans* **2014**, *119*, 805–823. [CrossRef]
265. Mogensen, K.S.; Magnusson, L.; Bidlot, J. Tropical cyclone sensitivity to ocean coupling in the ECMWF coupled model. *J. Geophys. Res. Oceans* **2017**, *122*, 4392–4412. [CrossRef]
266. Salisbury D.; Mogensen K.; Balsamo G. Use of in situ observations to verify the diurnal cycle of sea surface temperature in ECMWF coupled model forecasts. *ECMWF Tech. Memo.* **2018**, *826*, 1–19.
267. Danabasoglu, G.; Large, W.G.; Tribbia, J.J.; Gent, P.R.; Briegleb, B.P.; McWilliams, J.C. Diurnal coupling in the tropical oceans of CCSM3. *J. Clim.* **2006**, *19*, 2347–2365. [CrossRef]
268. Ham, Y.G.; Kug, J.S.; Kang, I.S.; Jin, F.F.; Timmermann, A. Impact of diurnal atmosphere–ocean coupling on tropical climate simulations using a coupled GCM. *Clim. Dyn.* **2010**, *34*, 905–917. [CrossRef]
269. Tian, F.; von Storch, J.S.; Hertwig, E. Air–sea fluxes in a climate model using hourly coupling between the atmospheric and the oceanic components. *Clim. Dyn.* **2017**, *48*, 2819–2836. [CrossRef]
270. Li, G.; Xie, S.P. Tropical biases in CMIP5 multimodel ensemble: The excessive equatorial Pacific cold tongue and double ITCZ problems. *J. Clim.* **2014**, *27*, 1765–1780. [CrossRef]
271. Slingo, J.; Inness, P.; Neale, R.; Woolnough, S.; Yang, G. Scale interactions on diurnal toseasonal timescales and their relevanceto model systematic errors. *Ann. Geophys.* **2003**, *45*, 139–155.
272. Seo, H.; Subramanian, A.C.; Miller, A.J.; Cavanaugh, N.R. Coupled impacts of the diurnal cycle of sea surface temperature on the Madden-Julian oscillation. *J. Clim.* **2014**, *27*, 8422–8443. [CrossRef]
273. Masson, S.; Terray, P.; Madec, G.; Luo, J.J.; Yamagata, T.; Takahashi, K. Impact of intra-daily SST variability on ENSO characteristics in a coupled model. *Clim. Dyn.* **2012**, *39*, 681–707. [CrossRef]
274. Fairall, C.; Bradley, E.F.; Hare, J.; Grachev, A.; Edson, J. Bulk parameterization of air–sea fluxes: Updates and verification for the COARE algorithm. *J. Clim.* **2003**, *16*, 571–591. [CrossRef]
275. Beljaars, A.C. The parametrization of surface fluxes in large-scale models under free convection. *Q. J. R. Meteorol. Soc.* **1995**, *121*, 255–270. [CrossRef]
276. Cronin, M.F.; Fairall, C.W.; McPhaden, M.J. An assessment of buoy-derived and numerical weather prediction surface heat fluxes in the tropical Pacific. *J. Geophys. Res. Oceans* **2006**, *111*. Available online: https://agupubs.onlinelibrary.wiley.com/doi/full/10.1029/2005JC003324 (accessed on 28 June 2006). [CrossRef]

277. Janssen, P.A.; Bidlot, J.R. Progress in Operational Wave Forecasting. *Procedia {IUTAM}* **2018**, *26*, 14–29. [CrossRef]

278. Cavaleri, L.; Fox-Kemper, B.; Hemer, M. Wind Waves in the Coupled Climate System. *Bull. Am. Meteorol. Soc.* **2012**, *93*, 1651–1661. [CrossRef]

279. Uttal, T.; Curry, J.A.; McPhee, M.G.; Perovich, D.K.; Moritz, R.E.; Maslanik, J.A.; Guest, P.S.; Stern, H.L.; Moore, J.A.; Turenne, R.; et al. Surface heat budget of the Arctic Ocean. *Bull. Am. Meteorol. Soc.* **2002**, *83*, 255–276. [CrossRef]

280. Zuo, H.; Balmaseda, M.A.; Mogensen, K.; Tietsche, S. 0CEAN5: The ECMWF Ocean Reanalysis System ORAS5 and its Real-Time analysis component. *ECMWF Tech. Memo* **2018**. Available online: https://www.ecmwf.int/en/elibrary/18519-ocean5-ecmwf-ocean-reanalysis-system-and-its-real-time-analysis-component (accessed on 1 August 2018).

281. Stark, J.D.; Jeff, R.; Matthew, M.; Adrian, H. OCEAN5: The ECMWF Ocean Reanalysis System and Its Real-Time Analysis Component. Available online: https://www.ecmwf.int/sites/default/files/elibrary/2018/18519-ocean5-ecmwf-ocean-renalysis-system-and-its-real-time-analysis-component.pdf (accessed on 1 August 2018).

282. Chevallier, M.; Smith, G.C.; Dupont, F.; Lemieux, J.F.; Forget, G.; Fujii, Y.; Hernandez, F.; Msadek, R.; Peterson, K.A.; Storto, A.; et al. Intercomparison of the Arctic sea ice cover in global ocean–sea ice reanalyses from the ORA-IP project. *Clim. Dyn.* **2017**, *49*, 1107–1136. [CrossRef]

283. Peterson, K.A.; Arribas, A.; Hewitt, H.T.; Keen, A.B.; Lea, D.J.; McLaren, A.J. Assessing the forecast skill of Arctic sea ice extent in the GloSea4 seasonal prediction system. *Clim. Dyn.* **2015**, *44*, 147–162. [CrossRef]

284. Lemieux, J.F.; Beaudoin, C.; Dupont, F.; Roy, F.; Smith, G.C.; Shlyaeva, A.; Buehner, M.; Caya, A.; Chen, J.; Carrieres, T.; et al. The Regional Ice Prediction System (RIPS): Verification of forecast sea ice concentration. *Q. J. R. Meteorol. Soc.* **2016**, *142*, 632–643.

285. Takaya, Y.; Hirahara, S.; Yasuda, T.; Matsueda, S.; Toyoda, T.; Fujii, Y.; Sugimoto, H.; Matsukawa, C.; Ishikawa, I.; Mori, H.; et al. Japan Meteorological Agency/Meteorological Research Institute-Coupled Prediction System version 2 (JMA/MRI-CPS2): Atmosphere–land–ocean–sea ice coupled prediction system for operational seasonal forecasting. *Clim. Dyn.* **2018**, *50*, 751–765. [CrossRef]

286. Tietsche, S.; Day, J.J.; Guemas, V.; Hurlin, W.J.; Keeley, S.P.E.; Matei, D.; Hawkins, E. Seasonal to interannual Arctic sea ice predictability in current global climate models. *Geophys. Res. Lett.* **2014**, *41*, 1035–1043. [CrossRef]

287. Wang, W.; Chen, M.; Kumar, A. Seasonal Prediction of Arctic Sea Ice Extent from a Coupled Dynamical Forecast System. *Mon. Weather Rev.* **2013**, *141*, 1375–1394. [CrossRef]

288. Chevallier, M.; Salas-Mélia, D. The role of sea ice thickness distribution in the Arctic sea ice potential predictability: A diagnostic approach with a coupled GCM. *J. Clim.* **2012**, *25*, 3025–3038. [CrossRef]

289. Day, J.; Hawkins, E.; Tietsche, S. Will Arctic sea ice thickness initialization improve seasonal forecast skill? *Geophys. Res. Lett.* **2014**, *41*, 7566–7575. [CrossRef]

290. Simmons, A.J.; Poli, P. Arctic warming in ERA-Interim and other analyses. *Q. J. R. Meteorol. Soc.* **2015**, *141*, 1147–1162. [CrossRef]

291. Tian-Kunze, X.; Kaleschke, L.; Maaß, N.; Mäkynen, M.; Serra, N.; Drusch, M.; Krumpen, T. SMOS-derived thin sea ice thickness: Algorithm baseline, product specifications and initial verification. *Cryosphere* **2014**, *8*, 997–1018. [CrossRef]

292. Tietsche, S.; Alonso-Balmaseda, M.; Rosnay, P.; Zuo, H.; Tian-Kunze, X.; Kaleschke, L. Thin Arctic sea ice in L-band observations and an ocean reanalysis. *Cryosphere* **2018**, *12*, 2051–2072. [CrossRef]

293. Richter, F.; Drusch, M.; Kaleschke, L.; Maaß, N.; Tian-Kunze, X.; Mecklenburg, S. Arctic sea ice signatures: L-band brightness temperature sensitivity comparison using two radiation transfer models. *Cryosphere* **2018**, *12*, 921–933. [CrossRef]

294. Florence, F.; Norbert, U. Observations of melt ponds on Arctic sea ice. *J. Geophys. Res. Oceans* **1998**, *103*, 24821–24835.

295. Rösel, A.; Kaleschke, L.; Birnbaum, G. Melt ponds on Arctic sea ice determined from MODIS satellite data using an artificial neural network. *Cryosphere* **2012**, *6*, 431–446. [CrossRef]

296. Lecomte, O.; Fichefet, T.; Flocco, D.; Schroeder, D.; Vancoppenolle, M. Interactions between wind-blown snow redistribution and melt ponds in a coupled ocean–sea ice model. *Ocean Modell.* **2015**, *87*, 67–80. [CrossRef]

297. Erko, J.; Timo, V.; Timo, P.; Liisi, J.; Hannes, K.; Jaak, J. Validation of atmospheric reanalyses over the central Arctic Ocean. *Geophys. Res. Lett.* **2012**, *39*. Available online: https://agupubs.onlinelibrary.wiley.com/doi/full/10.1029/2012GL051591 (accessed on 16 May 2012).

298. Lindsay, R.; Wensnahan, M.; Schweiger, A.; Zhang, J. Evaluation of seven different atmospheric reanalysis products in the Arctic. *J. Clim.* **2014**, *27*, 2588–2606. [CrossRef]

299. Rasmussen, T.A.; Høyer, J.L.; Ghent, D.; Bulgin, C.E.; Dybkjær, G.; Ribergaard, M.H.; Nielsen-Englyst, P.; Madsen, K.S. Impact of Assimilation of Sea-Ice Surface Temperatures on a Coupled Ocean and Sea-Ice Model. *J. Geophys. Res. Oceans* **2018**, *123*, 2440–2460. [CrossRef]

300. Yang, Q.; Wang, M.; Overland, J.E.; Wang, W.; Collow, T.W. Impact of Model Physics on Seasonal Forecasts of Surface Air Temperature in the Arctic. *Mon. Weather Rev.* **2017**, *145*, 773–782. [CrossRef]

301. Oleson, K.W.; Bonan, G.B.; Feddema, J.; Vertenstein, M.; Grimmond, C. An urban parameterization for a global climate model. Part I: Formulation and evaluation for two cities. *J. Appl. Meteorol. Climatol.* **2008**, *47*, 1038–1060. [CrossRef]

302. Grimmond, C.; Blackett, M.; Best, M.; Barlow, J.; Baik, J.; Belcher, S.; Bohnenstengel, S.; Calmet, I.; Chen, F.; Dandou, A.; et al. The international urban energy balance models comparison project: first results from phase 1. *J. Appl. Meteorol. Climatol.* **2010**, *49*, 1268–1292. [CrossRef]

303. Grimmond, C.; Blackett, M.; Best, M.; Baik, J.J.; Belcher, S.; Beringer, J.; Bohnenstengel, S.; Calmet, I.; Chen, F.; Coutts, A.; et al. Initial results from Phase 2 of the international urban energy balance model comparison. *Int. J. Climatol.* **2011**, *31*, 244–272. [CrossRef]

304. Stewart, I.D.; Oke, T.R. Local Climate Zones for Urban Temperature Studies. *Bull. Am. Meteorol. Soc.* **2012**, *93*, 1879–1900. [CrossRef]

305. Trusilova, K.; Schubert, S.; Wouters, H.; Fruh, B.; Grossman-Clarke, S.; Demuzere, M.; Becker, P. The urban land use in the COSMO-CLM model: A comparison of three parameterizations for Berlin. *Meteorol. Z.* **2016**, *25*, 231–244. [CrossRef]

306. Chrysoulakis, N.; Grimmond, S.; Feigenwinter, C.; Lindberg, F.; Gastellu-Etchegorry, J.P.; Marconcini, M.; Mitraka, Z.; Stagakis, S.; Crawford, B.; Olofson, F.; et al. Urban energy exchanges monitoring from space. *Sci. Rep.* **2018**, *8*, 11498. [CrossRef] [PubMed]

307. Pekel, J.F.; Cottam, A.; Gorelick, N.; Belward, A.S. High-resolution mapping of global surface water and its long-term changes. *Nature* **2016**, *540*, 418. [CrossRef] [PubMed]

308. Rodell, M.; Beaudoing, H.K.; L'Ecuyer, T.S.; Olson, W.S.; Famiglietti, J.S.; Houser, P.R.; Adler, R.; Bosilovich, M.G.; Clayson, C.A.; Chambers, D.; et al. The observed state of the water cycle in the early twenty-first century. *J. Clim.* **2015**, *28*, 8289–8318. [CrossRef]

309. Duhovny, V.; Avakyan, I.; Zholdasova, I.; Mirabdullaev, I.; Muminov, S.; Roshenko, E.; Ruziev, I.; Ruziev, M.; Stulina, G.; Sorokin, A.; et al. *Aral Sea and Its Surrounding*; UNESCO Office in Uzbekistan and Baktria Press, 2017; pp. 1–120. Available online: https://unesdoc.unesco.org/ark:/48223/pf0000260741 (accessed on 1 January 2017).

310. Eerola, K.; Rontu, L.; Kourzeneva, E.; Kheyrollah Pour, H.; Duguay, C. Impact of partly ice-free Lake Ladoga on temperature and cloudiness in an anticyclonic winter situation—A case study using a limited area model. *Tellus A Dyn. Meteorol. Oceanogr.* **2014**, *66*, 23929. [CrossRef]

311. Wisser, D.; Frolking, S.; Douglas, E.M.; Fekete, B.M.; Vörösmarty, C.J.; Schumann, A.H. Global irrigation water demand: Variability and uncertainties arising from agricultural and climate data sets. *Geophys. Res. Lett.* **2008**, *35*. Available online: https://agupubs.onlinelibrary.wiley.com/doi/10.1029/2008GL035296 (accessed on 31 December 2008). [CrossRef]

312. De Rosnay, P.; Polcher, J.; Laval, K.; Sabre, M. Integrated parameterization of irrigation in the land surface model ORCHIDEE. Validation over Indian Peninsula. *Geophys. Res. Lett.* **2003**, *30*. Available online: https://agupubs.onlinelibrary.wiley.com/doi/full/10.1029/2003GL018024 (accessed on 7 October 2003). [CrossRef]

313. Puma, M.; Cook, B. Effects of irrigation on global climate during the 20th century. *J. Geophys. Res. Atmos.* **2010**, *115*. Available online: https://agupubs.onlinelibrary.wiley.com/doi/full/10.1029/2010JD014122 (accessed on 28 August 2010). [CrossRef]

314. Miyazaki, K.; Eskes, H.; Sudo, K.; Boersma, K.F.; Bowman, K.; Kanaya, Y. Decadal changes in global surface NOx emissions from multi-constituent satellite data assimilation. *Atmos. Chem. Phys* **2017**, *17*, 807–837. [CrossRef]

315. Van der Werf, G.R.; Randerson, J.T.; Giglio, L.; Collatz, G.J.; Kasibhatla, P.S.; Arellano, A.F., Jr. Interannual variability in global biomass burning emissions from 1997 to 2004. *Atmos. Chem. Phys.* **2006**, *6*, 3423–3441. [CrossRef]

316. Fortems-Cheiney, A.; Dufour, G.; Hamaoui-Laguel, L.; Foret, G.; Siour, G.; Van Damme, M.; Meleux, F.; Coheur, P.F.; Clerbaux, C.; Clarisse, L.; et al. Unaccounted variability in NH3 agricultural sources detected by IASI contributing to European spring haze episode. *Geophys. Res. Lett.* **2016**, *43*, 5475–5482. [CrossRef]

317. Pandey, S.; Houweling, S.; Krol, M.; Aben, I.; Monteil, G.; Nechita-Banda, N.; Dlugokencky, E.J.; Detmers, R.; Hasekamp, O.; Xu, X.; et al. Enhanced methane emissions from tropical wetlands during the 2011 La Niña. *Sci. Rep.* **2017**, *7*, 45759. [CrossRef] [PubMed]

318. Yin, Y.; Ciais, P.; Chevallier, F.; Li, W.; Bastos, A.; Piao, S.; Wang, T.; Liu, H. Changes in the response of the Northern Hemisphere carbon uptake to temperature over the last three decades. *Geophys. Res. Lett.* **2018**, *45*, 4371–4380. [CrossRef]

319. Escribano, J.; Boucher, O.; Chevallier, F.; Huneeus, N. Subregional inversion of North African dust sources. *J. Geophys. Res. Atmos.* **2016**, *121*, 8549–8566. [CrossRef]

320. Noilhan, J.; Mahfouf, J.F. The ISBA land surface parameterisation scheme. *Glob. Planet. Chang.* **1996**, *13*, 145–159. [CrossRef]

321. Mahfouf, J.F. Assimilation of satellite-derived soil moisture from ASCAT in a limited-area NWP model. *Q. J. R. Meteorol. Soc.* **2010**, *136*, 784–798. [CrossRef]

322. Kumar, S.V.; Reichle, R.H.; Koster, R.D.; Crow, W.T.; Peters-Lidard, C.D. Role of subsurface physics in the assimilation of surface soil moisture observations. *J. Hydrol.* **2009**, *10*, 1534–1547. [CrossRef]

323. Parrens, M.; Mahfouf, J.F.; Barbu, A.; Calvet, J.C. Assimilation of surface soil moisture into a multilayer soil model: Design and evaluation at local scale. *Hydrol. Earth Syst. Sci.* **2014**, *18*, 673–689. [CrossRef]

324. Penny, S.G.; Hamill, T.M. Coupled data assimilation for integrated earth system analysis and prediction. *Bull. Am. Meteorol. Soc.* **2017**, *97*, ES169–ES172. [CrossRef]

325. Orth, R.; Dutra, E.; Pappenberger, F. Improving Weather Predictability by Including Land Surface Model Parameter Uncertainty. *Mon. Weather Rev.* **2016**, *144*, 1551–1569. [CrossRef]

326. Buizza, R.; Anderson, E.; Forbes, R.; Sleigh, M. The ECMWF Research to Operations (R2O) process. *ECMWF Res. Dep. Tech. Memo.* **2017**, *806*, 1–16.

327. Buizza, R.; Alonso-Balmaseda, M.; Brown, A.; English, S.; Forbes, R.; Geer, A.; Haiden, T.; Leutbecher, M.; Magnusson, L.; Rodwell, M.; et al. The development and evaluation process followed at ECMWF to upgrade the Integrated Forecasting System (IFS). *ECMWF Res. Dep. Tech. Memo.* **2018**, *829*, 1–47.

328. Parsons, D.; Beland, M.; Burridge, D.; Bougeault, P.; Brunet, G.; Caughey, J.; Cavallo, S.; Charron, M.; Davies, H.; Niang, A.D.; et al. THORPEX Research and the Science of Prediction. *Bull. Am. Meteorol. Soc.* **2017**, *98*, 807–830. [CrossRef]

329. Rabier, F.; Cohn, S.; Cocquerez, P.; Hertzog, A.; Avallone, L.; Deshler, T.; Haase, J.; Hock, T.; Doerenbecher, A.; Wang, J.; et al. The Concordiasi field experiment over Antarctica: First results from innovative atmospheric measurements. *Bull. Am. Meteorol. Soc.* **2013**, *94*, ES17–ES20. [CrossRef]

330. Cohn, S.A.; Hock, T.; Cocquerez, P.; Wang, J.; Rabier, F.; Parsons, D.; Harr, P.; Wu, C.C.; Drobinski, P.; Karbou, F.; et al. Driftsondes: Providing in situ long-duration dropsonde observations over remote regions. *Bull. Am. Meteorol. Soc.* **2013**, *94*, 1661–1674. [CrossRef]

331. Rabier, F.; Bouchard, A.; Brun, E.; Doerenbecher, A.; Guedj, S.; Guidard, V.; Karbou, F.; Peuch, V.H.; El Amraoui, L.; Puech, D.; et al. The CONCORDIASI project in Antarctica. *Bull. Am. Meteorol. Soc.* **2010**, *91*, 69–86. [CrossRef]

332. Brun, E.; Six, D.; Picard, G.; Vionnet, V.; Arnaud, L.; Bazile, E.; Boone, A.; Bouchard, A.; Genthon, C.; Guidard, V.; et al. Snow/atmosphere coupled simulation at Dome C, Antarctica. *J. Glaciol.* **2011**, *57*, 721–736. [CrossRef]

333. GEWEX. GEWEX plans for 2013 and beyond—GEWEX Science Questions. *GEWEX Doc.* **2012**, *2*, 1–21.

334. Seneviratne, S.I.; Lüthi, D.; Litschi, M.; Schär, C. Land–atmosphere coupling and climate change in Europe. *Nature* **2006**, *443*, 205. [CrossRef]

335. Fischer, E.M.; Seneviratne, S.I.; Lüthi, D.; Schär, C. Contribution of land-atmosphere coupling to recent European summer heat waves. *Geophys. Res. Lett.* **2007**, *34*. Available online: https://agupubs.onlinelibrary.wiley.com/doi/10.1029/2006GL029068 (accessed on 24 March 2007). [CrossRef]

336. Hirschi, M.; Seneviratne, S.I.; Alexandrov, V.; Boberg, F.; Boroneant, C.; Christensen, O.B.; Formayer, H.; Orlowsky, B.; Stepanek, P. Observational evidence for soil-moisture impact on hot extremes in southeastern Europe. *Nat. Geosci.* **2011**, *4*, 17–21. [CrossRef]

337. Mueller, B.; Seneviratne, S.I. Hot days induced by precipitation deficits at the global scale. *Proc. Natl. Acad. Sci. USA* **2012**, *109*, 12398–12403. [CrossRef]

338. Quesada, B.; Vautard, R.; Yiou, P.; Hirschi, M.; Seneviratne, S.I. Asymmetric European summer heat predictability from wet and dry southern winters and springs. *Nat. Clim. Chang.* **2012**, *2*, 736–741. [CrossRef]

339. Seneviratne, S.I.; Wilhelm, M.; Stanelle, T.; Hurk, B.; Hagemann, S.; Berg, A.; Cheruy, F.; Higgins, M.E.; Meier, A.; Brovkin, V.; et al. Impact of soil moisture-climate feedback on CMIP5 projections: First results from the GLACE-CMIP5 experiment. *Geophys. Res. Lett.* **2013**, *40*, 5212–5217. [CrossRef]

340. Miralles, D.G.; Teuling, A.J.; van Heerwaarden, C.C.; Vilà-Guerau de Arellano, J. Mega-heatwave temperatures due to combined soil desiccation and atmospheric heat accumulation. *Nat. Geosci.* **2014**, *7*, 345–349. [CrossRef]

341. Seneviratne, S.I.; Donat, M.G.; Pitman, A.J.; Knutti, R.; Wilby, R.L. Allowable CO_2 emissions based on regional and impact-related climate targets. *Nature* **2016**, *529*, 477. [CrossRef] [PubMed]

342. Davin, E.L.; Seneviratne, S.I.; Ciais, P.; Olioso, A.; Wang, T. Preferential cooling of hot extremes from cropland albedo management. *Proc. Natl. Acad. Sci. USA* **2014**, *111*, 9757–9761. [CrossRef] [PubMed]

343. Wilhelm, M.; Davin, E.L.; Seneviratne, S.I. Climate engineering of vegetated land for hot extremes mitigation: An Earth system model sensitivity study. *J. Geophys. Res. Atmos.* **2015**, *120*, 2612–2623. [CrossRef]

344. Taylor, C.M.; de Jeu, R.A.; Guichard, F.; Harris, P.P.; Dorigo, W.A. Afternoon rain more likely over drier soils. *Nature* **2012**, *489*, 423. [CrossRef]

345. Martius, O.; Sodemann, H.; Joos, H.; Pfahl, S.; Winschall, A.; Croci-Maspoli, M.; Graf, M.; Madonna, E.; Mueller, B.; Schemm, S.; et al. The role of upper-level dynamics and surface processes for the Pakistan flood of July 2010. *Q. J. R. Meteorol. Soc.* **2013**, *139*, 1780–1797. [CrossRef]

346. Lorenz, R.; Argüeso, D.; Donat, M.G.; Pitman, A.J.; Hurk, B.; Berg, A.; Lawrence, D.M.; Chéruy, F.; Ducharne, A.; Hagemann, S.; et al. Influence of land-atmosphere feedback on temperature and precipitation extremes in the GLACE-CMIP5 ensemble. *J. Geophys. Res. Atmos.* **2016**, *121*, 607–623. [CrossRef]

347. Greve, P.; Orlowsky, B.; Mueller, B.; Sheffield, J.; Reichstein, M.; Seneviratne, S.I. Global assessment of trends in wetting and drying over land. *Nat. Geosci.* **2014**, *7*, 716–721. [CrossRef]

348. Malbéteau, Y.; Merlin, O.; Balsamo, G.; Er-Raki, S.; Khabba, S.; Walker, J.; Jarlan, L. Toward a Surface Soil Moisture Product at High Spatiotemporal Resolution: Temporally Interpolated, Spatially Disaggregated SMOS Data. *J. Hydrometeorol.* **2018**, *19*, 183–200. [CrossRef]

349. Tabatabaeenejad, A.; Burgin, M.; Duan, X.; Moghaddam, M. P-Band Radar Retrieval of Subsurface Soil Moisture Profile as a Second-Order Polynomial: First AirMOSS Results. *IEEE Trans. Geosci. Remote Sens.* **2015**, *53*, 645–658. [CrossRef]

350. Dee, D.P.; Uppala, S. Variational bias correction of satellite radiance data in the ERA-Interim reanalysis. *Q. J. R. Meteorol. Soc.* **2009**, *135*, 1830–1841. [CrossRef]

Correction

Correction: Balsamo, G., *et al.* Satellite and In Situ Observations for Advancing Global Earth Surface Modelling: A Review. *Remote Sensing* 2018, *10*, 2038

Gianpaolo Balsamo [1], Anna Agusti-Panareda [1], Clement Albergel [2], Gabriele Arduini [1], Anton Beljaars [1], Jean Bidlot [1], Eleanor Blyth [3], Nicolas Bousserez [1], Souhail Boussetta [1], Andy Brown [1], Roberto Buizza [1,4], Carlo Buontempo [1], Frédéric Chevallier [5], Margarita Choulga [1], Hannah Cloke [6], Meghan F. Cronin [7], Mohamed Dahoui [1], Patricia De Rosnay [1], Paul A. Dirmeyer [8], Matthias Drusch [9], Emanuel Dutra [10], Michael B. Ek [11], Pierre Gentine [12], Helene Hewitt [13], Sarah P.E. Keeley [1], Yann Kerr [14], Sujay Kumar [15], Cristina Lupu [1], Jean-François Mahfouf [2], Joe McNorton [1], Susanne Mecklenburg [9], Kristian Mogensen [1], Joaquín Muñoz-Sabater [1], Rene Orth [16], Florence Rabier [1], Rolf Reichle [15], Ben Ruston [17], Florian Pappenberger [1], Irina Sandu [1], Sonia I. Seneviratne [18], Steffen Tietsche [1], Isabel F. Trigo [19], Remko Uijlenhoet [20], Nils Wedi [1], R. Iestyn Woolway [6] and Xubin Zeng [21]

[1] European Centre for Medium-range Weather Forecasts (ECMWF), Reading RG2 9AX, UK; Anna.Agusti-Panareda@ecmwf.int (A.A.-P.); Gabriele.Arduini@ecmwf.int (G.A.); Anton.Beljaars@ecmwf.int (A.B.); Jean.Bidlot@ecmwf.int (J.B.); Nicolas.Bousserez@ecmwf.int (N.B.); Souhail.Boussetta@ecmwf.int (S.B.); Andy.Brown@ecmwf.int (A.B.); Carlo.Buontempo@ecmwf.int (C.B.); Margarita.Choulga@ecmwf.int (M.C.); Mohamed.Dahoui@ecmwf.int (M.D.); Patricia.Rosnay@ecmwf.int (P.D.R.); Sarah.Keeley@ecmwf.int (S.P.E.K.); Cristina.Lupu@ecmwf.int (C.L.); Joe.McNorton@ecmwf.int (J.M.); Kristian.Mogensen@ecmwf.int (K.M.); Joaquin.Munoz@ecmwf.int (J.M.-S.); Florence.Rabier@ecmwf.int (F.R.); Florian.Pappenberger@ecmwf.int (F.P.); Irina.Sandu@ecmwf.int (I.S.); Steffen.Tietsche@ecmwf.int (S.T.); Nils.Wedi@ecmwf.int (N.W.)

[2] Météo-France, Centre National de Recherches Météorologique, 31000 Toulouse, France; clement.albergel@meteo.fr (C.A.); jean-francois.mahfouf@meteo.fr (J.F.M.)

[3] Centre for Ecology & Hydrology, Wallingford OX10 8BB, UK; emb@ceh.ac.uk

[4] Scuola Superiore Sant'Anna, 56127 Pisa, Italy; roberto.buizza@santannapisa.it

[5] Laboratoire des Sciences du Climat et de l'Environnement, Institut Pierre-Simon-Laplace, Commissariat à l'énergie atomique et aux énergies alternatives, LSCE/IPSL/CEA, 91190 Gif sur Yvette, France; frederic.chevallier@lsce.ipsl.fr

[6] Meteorology Depart., University of Reading, Reading RG6 7BE, UK; h.l.cloke@reading.ac.uk (H.C.); r.i.woolway@reading.ac.uk (R.I.W.)

[7] National Oceanic and Atmospheric Administration, Pacific Marine Environmental Laboratory, Seattle, WA 98115, USA; meghan.f.cronin@noaa.gov

[8] Center for Ocean-Land-Atmosphere Studies, George Mason University, Fairfax, VA 22030, USA; pdirmeye@gmu.edu

[9] European Space Agency (ESA), European Space Research and Technology Centre (ESTEC), 2201AZ Noordwijk, The Netherlands; matthias.drusch@esa.int (M.D.); Susanne.Mecklenburg@esa.int (S.M.)

[10] Instituto Dom Luiz, University of Lisbon, 1749-016 Lisbon, Portugal; endutra@fc.ul.pt

[11] National Center for Atmospheric Research, Boulder, CO 80305, USA; ek@ucar.edu

[12] Department of Earth and Environmental Engineering, Columbia University, New York, NY 10027, USA; pg2328@columbia.edu

[13] UK MetOffice, Exeter EX1 3PB, UK; helene.hewitt@metoffice.gov.uk

[14] Centre National d'Etudes Spatiales, CESBIO, 31401 Toulouse, France; yann.kerr@cesbio.cnes.fr

[15] National Aeronautics and Space Administration (NASA), Goddard Space Flight Center (GSFC), Greenbelt, MD 20771, USA; sujay.v.kumar@nasa.gov (S.K.); rolf.reichle@nasa.gov (R.R.)

[16] Max Planck Institute for Biogeochemistry, 07745 Jena, Germany; rene.orth@bgc-jena.mpg.de

[17] Naval Research Laboratory (NRL), Monterey, CA 93943, USA; Ben.Ruston@nrlmry.navy.mil

[18] Eidgenössische Technische Hochschule (ETH), 8092 Zürich, Switzerland; sonia.seneviratne@ethz.ch

[19] Instituto Português do Mar e da Amosfera (IPMA), 1749-077 Lisbon, Portugal; isabel.trigo@ipma.pt

[20] Department of Environmental Sciences, Wageningen University and Research, 6708 PB Wageningen, The Netherlands; remko.uijlenhoet@wur.nl

[21] Department of Hydrology and Atmospheric Sciences, University of Arizona, Tucson, AZ 85721, USA; xubin@atmo.arizona.edu

* Correspondence: gianpaolo.balsamo@ecmwf.int; Tel.: +44-759-5331517

† Current address: European Centre for Medium-Range Weather Forecasts (ECMWF), Shinfield Park, Reading RG2 9AX, UK.

Received: 4 April 2019; Accepted: 7 April 2019; Published: 18 April 2019

The authors wish to make the following corrections to this paper [1]:
Update of Figure 1 and correct authorship to include Dr. Eleanor Blyth (CEH).

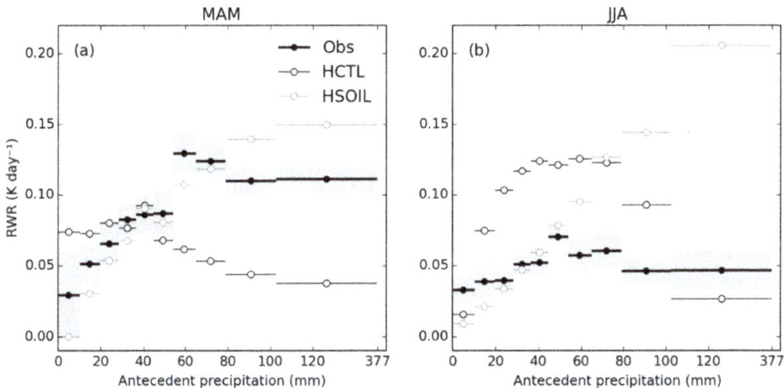

Figure 1. Example of usage of land surface temperature during dry episodes after a precipitation event. The plot shows a composite of the so-called Relative Warming Rate (RWR) as a function of the amount of precipitation during the preceding event for March–April–May (MAM, **left**) and June–July–August (JJA, **right**). RWR quantifies the increase in dry spell land surface temperature relative to air temperature, and is a measure for the evaporation regime of the land surface.

The authors wish to make the following corrections to this paper [2]:
Correction to the legend of Figure 15 to mention this is adapted from Rodriguez-Fernandez et al., 2018 ([2] and referenced as [182] in [1]).
The authors would like to apologize for any inconvenience caused to the readers by these changes.

References

1. Balsamo, G.; Agustì-Parareda, A.; Albergel, C.; Arduini, G.; Beljaars, A.; Bidlot, J.; Bousserez, N.; Boussetta, S.; Brown, A.; Buizza, R.; et al. Satellite and In Situ Observations for Advancing Global Earth Surface Modelling: A Review. *Remote Sens.* **2018**, *10*, 2038. [CrossRef]

2. Rodríguez-Fernández, N.J.; Mialon, A.; Mermoz, S.; Bouvet, A.; Richaume, P.; Al Bitar, A.; Al-Yaari, A.; Brandt, M.; Kaminski, T.; Le Toan, T.; et al. An evaluation of SMOS L-band vegetation optical depth (L-VOD) data sets: high sensitivity of L-VOD to above-ground biomass in Africa. *Biogeosciences* **2018**, *15*, 4627–4645. [CrossRef]

MDPI
St. Alban-Anlage 66
4052 Basel
Switzerland
Tel. +41 61 683 77 34
Fax +41 61 302 89 18
www.mdpi.com

Remote Sensing Editorial Office
E-mail: remotesensing@mdpi.com
www.mdpi.com/journal/remotesensing